AF411173

Unmanned Aircraft Systems with Autonomous Navigation

Unmanned Aircraft Systems with Autonomous Navigation

Editors

Umberto Papa
Marcello Rosario Napolitano
Giuseppe Del Core
Salvatore Ponte

MDPI • Basel • Beijing • Wuhan • Barcelona • Belgrade • Manchester • Tokyo • Cluj • Tianjin

Editors

Umberto Papa
Department of Science
and Technology,
University of
Naples "Parthenope",
Naples, Italy

Marcello Rosario Napolitano
Department of Mechanical
and Aerospace Engineering,
West Virginia University,
Morgantown, WV, USA

Giuseppe Del Core
Department of Science
and Technology,
University of
Naples "Parthenope",
Naples, Italy

Salvatore Ponte
Department of Engineering,
University of Campania
"L. Vanvitelli",
Aversa, Italy

Editorial Office
MDPI
St. Alban-Anlage 66
4052 Basel, Switzerland

This is a reprint of articles from the Special Issue published online in the open access journal *Electronics* (ISSN 2079-9292) (available at: https://www.mdpi.com/journal/electronics/special_issues/UAS_electronics).

For citation purposes, cite each article independently as indicated on the article page online and as indicated below:

LastName, A.A.; LastName, B.B.; LastName, C.C. Article Title. *Journal Name* **Year**, *Volume Number*, Page Range.

ISBN 978-3-0365-8052-4 (Hbk)
ISBN 978-3-0365-8053-1 (PDF)

Cover image courtesy of Umberto Papa.

Contents

About the Editors

Umberto Papa

Umberto Papa was born in Caserta, Italy, in 1986. He received his bachelor's and master's degrees in aerospace engineering from the Parthenope University of Naples, Naples, Italy, in 2008 and 2012, respectively. He received his Ph.D. degree in applied science on sea, environment, and territory from Parthenope University of Naples in 2017. In 2014, he joined the University of Naples Parthenope as a Research Fellow. Since 2016, he has been an Honorary Fellow in Flight Mechanics at Department of Science and Technology of Parthenope University of Naples. He is currently an Aerospace Engineer with Leonardo Company—Aircraft Division, Pomigliano d'Arco, Naples. His current research interests include flight mechanics, navigation, and guidance of small UASs, Dr. Papa was a recipient of the "Leonardo Innovation Award" in the Ph.D. student category in 2015.

Marcello Rosario Napolitano

Marcello Rosario Napolitano received a Ph.D. in Mechanical Engineering from Oklahoma State University, Stillwater, Oklahoma (1989), and an M.S. in Aeronautical Engineering from University of Naples, Naples, Italy (1985). He is a Professor at the Dept. of Mechanical and Aerospace Eng. (MAE), WVU (2001–Present). Previously, he held the positions of Associate Professor (1996–2001) and Assistant Professor (1990–1996) in the same department. He was PI or Co-PI for approx. 60 projects sponsored by NASA, DOD, and other agencies. He is the author of the book "Aircraft Dynamics: From Modeling to Simulation".

Giuseppe Del Core

Giuseppe Del Core was born in Naples, Italy. He is currently an Associate Professor of flight mechanics with the Department of Sciences and Technologies, University of Naples Parthenope, Naples, Italy. He has taught in the area of flight mechanics, airplane design, flight testing, and aircraft systems at the Universities of Palermo, Caserta "Luigi Vanvitelli", and Salento, Italy. His current research interests include experimental aerodynamics, flight performance, airplane design optimization, UAVs design, and fault diagnosis.

Salvatore Ponte

Salvatore Ponte received a Ph.D. in Aerospace Engineering from University of Naples Federico II, Naples, Italy (1995). Since 1995, he has been an Assistant Professor at Department of Aerospace and Mechanical Engineering at Università della Campania, Aversa.

 *electronics*

Editorial

Unmanned Aircraft Systems with Autonomous Navigation

Umberto Papa

Department of Science and Technology, Parthenope University of Naples, 80143 Naples, Italy; umberto.papa@collaboratore.uniparthenope.it

Citation: Papa, U. Unmanned Aircraft Systems with Autonomous Navigation. *Electronics* **2023**, *12*, 1591. https://doi.org/10.3390/ electronics12071591

Received: 14 March 2023
Accepted: 18 March 2023
Published: 28 March 2023

1. Introduction

Unmanned aerial systems play an increasingly remarkable role in widely diffused application fields, from military defense programs and strategies to civil and commercial utilization. UAS are usually involved in dull, dirty, and dangerous (DDD) scenarios, which require reliable, extended-capability, easy-to-use, and cost-effective fixed-wing or rotary-wing platforms. Therefore, it is important to provide onboard systems capable of recognizing the environment around the aerial vehicle, detecting and avoiding obstacles, implementing path planning and management strategies, defining safe landing areas, and achieving full autonomy, especially for BVLOS (beyond visual line-of-sight) missions. The technical and economic challenges implied by the issues related to autonomous navigation range from hardware (sensors, platforms, controllers, etc.) to software (data processing and filtering techniques, optimal control, state estimation, innovative algorithms, etc.), and from modeling to practical realizations.

The aim of this Special Issue is to seek high-quality contributions that highlight novel research results and emerging applications, addressing recent breakthroughs in UAS autonomous navigation and related fields, such as flight mechanics and control, structural design, sensor design, etc.

The topics of interest are as follows:

- Two-dimensional and three-dimensional mapping, target detection, and obstacle avoidance;
- The active perception of targets in cluttered environments (foliage, forests, etc.);
- Vision-based and optical flow techniques;
- Sensors and sensor fusion techniques;
- Design models for guidance and controlled flight;
- State estimation, data analysis and filtering techniques (KF, EKF, particle filtering, fuzzy logic, etc.);
- Path planning and path management;
- Optimal control and strategies (neural networks, fuzzy logic, reinforcement learning, evolutionary and genetic algorithms, AI, etc.);
- Navigation in GPS-denied environments;
- Autolanding and safe landing area definition (SLAD);
- Environmental effects on UAVs (wind, etc.);
- Autonomous UAV or MAV swarms, and distributed architectures;
- BVLOS autonomous navigation.

2. Review Papers

Isaac S. Leal et al. [1] present a comprehensive approach for the dynamic modeling, control system design, simulation, and optimization of a quadcopter. The main objective is to study the behavior of different controllers when the model is working under linear and/or non-linear conditions, and, therefore, to define the possible limitations of the controllers. Five different control systems are proposed to improve the control performance, mainly the stability of the system. Additionally, a path simulator was also developed with the intention of describing the vehicle's movements and hence to detect faults intuitively. The proposed PID and fuzzy-PD control systems showed promising responses to the tests which were carried out. The results indicated the limits of the PID controller over non-linear conditions, and the effectiveness of the controllers was enhanced by the implementation of a genetic algorithm to autotune the controllers in order to adapt to changing conditions.

Ariante G. et al. [2] propose a practical method for the estimation of true and calibrated airspeed, angle of attack (AOA), and angle of sideslip (AOS) for small unmanned aerial vehicles (UAVs, up to 20 kg mass, 1200 ft altitude above ground level, and airspeed of up to 100 knots) or light aircraft, for which weight, size, cost, and power-consumption requirements do not allow solutions used in large airplanes (typically, arrays of multi-hole Pitot probes). The sensors used in this research were a static and dynamic pressure sensor ("micro-Pitot tube" MPX2010DP differential pressure sensor) and a 10 degrees of freedom (DoF) inertial measurement unit (IMU) for attitude determination. Kalman and complementary filtering were applied for measurement noise removal and data fusion, respectively, achieving the global exponential stability of the estimation error. The methodology was tested using experimental data from a prototype of the devised sensor suite, in various indoor-acquisition campaigns and laboratory tests under controlled conditions. AOA and AOS estimates were validated via a correlation between the AOA measured by the micro-Pitot and vertical accelerometer measurements, since lift force can be modeled as a linear function of AOA in normal flight. The results confirmed the validity of the proposed approach, which could have interesting applications in energy-harvesting techniques.

Zhang R. et al. [3] propose in their paper a novel hybrid-driven fixed-wing UAV maneuver optimization framework, inspired by apprenticeship learning and nonlinear programing approaches. The work consists of two main aspects: (1) identifying the model parameters for a certain fixed-wing UAV based on the demonstrated flight data performed by human pilot. Then, the features of the maneuvers can be described by the positional/attitude/compound key frames. Eventually, each of the maneuvers can be decomposed into several motion primitives. Formulating the maneuver planning issue into a minimum-time optimization problem, a novel nonlinear programming algorithm was developed, which was unnecessary to determine the exact time for the UAV to pass by the key frames. The simulation results illustrate the effectiveness of the proposed framework in several scenarios, as both the preservation of geometric features and the minimization of maneuver times were ensured.

Nguyen L.V. et al. [4] present a novel iterative learning sliding mode controller (ILSMC) that can be applied to the trajectory tracking of quadrotor unmanned aerial vehicles (UAVs) subject to model uncertainties and external disturbances. Here, the proposed ILSMC is integrated in the outer loop of a controlled system. The control development, conducted in the discrete-time domain, does not require a priori information of the disturbance to be bound as with conventional SMC techniques. It only involves an equivalent control term for the desired dynamics in the closed loop and an iterative learning term to drive the system state toward the sliding surface to maintain a robust performance. By learning from previous iterations, the ILSMC can yield a very accurate tracking performance when a sliding mode is induced without control chattering. The design is then applied to the attitude control of a 3DR Solo UAV with a built-in PID controller. The simulation results and experimental validation with real-time data demonstrate the advantages of the proposed control scheme over the existing techniques.

Lerro A. et al. [5] in their paper aimed to describe the verification in a relevant environment of a physics-based approach using a dedicated technological demonstrator. The flow angle synthetic solution is based on a model-free, or physics-based, scheme and, therefore, it is applicable to any flying body. The demonstrator also encompasses physical sensors that provide all the necessary inputs to the synthetic sensors to estimate the angle-of-attack and the angle-of-sideslip. The uncertainty budgets of the physical sensors are evaluated to corrupt the flight simulator data with the aim of reproducing a realistic scenario to verify the synthetic sensors. The proposed approach for the flow angle estimation is suitable for modern and future aircraft, such as drones and urban mobility air vehicles. The results presented in this work show that the proposed approach can be effective in relevant scenarios, even though some limitations can arise.

Song Y. et al. [6] propose a two-way neighbor discovery algorithm based on a spatial multi-channel through cross-layer optimization. First, they provide two boundary conditions of the physical (PHY) layer and media access control (MAC) layer for a successful link establishment of mmWave neighbor discovery and to provide the optimal pairing of antenna beamwidth in different stages and scenarios using cross-layer optimization. Then, a mmWave neighbor discovery algorithm based on a spatial multi-channel is proposed, which greatly reduces the convergence time by increasing the discovery probability of nodes in the network. Finally, a random reply algorithm is proposed based on dynamic reserved time slots. By adjusting the probability of reply and the number of reserved time slots, the neighbor discovery time can be further reduced when the number of nodes is larger. Simulations show that as the network scale is 100 to 500 nodes, the convergence time is 10 times higher than that of the single channel algorithm.

Amphawan A. et al. [7] in their paper discuss the deployment of unmanned aerial vehicles (UAVs) for free space optical communications. In particular, a critical challenge to this is maintaining an acceptable signal quality between the ground base station and UAV-based free space optics relay. This is largely unattainable due to rapid UAV propeller and body movements, which result in fluctuations in the beam alignment and frequent link failures. To address this issue, linearly polarized Laguerre–Gaussian modes were leveraged for spatial mode diversity to prevent link failures over a 400 m link. Spatial mode diversity successfully improved the bit error rate by 38% to 55%. This was due to a 10% to 19% increase in the predominant mode power from spatial mode diversity. The time-varying channel matrix indicated the presence of nonlinear deterministic chaos. This opens up new possibilities for research on state space reconstruction of the channel matrix.

Pfeiffer R. et al. [8] present important foundational blocks that can be expanded into an autonomous monitoring-and-retrieval pipeline based on drone surveys and object detection using deep learning. Drone footage collected on the islands of Malta and Gozo in Sicily (Italy) and the Red Sea coast was combined with publicly available litter datasets and used to train an object detection algorithm (YOLOv5) to detect litter objects in footage recorded during drone surveys. Across all classes of litter objects, the 50–95% mean average precision (mAP50-95) was 0.252, with the performance on single well-represented classes reaching up to 0.674. We also present an approach to geolocate objects detected by the algorithm, assigning latitude and longitude coordinates to each detection. In combination with beach morphology information derived from digital elevation models (DEMs) for path finding and identifying inaccessible areas for an autonomous litter retrieval robot, this research provides important building blocks for an automated monitoring-and-retrieval pipeline.

Firdaus A.R. et al. [9] in their paper consider indoor navigation applications using the ZED2 stereo camera for the quadcopter. To use the ZED2 camera as a navigation sensor, we first transformed its coordinates into the north, east, and down (NED) system to enable the drone to understand its position and maintain stability in a particular position. The experiment was performed using a real-time application to confirm the feasibility of this approach for indoor localization. In the real-time application, we commanded the quadcopter to follow triangular and rectangular paths. The results indicated that the quadcopter was able to follow the paths and maintain its stability in specific coordinate positions.

Koukiou G. et al. [10] propose a sensor autonomous integrity monitoring (SAIM) methodology to enhance the sensors' failures in a collision avoidance system (CAS) field. The configuration of the sensors and their interaction is based on a fusion procedure that involves a total of five sensors. Accordingly, the performance of each one of the sensors is continuously checked against the combined (fused) operation of the other four. A complementary experiment with a total of four sensors, one of which had a low performance, was also conducted. The experimental results reveal a reliable approach for sensor autonomous integrity monitoring (SAIM). The method can be easily extended to a larger number of sensors.

Funding: This research received no external funding.

Acknowledgments: We would like to take this opportunity to appreciate and thank all authors for their outstanding contributions and the reviewers for their fruitful comments and feedback. Special appreciation should also be paid to Ariante Gennaro, Alberto Greco, and to the *Electronics* Editorial Office staff for their hard and precise work in maintaining a rigorous peer-review schedule and timely publication. A special thanks also to the Editorial Board of MDPI's *Electronics* journal for the opportunity to guest edit this Special Issue.

Conflicts of Interest: The authors declare no conflict of interest.

References

1. Leal, I.S.; Abeykoon, C.; Perera, Y.S. Design, simulation, analysis and optimization of PID and fuzzy based control systems for a quadcopter. *Electronics* **2021**, *10*, 2218. [CrossRef]
2. Ariante, G.; Ponte, S.; Papa, U.; Del Core, G. Estimation of airspeed, angle of attack, and sideslip for small unmanned aerial vehicles (UAVs) using a micro-pitot tube. *Electronics* **2021**, *10*, 2325. [CrossRef]
3. Zhang, R.; Cao, S.; Zhao, K.; Yu, H.; Hu, Y. A hybrid-driven optimization framework for fixed-wing uav maneuvering flight planning. *Electronics* **2021**, *10*, 2330. [CrossRef]
4. Nguyen, L.V.; Phung, M.D.; Ha, Q.P. Iterative Learning Sliding Mode Control for UAV Trajectory Tracking. *Electronics* **2021**, *10*, 2474. [CrossRef]
5. Lerro, A.; Gili, P.; Pisani, M. Verification in Relevant Environment of a Physics-Based Synthetic Sensor for Flow Angle Estimation. *Electronics* **2022**, *11*, 165. [CrossRef]
6. Song, Y.; Zeng, L.; Liu, Z.; Song, Z.; Zeng, J.; An, J. Cross-Layer Optimization Spatial Multi-Channel Directional Neighbor Discovery with Random Reply in mmWave FANET. *Electronics* **2022**, *11*, 1566. [CrossRef]
7. Amphawan, A.; Arsad, N.; Neo, T.-K.; Jasser, M.B.; Mohd Ramly, A. Post-Flood UAV-Based Free Space Optics Recovery Communications with Spatial Mode Diversity. *Electronics* **2022**, *11*, 2257. [CrossRef]
8. Pfeiffer, R.; Valentino, G.; D'Amico, S.; Piroddi, L.; Galone, L.; Calleja, S.; Farrugia, R.A.; Colica, E. Use of UAVs and Deep Learning for Beach Litter Monitoring. *Electronics* **2022**, *12*, 198. [CrossRef]
9. Firdaus, A.R.; Hutagalung, A.; Syahputra, A.; Analia, R. Indoor Localization Using Positional Tracking Feature of Stereo Camera on Quadcopter. *Electronics* **2023**, *12*, 406. [CrossRef]
10. Koukiou, G.; Anastassopoulos, V. UAV Sensors Autonomous Integrity Monitoring—SAIM. *Electronics* **2023**, *12*, 746. [CrossRef]

Article

UAV Sensors Autonomous Integrity Monitoring—SAIM

Georgia Koukiou * and Vassilis Anastassopoulos *

Electronics Lab., Physics Department, University of Patras, 26504 Patras, Greece
* Correspondence: gkoukiou@upatras.gr (G.K.); vassilis@upatras.gr (V.A.)

Abstract: For Unmanned Aerial Vehicles (UAVs), it is of crucial importance to develop a technically advanced Collision Avoidance System (CAS). Such a system must necessarily consist of many sensors of various types, each one having special characteristics and performance. The poor performance of one of the sensors can lead to a total failure in collision avoidance if there is no provision for the performance of each separate sensor to be continuously monitored. In this work, a Sensor Autonomous Integrity Monitoring (SAIM) methodology is proposed. The configuration of the sensors and their interaction is based on a fusion procedure that involves a total of five sensors. Accordingly, the performance of each one of the sensors is continuously checked against the combined (fused) operation of the other four. A complementary experiment with a total of four sensors, one of which had low performance, was also conducted. Experimental results reveal a reliable approach for Sensor Autonomous Integrity Monitoring (SAIM). The method can be easily extended to a larger number of sensors.

Keywords: sensor integrity; collision avoidance; sensor fusion; unmanned vehicles

Citation: Koukiou, G.; Anastassopoulos, V. UAV Sensors Autonomous Integrity Monitoring—SAIM. *Electronics* **2023**, *12*, 746. https://doi.org/10.3390/electronics12030746

Academic Editor: Cecilio Angulo

Received: 19 December 2022
Revised: 13 January 2023
Accepted: 30 January 2023
Published: 2 February 2023

1. Introduction

Unmanned Aerial Vehicles (UAVs) are employed for operations where pilots' lives may be at risk [1,2]. Research on UAVs for numerous applications is currently carried out [3–5]. UAVs embody various sensors for monitoring the environment and performing real-time decision-making [6]. Due to the fact that UAVs are intended to operate in hazardous environments, the most important issue is planning the path of their mission [7]. Accordingly, the need arises for on-board sensors to avoid collisions with any type of obstacle even with other UAVs [8,9]. Thus, sophisticated Collision Avoidance Systems (CASs) are of primary importance for the efficient performance of a UAV [10].

The collision avoidance procedure incorporates two stages [1]: the perception of the environment and, after that, the probable required action. Perception is carried out using various sensors, passive or active, while the required actions for collision avoidance are categorized as [1]:

- Geometric;
- Force-field;
- Optimized;
- Sense and Avoid.

The capability of having many sensors on board increases the possibility of improving the performance of a CAS system, but, on the other hand, it makes the UAV heavier and the required data processing time-consuming [11]. A high-performance CAS system should incorporate the monitoring of each of the sensors for optimal performance. Thus, a kind of Sensor Autonomous Integrity Monitoring (SAIM) is required in correspondence with the Receiver Autonomous Integrity Monitoring (RAIM) systems [12].

During the last decades, a procedure has been developed to assess the integrity of the global positioning system (GPS) signals recorded with a GPS receiver system. The technique is called Receiver Autonomous Integrity Monitoring (RAIM) [12–15]. The integrity of the

GPS receiver signals is of crucial importance in GPS applications, such as in aviation or marine navigation, which are characterized as safety critical. The GPS system does not incorporate any technical procedure to correct possible inaccurate information transmitted by its systems. Simultaneously, a receiver of the GPS signals cannot be aware of this inaccurate information and will not try to correct its navigational capabilities. For GPS receivers, RAIM availability usually is tested when fewer than 24 GPS satellites are available by employing mathematical prediction approaches implemented at ground stations.

Currently, Time-Receiver Autonomous Integrity Monitoring (T-RAIM) algorithms assess the reliability of the timing solution provided by a Global Navigation Satellite System (GNSS) timing receiver. A potential T-RAIM approach developed in a multi-constellation context is described in [12]. In this work, measurements from several GNSS constellations provide increased redundancy at the cost of increased system complexity. Integrity requirements for the airborne use of GPS are reviewed in [13]. This is followed by the description of a baseline fault detection algorithm which is shown to be capable of satisfying tentative integrity requirements. Preliminary performance results for the baseline fault detection algorithm are presented in [14] along with the potential of RAIM techniques for achieving GPS integrity. The focus of the paper in [15] is to implement a fault detection and exclusion algorithm in a software GPS receiver in order to provide timely warnings to the user when it is not advisable to use the GPS system for navigation. Several GPS-related systems also provide integrity signals separate from the GPS. Among these is the Wide Area Augmentation System (WAAS) [16], which is an air navigation aid developed by the Federal Aviation Administration to augment the GPS with the goal of improving its accuracy, integrity, and availability.

A cooperative integrity monitoring (CIM) algorithm is proposed in [17] which can be applied to many existing multi-sensor cooperative positioning algorithms. In [18], an integrity monitoring framework is proposed that can assess the quality of multimodal data from exteroceptive sensors. The problem of fault detection, isolation, and adaptation (FDIA) is addressed in [19] for navigation systems on board passenger vehicles.

In this paper, a method for monitoring the operational integrity of each one of the sensors on board the UAV is proposed. It is assumed that five sensors of the same type are employed on the UAV in order to combine their decisions regarding the physical quantities such as distance and angle required to assess the collision parameters. The method is based on comparing the decision of the sensor under inspection with the decision obtained after fusing the decisions of the other four sensors, the operation of which is assumed to be normal. The proposed Sensor Autonomous Integrity Monitoring (SAIM) approach in fact employs the property which emerges from decision fusion methods [20,21], which states that at least three sensors give a decision that is more reliable than the most reliable sensor of the three used in the fusion procedure. If the inspected sensor is assessed to convey poor or unreliable information, its decision is ignored or otherwise, the sensor is gated. A complementary experiment with a total of four sensors, one of which has low performance, was also conducted.

This manuscript is organized as follows. In Section 2, the basics of UAV collision avoidance sensors are given. The RAIM essentials are presented in Section 3. The proposed SAIM method for UAVs is analyzed in Section 4. The experimental results are presented in Section 5, while the conclusions are drawn in Section 6.

2. Basics of UAV Collision Avoidance Sensors

In order for an unmanned vehicle to be able to navigate autonomously avoiding obstacles, a series of procedures are necessary such as the detection of the obstacle, its avoidance, planning continuously the path to be followed, localization, and control systems management [22]. In general, the environment in which a UAV is operating is dynamic, and there are limitations for the on-board payload, which, however, is very crucial for the operation of the UAV. Furthermore, severe weather conditions can dramatically reduce the operational capabilities of the UAV. Obstacle detection and collision avoidance become more challenging tasks if multiple UAVs or multiple moving obstacles are considered [11].

The collision radius R_c determines the minimum distance from a UAV which is adequate to avoid a collision (Figure 1). The detection range is much larger (Figure 1), and the larger its length is, the better the UAV can resolve, detect, and avoid a collision.

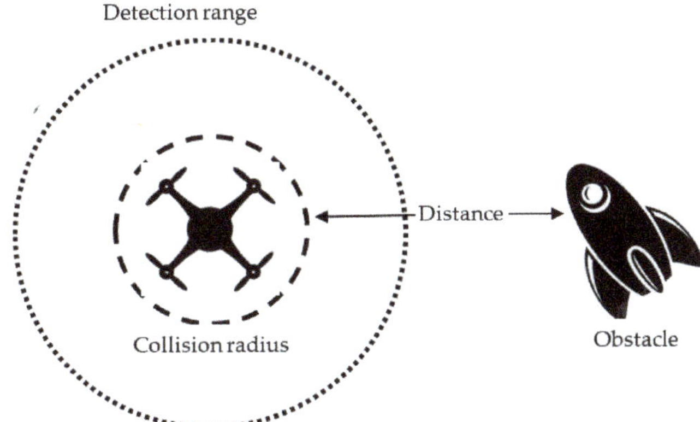

Figure 1. Detection range and collision radius.

A collision avoidance system (CAS) for a UAV is responsible for ensuring that no collisions with any obstacle will happen whether moving or stationary. In this work, we take advantage of the available information given for each sensor to sense the obstacle, and then we improve the detection phase of the system trying to assess the risk. Based on this, the collision avoidance module performs the necessary calculations to compute the amount of deviation needed from the original path to avoid the potential collision. Collision avoidance algorithms are categorized in [1] into the following major methods:

(1) *Geometric* methods, which work by computing the distance between the UAV and the obstacle.
(2) *Force field* methods, in which the main idea is inspired by attractive or repulsive electric forces that exist among charged objects. In a swarm of drones, each UAV node is considered a charged particle.
(3) *Optimization-based* methods, which aim at finding the optimal or near-optimal solutions for path planning.
(4) *Sense-and-avoid* methods, which mainly focus on reducing the computational cost with a short response time.

In this work, the geometric approach is followed. Geometric approaches rely on computing the time to collision by utilizing the distances between the UAVs and their velocities. In [23], the authors studied geometry-based collision avoidance strategies for a swarm of UAVs. The proposed approach uses line-of-sight vectors in combination with relative velocity vectors while considering the dynamic constraints of a formation. By calculating a collision envelope, each UAV can determine the available direction for avoiding a collision and decide whether the formation can be kept while avoiding collisions. In [24], the authors presented a new methodology of the Fast Geometric Avoidance Algorithm (FGA) based on kinematics, the probability of collisions, and navigational limitations by combining geometric avoidance and the selection of start time from critical avoidance. In a multiple obstacles scenario, instead of avoiding the obstacles simultaneously, FGA can assign different threat levels to obstacles based on the critical time for avoidance and avoid them sequentially and, hence, increase the avoidance success rate.

In the scenario of this work, the sensors are considered to be distributed in a very small space in the UAV. Usually, each sensor is of a different type from the others and monitors a different and complementary region. In our scenario, we assume that a specific region outside the UAV is monitored periodically by four or five active sensors simultaneously of different types (Radar, Lidar, Near-Infrared). Accordingly, all the sensors can be considered active, so that they will operate in any environmental and weather conditions. Furthermore, it is assumed that their quality characteristics are similar. The variation in the quality characteristics is mainly due to the different types of sensors (Radar, Lidar, Near-Infrared, etc.).

3. RAIM Essentials

Receiver Autonomous Integrity Monitoring (RAIM) approaches provide monitoring for the integrity of GPS signals which are employed in numerous aviation applications. In order to achieve integrity monitoring of the signals of a sensor, a minimum of five satellites visible to the sensor with satisfactory geometry must be utilized. The sensor or receiver sends a signal to the pilot regarding its integrity. It is very important for aviation receivers to possess RAIM capabilities to support safety issues. First of all, the receiver availability is a function of the geometry of the satellite constellation as well as the environmental conditions.

An enhanced version of RAIM employed in some receivers is known as fault detection and exclusion (FDE). It uses a minimum of six measurements which can be achieved with six satellites or five satellites with baro-aiding to not only detect a possible faulty satellite. The goal of fault detection is to detect the presence of a positioning failure. GNSS differs from traditional navigation systems because the satellites and areas of degraded coverage are in constant motion. Therefore, if a satellite fails or is taken out of service for maintenance, it is not immediately clear which areas of the airspace will be affected, if any. The location and duration of these deficiencies can be predicted with the aid of computer analysis and reported to pilots during the pre-flight planning process.

Because RAIM operates autonomously, that is without the assistance of external signals, it requires redundant pseudo-range measurements. To obtain a 3D position solution, at least four measurements are required. To detect a fault, at least five measurements are required, and to isolate and exclude a fault, at least six measurements are required. However, often more measurements are needed depending on the satellite geometry. Typically, there are seven to twelve satellites in view. The test statistic used is a function of the pseudo-range measurement residual (the difference between the expected measurement and the observed measurement) and the amount of redundancy. The test statistic is compared with a threshold value, which is determined based on the requirements for the probability of false alarm (P_{fa}) and the expected measurement noise. In aviation systems, the P_{fa} is fixed at $1/15,000$.

Various attempts have been made in the past to support autonomous integrity monitoring of sensors employed for collision avoidance. The cooperative integrity monitoring (CIM) algorithm proposed in [17] can fully exploit the global navigation satellite system (GNSS) data and inter-vehicle measurements data to improve the detection and isolation of faulty measurements due to multipath or non-line of sight (NLOS). An integrity monitoring framework that can assess the quality of multimodal data from exteroceptive sensors has been proposed in [18]. The proposed multisource coherence-based integrity assessment framework is capable of handling highway as well as complex semi-urban and urban scenarios. The work in [19] addresses the problem of fault detection, isolation, and adaptation (FDIA) in navigation systems on board passenger vehicles. It succeeds to prevent malfunctions in systems such as advanced driving assistance systems and autonomous driving functions that use data provided by the navigation system. The integrity of the estimation of the vehicle position provided by the navigation system is continuously monitored and assessed.

4. Proposed SAIM in UAVs

Sensor Autonomous Integrative Monitoring is crucial for the effective operation of the collision avoidance system of the UAV. In a similar manner as in RAIM, each sensor in the

UAV must check itself for the reliability of the decisions taken. If the decisions taken by a sensor are not reliable, then the specific sensor is to be deactivated. The reliability of the sensor is accessed by comparing its characteristics P_d and P_f with the performance at the fusion center.

The Sensor Autonomous Integrity Monitoring (SAIM) method proposed in this work is explained in four distinctive stages. It is assumed that five different sensors are on board to assess the distance and the relative movement of the obstacle with respect to the UAV. A complementary simulation experiment with a total of four sensors, one of which has low performance, was also conducted.

4.1. Performance Characteristics of Each Separate Sensor

Each separate sensor is measuring the geometric attributes of speed and distance of the obstacle while it is operating under specific quality characteristics. These characteristics are uniquely selected in this work to be the probability of detection P_{d_i} and the probability of false alarm P_{f_i} for the measured quantities. Consequently, we consider that we have a two-hypothesis testing problem with H_1 corresponding to obstacle presence and H_0 to obstacle absence (free space). There is no knowledge of a priori probabilities, so they are not taken into consideration, or otherwise, they are assumed equal.

4.2. Fusion Performance in Parallel Configuration

A Fusion Center (FC) operates on board and is responsible for two different tasks:

a.   Implement a Neyman–Pearson (N-P) test using all five decisions except one (the k-th, $k = 1$ to 5).

b.   Decide for gating the k-th sensor if a specific condition reports the k-th sensor as being unreliable.

Let u_j be the decision of the j-th sensor having considered all the observations regarding the obstacle distance and velocity. If the decision of the j-th sensor favors hypothesis H_1, then $u_j = +1$, otherwise, $u_j = 0$. All five decisions are transmitted to the Fusion Center and are available for processing. Let $(P_{f_j},\ P_{d_j})$ be the operating characteristics or otherwise the quality of the decision of the j-th sensor. The Fusion Center implements the N-P test using all four of the five decisions based on the Likelihood Ratio (LR) test:

$$\Lambda(u) = \frac{P(u_1, u_2,\ u_3, u_4 | H_1)}{P(u_1, u_2,\ u_3, u_4 | H_0)} \underset{H_0}{\overset{H_1}{\gtrless}} t \tag{1}$$

where $u = (u_1, u_2, u_3, u_4)$ is a 1×4 row vector with entries being the decisions of the four individual sensors (one is excluded), and t the threshold to be determined with the desirable probability of false alarm at the Fusion Center P_F^{fc}, [20], i.e.,

$$\sum_{\Lambda(u) > t^*} P(\Lambda(u) | H_0) = P_F^{fc} \tag{2}$$

Since the decisions of each sensor are independent from those of the other sensors, the LR test from (1) gives

$$\Lambda(u) = \prod_{i=1}^{N} \frac{P(u_1 | H_1)}{P(u_1 | H_0)} \underset{H_0}{\overset{H_1}{\gtrless}} t \tag{3}$$

For the implementation of the N-P, the computation of $P(\Lambda(u) | H_0)$ is required. Taking into consideration the independence of the sensors, we obtain, for easier evaluation, the distribution $P(log\Lambda(u) | H_0)$ which can be expressed as the convolution of the individual $P(log\Lambda(u_i) | H_0)$.

Accordingly, the LR $\Lambda(P_{d_i})$ assumes two values. Either $\frac{1-P_{d_i}}{1-P_{f_i}}$, when $u_i = 0$ with probability $1 - P_{f_i}$ under hypothesis H_0 and probability $1 - P_{d_i}$ under hypothesis H_1, or $\frac{P_{d_i}}{P_{f_i}}$, when $u_i = 1$ with probability P_{f_i} under hypothesis H_0 and probability P_{d_i} under hypothesis H_1.

At the Fusion Center, the probability of false alarm is obtained as

$$P_F^{fc} = \sum_{\Lambda(u)>t} P(\Lambda(u)|H_0) \tag{4}$$

where t is a threshold chosen to satisfy (4) for a given P_F^{fc}. Similarly, the probability of detection at the Fusion Center is

$$P_D^{fc} = \sum_{\Lambda(u)>t} P(\Lambda(u)|H_1) \tag{5}$$

We are interested to know if a configuration of four sensors can provide, by means of the N-P test, a $\left(P_F^{fc},\, P_D^{fc}\right)$ pair such that

$$P_F^{fc} \le \min_{i\in 1,\dots,4}\left\{P_{f_i}\right\} \text{ and } P_D^{fc} > \max_{i\in 1,\dots,4}\left\{P_{d_i}\right\} \tag{6}$$

In [20], it is proved, in the form of a theorem, that the condition (6) can be satisfied if the number of sensors N is greater than two, and all the sensors are characterized by the same $\left(P_f,\, P_d\right)$ pair.

Theorem 1 ([20]). *In a configuration of N similar sensors, all operating at the same $\left(P_f,\, P_d\right) = (p,\, q)$, the randomized N-P test at the Fusion Center can provide a$(P_F^{fc},\, P_D^{fc})$ satisfying (15) if $N \ge 3$.*

More precisely, for $N \ge 3$, the randomized N-P test can be fixed so that

$$P_F^{fc} = P_f = p \text{ and } P_D^{fc} > P_d = q \tag{7}$$

4.3. Conditions for a Sensor Gating

According to the previous theorem, if the Fusion Center is employing four sensors, it can give a probability of false alarm at the Fusion Center P_F^{fc} smaller than that of the probability of false alarm P_{f_5} at the fifth sensor under testing, and simultaneously a P_D^{fc} larger than that of the probability of detection P_{d_5} at the fifth sensor under testing. In this case, this fifth sensor can be considered as unreliable and be gated (excluded) for evaluating the Fusion Center performance.

4.4. Total Configuration of the SAIM Method

The configuration approach for applying the proposed SAIM method for each of the available sensors is as follows:

a. Each of the five sensors is excluded, and the performance of the Fusion Center is evaluated by employing the other four sensors. It is required that:

$$P_F^{fc} = min\, P_{f_i} \text{ and } P_D^{fc} > max\, P_{d_i}\ i = 1,\dots,4$$

b. Procedure a. is repeated for all five sensors.
c. In the case of dissimilar sensors, we have to record the performance of the Fusion Center when the k-th sensor is excluded.

d.   The SAIM configuration leads the specific k-th sensor among the five sensors to be autonomously disabled with a special signal for its participation in the decision of the Fusion Center when this sensor finds out that:

$$P_F^{fc} = P_{f_k}, \text{ and the corresponding } P_{d_k} << P_D^{fc}, \ i = 1, \dots 4 \tag{8}$$

This means that the k-th sensor cannot contribute significant information to the Fusion Center if its characteristic probability of false alarm is larger or equal to that achieved by the Fusion Center with four sensors while its probability of detection is well below the achieved corresponding probability of detection at the Fusion Center of the UAV. If condition (8) is not valid for any one of the five sensors, then all available sensors are employed to contribute to the Fusion Center for its final decision.

5. Experimental Results

The SAIM procedure is demonstrated experimentally by means of the numerical evaluation of the performance at the Fusion Center. For this purpose, five sensors are considered on board the UAV having the quality characteristics (P_{f_i} and P_{d_i}) as depicted in Table 1. As it was previously stated, the evaluation of the distribution $P(log\Lambda(u)|H_0)$, given the independence of the sensors, can be expressed as the convolution of the individual $P(log\Lambda(u_i)|H_0)$. Accordingly, the distribution $P(log\Lambda(u)|H_0)$ is graphically depicted in Figure 2. Only the first to fourth sensors have been considered for evaluating $P(log\Lambda(u)|H_0)$. In a similar way, the distribution of $P(log\Lambda(u)|H_1)$ can be expressed as the convolution of the individual $P(log\Lambda(u_i)|H_1)$, and this is graphically depicted in Figure 3. In this case as well, only the first to fourth sensors have been considered for evaluating $P(log\Lambda(u)|H_1)$. In Table 2 are given analytically all possible threshold positions at the Fusion Center and the achieved P_F^{fc} and P_D^{fc}.

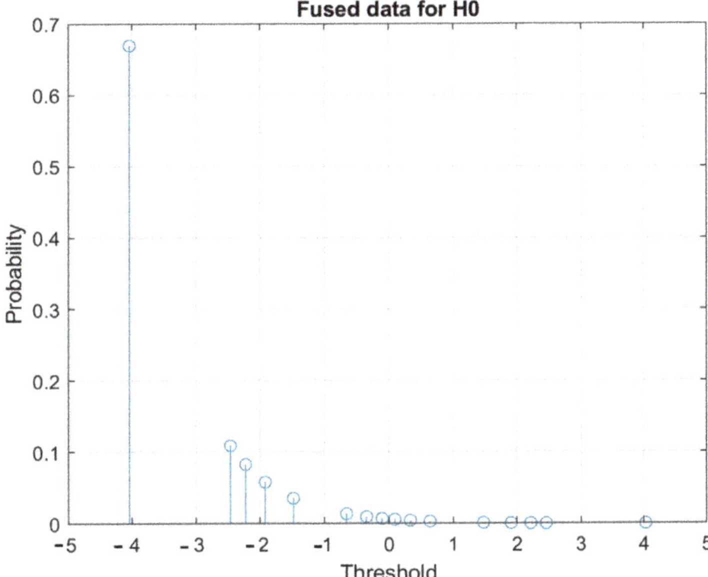

Figure 2. The discreet probability of false alarm $P(log\Lambda(u)|H_0)$ at the Fusion Center. Depending on the threshold abscissa, given in the second column of Table 2, the various values of P_F^{fc} in the third column of Table 2 are obtained.

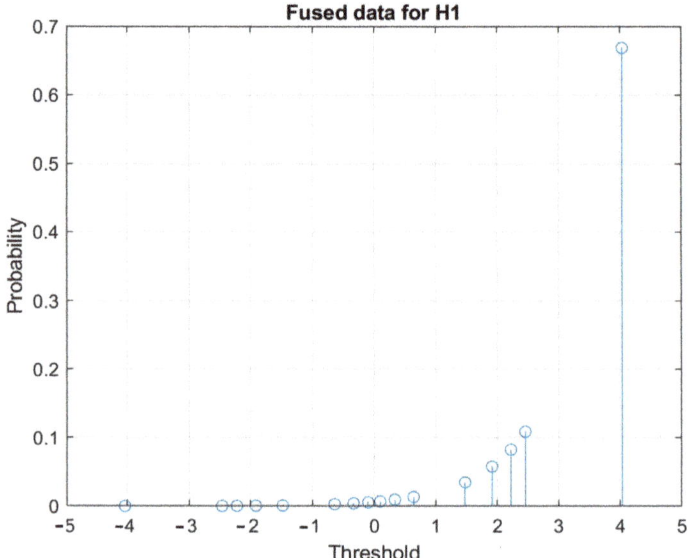

Figure 3. The discreet probability of detection $P(log\Lambda(u)|H_1)$ at the Fusion Center. Depending on the threshold abscissa, given in the second column of Table 2, the various values of P_D^{fc} in the fourth column of Table 2 are obtained.

Table 1. Quality characteristics (P_{f_i} and P_{d_i}) for each of the five sensors on board a single UAV.

Sensor	Sensor Probability of Detection	Sensor Probability of False Alarm
1st	0.95	0.05
2nd	0.92	0.08
3rd	0.89	0.11
4th	0.86	0.14
5th	0.83	0.17

Table 2. Detection performance at the Fusion Center on board the specific UAV with only four of the five sensors in Table 1 being active. The fifth sensor was excluded.

Threshold Number	Threshold Abscissa	Threshold Abscissa (log)	Cumulative Probability of Detection	Cumulative Probability of False Alarm
1	10859.733	4.035819	0.6689596	6.1600×10^{-5}
2	287.79175	2.459078	0.77786 00	0.0004400
3	165.89165	2.219825	0.8605404	0.0009384
4	82.115189	1.914423	0.9187108	0.0016468
5	30.082365	1.478312	0.9539192	0.0028172
6	4.3962633	0.643084	0.9673788	0.0058788
7	2.1761191	0.337683	0.9768484	0.0102304
8	1.2543792	0.098429	0.9840380	0.0159620
9	0.7972070	-0.09843	0.9897696	0.0231516
10	0.4595336	-0.33768	0.9941212	0.0326212
11	0.2274658	-0.64308	0.9971828	0.0460808
12	0.0332420	-1.47831	0.9983532	0.0812892
13	0.0121780	-1.91442	0.9990616	0.1394596
14	0.0060280	-2.21983	0.9995600	0.2221400
15	0.0034747	-2.45908	0.9999384	0.3310404
16	9.2083×10^{-5}	-4.03582	1.0000000	1.0000000

The achieved P_F^{fc} and P_D^{fc} for each threshold are evaluated by summing all the terms from the threshold position to the right of the distributions in Figures 2 and 3, respectively. According to the results in Tables 1 and 2, the quality characteristics of the fifth sensor are far exceeded by the performance of the Fusion Center, which employs the other four sensors (first to fourth). The 13th threshold, which is equal to 0.012, gives a $P_F^{fc} = 0.139$, which is smaller than the $P_{f5} = 0.17$. Simultaneously, for the same threshold, the $P_D^{fc} = 0.999$, which is very high in contrast to the P_{d5}, which equals 0.85. The results in Table 2 were obtained considering the fifth sensor in Table 1 as inactive.

Mathematic evaluations are carried out using simulation software built in MATLAB.

In an attempt to assess the information contribution of the fifth sensor to the final decision of the Fusion Center, this sensor was considered together with the other four to evaluate the P_F^{fc} and the P_D^{fc}. As depicted in Table 3, the fifth sensor does not add any significant information to the Fusion Center since at the threshold No. 28, the $P_F^{fc} = P_{f5}$ while the achieve P_D^{fc}, which is 0.9997, was reached using only the first four sensors (Table 2, threshold 13). Thus, the fifth sensor must be ignored since the information it conveys regarding the specific incident is poor or unreliable.

Table 3. Detection performance at the Fusion Center on board the specific UAV based on all five sensors in Table 1. The fifth sensor does not contribute to the Fusion Center performance.

Threshold Number	Threshold Abscissa	Cumulative Probability of Detection	Cumulative Probability of False Alarm
25	0.0294309	0.9991558	0.0760108
26	0.0169648	0.9994699	0.0945239
27	0.0068086	0.9996689	0.1237469
28	0.0024942	0.9997893	0.1720283
29	0.0012346	0.9998740	0.2406530
30	0.0007116	0.9999384	0.3310404
31	0.0004495	0.9999895	0.4447635
32	1.8860432	1	1

In order to further strengthen the reliability of the proposed SAIM method, another example is presented proving that a sensor with low performance has to be ignored in the decision fusion procedure in the Fusion Center. In this example, a total of four sensors are employed assuming that one of them has quite poor performance concerning its P_f and P_d. Accordingly, in Table 4 are given the operating characteristics of the four sensors with the fourth one being poorly-operating. In Table 5, the only threshold suitable for operating the Fusion Center with performance better than the best of the sensors is threshold No. 4. As it is shown in Table 6, the fourth poorly-operating sensor cannot add any important threshold discrimination capabilities since the eighth threshold is the same as the fourth one in Table 5, and it is equivalent to the ninth threshold.

Table 4. Quality characteristics (P_{f_i} and P_{d_i}) for each of the four sensors on board a single UAV. The fourth is poorly operating.

Sensor	Sensor Probability of Detection	Sensor Probability of False Alarm
1st	0.95	0.05
2nd	0.92	0.08
3rd	0.89	0.11
4th	0.60	0.40

Table 5. Detection performance at the Fusion Center on board the specific UAV with only three of the four sensors in Table 4 being active. The fourth sensor was excluded.

Threshold Number	Threshold Abscissa	Cumulative Probability of Detection	Cumulative Probability of False Alarm
1	1767.86363	0.77786	0.00044
2	27.00561	0.87400	0.00400
3	13.36758	0.94164	0.00906
4	4.89712	0.98258	0.01742
5	0.20420	0.99094	0.05836
6	0.07480	0.99600	0.12600
7	0.03702	0.99956	0.22214
8	0.00056	1	1

Table 6. Detection performance at the Fusion Center on board the specific UAV based on all four sensors in Table 4. The fourth sensor does not contribute to the Fusion Center performance.

Threshold Number	Threshold Abscissa	Cumulative Probability of Detection	Cumulative Probability of False Alarm
1	2651.79545	0.46671	0.00017
2	1178.57575	0.77786	0.00044
3	40.50842	0.83554	0.00186
4	20.05138	0.87612	0.00388
5	18.00374	0.91458	0.00602
6	8.91172	0.94164	0.00906
7	7.34569	0.96620	0.01240
8	3.26475	0.98258	0.01742
9	0.30630	0.98759	0.03379
10	0.13613	0.99094	0.05836
11	0.11221	0.99397	0.08541
12	0.05554	0.99611	0.12387
13	0.04987	0.99813	0.16445
14	0.02468	0.99956	0.22214
15	0.00084	0.99982	0.53328
16	0.00037	1	1

6. Conclusions

In this paper, the SAIM method for monitoring the operation integrity of each one of the sensors on board the UAV was presented. The assumed navigation scenario included the involvement of five sensors for monitoring collision avoidance. The five sensors have similar quality characteristics, and their decisions regarding the physical quantities such as the distance and angle required to assess collision parameters are combined at the Fusion Center of the UAV. The quality characteristics of each sensor is monitored against the performance at the Fusion Center when it employs the decision of the other four sensors. Experimental results reveal a reliable approach for Sensor Autonomous Integrity Monitoring. The method can be easily extended to a larger number of sensors.

According to the analyzed results in Tables 1 and 2, the quality characteristics of the fifth sensor are far exceeded by the performance of the Fusion Center, which employs the other four sensors (first to fourth). The information contribution of the fifth sensor, if it is taken into consideration in the final decision of the Fusion Center, is proved to be negligible due to the poor or unreliable performance of this specific sensor. This is evident from the comparison of the results in Tables 2 and 3, where the fifth sensor does not add any significant information to the Fusion Center and thus it must be ignored. A complementary experiment with a total of four sensors, one of which has low performance, was also conducted and resulted in similar conclusions as demonstrated in Tables 4–6.

The important requirement for applying the proposed SAIM method is that the Fusion Center on board the UAV must be continuously updated with the quality characteristics (probability of false alarm and probability of detection—(P_{f_j}, P_{d_j})) of each separate sensor. Thus, each sensor must transmit this information frequently to the Fusion Center. The quality characteristics of a specific sensor can be estimated based on the results of the sensor's monitoring with respect to the Fusion Center decisions.

Author Contributions: G.K. and V.A. have contributed equally to the present work. All authors have read and agreed to the published version of the manuscript.

Funding: This research received no external funding.

Institutional Review Board Statement: Not applicable.

Informed Consent Statement: Not applicable.

Data Availability Statement: Not applicable.

Conflicts of Interest: The authors declare no conflict of interest.

References

1. Yasin, J.N.; Mohamed, S.A.S.; Haghbayan, M.H.; Heikkonen, J.; Tenhunen, H.; Plosila, J. Unmanned Aerial Vehicles (UAVs): Collision Avoidance Systems and Approaches. *IEEE Access* **2020**, *8*, 105139–105155. [CrossRef]
2. Mcfadyen, A.; Mejias, L. A survey of autonomous vision-based see and avoid for unmanned aircraft systems. *Prog. Aerosp. Sci.* **2016**, *80*, 1–17. [CrossRef]
3. Goerzen, C.; Kong, Z.; Mettler, B. A survey of motion planning algorithms from the perspective of autonomous UAV guidance. *J. Intell. Robot. Syst.* **2009**, *57*, 65. [CrossRef]
4. Gageik, N.; Benz, P.; Montenegro, S. Obstacle detection and collision avoidance for a UAV with complementary low-cost sensors. *IEEE Access* **2015**, *3*, 599–609. [CrossRef]
5. Senanayake, M.; Senthooran, I.; Barca, J.C.; Chung, H.; Kamruzzaman, J.; Murshed, M. Search and tracking algorithms for swarms of robots: A survey. *Robot. Auton. Syst.* **2016**, *75*, 422–434. [CrossRef]
6. Zhang, W.; Zelinsky, G.; Samaras, D. Real-time accurate object detection using multiple resolutions. In Proceedings of the 2007 IEEE 11th International Conference on Computer Vision, Rio de Janeiro, Brazil, 14–21 October 2007; pp. 1–8.
7. Zhuge, C.; Cai, Y.; Tang, Z. A novel dynamic obstacle avoidance algorithm based on collision time histogram. *Chin. J. Electron.* **2017**, *26*, 522–529. [CrossRef]
8. Chao, H.; Cao, Y.; Chen, Y. Autopilots for small fixed-wing unmanned air vehicles: A survey. In Proceedings of the 2007 International Conference on Mechatronics and Automation, Harbin, China, 5–8 August 2007; pp. 3144–3149.
9. Vijayavargiya, A.; Anirudh, A.S.; Kumar, A.; Yadav, A.; Sharma, A.; Jangid, A.; Dubey, A. Unmanned aerial vehicle. *Imp. J. Interdiscipl. Res.* **2016**, *2*, 1747–1750.
10. Liu, Z.; Zhang, Y.; Yuan, C.; Ciarletta, L.; Theilliol, D. Collision avoidance and path following control of unmanned aerial vehicle in hazardous environment. *J. Intell. Robot. Syst.* **2019**, *95*, 193–210. [CrossRef]
11. Huang, S.; Teo, R.S.H.; Tan, K.K. Collision avoidance of multi unmanned aerial vehicles: A review. *Annu. Rev. Control* **2019**, *48*, 147–164. [CrossRef]
12. Farrell, J.L.; Van Graas, F. Receiver Autonomous Integrity Monitoring (RAIM): Techniques, Performance & Potential. In Proceedings of the 47th Annual Meeting of The Institute of Navigation, Williamsburg, VA, USA, 10–12 June 1991; pp. 421–428.
13. Gioia, C.; Borio, D. Multi-Constellation T-RAIM: An Experimental Evaluation. In Proceedings of the 30th International Technical Meeting of The Satellite Division of the Institute of Navigation (ION GNSS+), Portland, OR, USA, 25–29 September 2017; pp. 4248–4256.
14. Rakipi, A.; Kamo, B.; Cakaj, S.; Kolici, V.; Lala, A.; Shinko, I. *Red Book of RAIM, Special ION Publication on RAIM, Global Positioning System*; NAVIGATION, V, Institute of Navigation: Alexandria, Egypt, 1998.
15. Rakipi, A.; Kamo, B.; Cakaj, S.; Kolici, V.; Lala, A.; Shinko, I. Integrity Monitoring in Navigation Systems: Fault Detection and Exclusion RAIM Algorithm Implementation. *J. Comput. Commun.* **2015**, *3*, 25–33. [CrossRef]
16. Todd, W. Introduction to the Wide Area Augmentation System. *J. Glob. Position. Syst.* **2002**, *1*, 151–153. [CrossRef]
17. Xiong, J.; Cheong, J.W.; Xiong, Z.; Dempster, A.G.; Tian, S.; Wang, R. Integrity for Multi-Sensor Cooperative Positioning. *IEEE Trans. Intell. Transp. Syst.* **2021**, *22*, 792–807. [CrossRef]
18. Balakrishnan, A.; Florez, S.R.; Reynaud, R. Integrity Monitoring of Multimodal Perception System for Vehicle Localization. *Sensors* **2020**, *20*, 4654. [CrossRef] [PubMed]
19. Zinoune, C.; Bonnifait, P.; Guzmán, J.I. Sequential FDIA for Autonomous Integrity Monitoring of Navigation Maps on Board Vehicles. *IEEE Trans. Intell. Transp. Syst.* **2016**, *17*, 143–155. [CrossRef]

20. Thomopoulos, S.C.A.; Viswanathan, R.; Bougoulias, D.C. Optimal Decision Fusion in Multiple Sensor Systems. *IEEE Trans. Aerosp. Electron. Syst.* **1987**, *AES-23*, 644–653. [CrossRef]
21. Lampropoulos, G.A.; Anastassopoulos, V.; Boulter, J.F. Constant false alarm rate detection of point targets using distributed sensors. *Opt. Eng. J. SPIE* **1998**, *37*, 401–416. [CrossRef]
22. Foka, A.; Trahanias, P. Real-time hierarchical POMDPs for autonomous robot navigation. *Proc. IJCAI Workshop Reason. Uncertain. Robot* **2007**, *55*, 561–571. [CrossRef]
23. Seo, J.; Kim, Y.; Kim, S.; Tsourdos, A. Collision avoidance strategies for unmanned aerial vehicles in formation flight. *IEEE Trans. Aerosp. Electron. Syst.* **2017**, *53*, 2718–2734. [CrossRef]
24. Lin, Z.; Castano, L.; Mortimer, E.; Xu, H. Fast 3D collision avoidance algorithm for fixed wing UAS. *J. Intell. Robot. Syst.* **2020**, *97*, 577–604. [CrossRef]

 *electronics*

Article

Indoor Localization Using Positional Tracking Feature of Stereo Camera on Quadcopter

Ahmad Riyad Firdaus, Andreas Hutagalung, Agus Syahputra and Riska Analia *

Department of Electrical Engineering, Politeknik Negeri Batam, Kota Batam 29461, Indonesia
* Correspondence: riskaanalia@polibatam.ac.id

Abstract: During the maneuvering of most unmanned aerial vehicles (UAVs), the GPS is one of the sensors used for navigation. However, this kind of sensor cannot handle indoor navigation applications well. Using a camera might be the answer to performing indoor navigation using its coordinate system. In this study, we considered indoor navigation applications using the ZED2 stereo camera for the quadcopter. To use the ZED 2 camera as a navigation sensor, we first transformed its coordinates into the North, East, down (NED) system to enable the drone to understand its position and maintain stability in a particular position. The experiment was performed using a real-time application to confirm the feasibility of this approach for indoor localization. In the real-time application, we commanded the quadcopter to follow triangular and rectangular paths. The results indicated that the quadcopter was able to follow the paths and maintain its stability in specific coordinate positions.

Keywords: quadcopter; ZED 2 stereo camera; indoor localization; real-time application

1. Introduction

Research into unmanned aerial vehicles (UAVs), for example, the quadcopter type, has various purposes, including military applications [1], wildlife monitoring [2], agriculture [3], civilian purposes, for example, for the delivery of payload [4], or the recording of unreachable scenery [5]. With respect to configuration, a quadcopter can be categorized as "+" and "x" [6]. Even if the configuration is different, the quadcopter's performance is the same. To control this kind of UAV, researchers have introduced different types of control methods, including Fuzzy+PID, to control the attitude of an octocopter with the same configuration as a quadcopter [7], sliding mode control-based interval type-2 fuzzy logic [8], linear quadratic regulators (LQR) [9], fuzzy logic [10], H∞ control [11], adaptive control [12], gain scheduling [13], backstepping control [14], and PID control [15].

When a quadcopter has been controlled for the attitude and altitude, it can be ordered to maneuver by following a path or self-localizing in indoor or outdoor applications. For example, [16] used a GPS sensor to decentralize the localization to detect multiple quadcopters. Another localization method is the SLAM, which utilizes a Lidar sensor to reconstruct a 3D scene [17], or uses a camera to extract feature-based direct tracking and mapping (FDTAM) information to reconstruct a 3D scene in a real-time application [18]. Moreover, for indoor localization, [19] introduced the method called the time of arrival (ToA) or the time difference of arrival (TDoA), eliminating noise using a Gauss—Newton approach. This method resulted in a good precision. However, it generated acoustic noise. Another method, described in [20], introduced visual odometry by utilizing stereo vision; however, the visual odometry employed was sensitive to lights.

In [21], a hybrid acoustic was used based on time-code division multiple access (T-CDMA) and an optical module of a time-of-flight (TOF) camera for indoor positioning; however, the method uses an acoustic module to perform a 3D multilateration to estimate the position of the drone, which involves a significant calculation. To reduce the calculation

Citation: Firdaus, A.R.; Hutagalung, A.; Syahputra, A.; Analia, R. Indoor Localization Using Positional Tracking Feature of Stereo Camera on Quadcopter. *Electronics* **2023**, *12*, 406. https://doi.org/10.3390/electronics12020406

Academic Editor: Arturo de la Escalera Hueso

Received: 2 November 2022
Revised: 4 January 2023
Accepted: 10 January 2023
Published: 13 January 2023

involved for estimating the object coordinate, a ZED 2 stereo camera [22] can be used as this camera is equipped with positional tracking coordinates. Based on this feature provided by the stereo camera, in this investigation, we used a stereo camera to perform indoor localization. We evaluated the method used through experiments using real-time applications with different path patterns.

2. Related Work

The development of UAV navigation systems has been investigated for both outdoor and indoor applications. For outdoor applications, the global positioning system (GPS) sensor can be relied on to understand the environment surrounding the UAV during navigation. In [23], a GPS sensor was used to track waypoints based on a robot operating system (ROS) along with autopilot sensors, and a dense optical flow algorithm which were integrated for hovering and tracking in an outdoor environment. In [24], inertial devices and a satellite navigation system were combined to improve the fusion positioning, accuracy, and robustness. Another investigation used the Global Navigation Satellite System (GNSS) for an outdoor navigation system [25,26]. With respect to indoor applications, the GPS sensors are not able to transmit their position signal through a building; therefore, many investigators have introduced indoor localization methods utilizing several sensors. In [27], the fusion of a Marvelmind ultrasonic sensor and a PX4Flow flow camera was used to measure the position and optical flow for indoor navigation based on a robotic operation system (ROS).

Furthermore, in [28], an optical flow and Kalman filter were used to estimate the camera's position, then semantic segmentation was performed based on deep learning to determine the wall position in front of the drone. The authors of [29] employed the RFID received signal strength and sonar value to perform a localization in indoor applications, in cooperation with a vision system for landing procedures. In [30], a LIDAR-based 2D SLAM was used to enable the drone to understand the environment surrounding it in a simulation using MATLAB. Other investigators have sought to combine a stereo camera and ultrasonic sensor to detect a surrounded object and extract three-dimensional (3D) clouds for path planning. However, this method involves a limited field of view when handling the UAV movement; the system needs to change the heading before moving side-to-side or backwards, and, in addition, potentially requires heavy computational resources [31].

In contrast to [31], the present investigation was performed using only a single ZED 2 stereo camera to perform an indoor localization or navigation based on the features given by the camera. To use the stereo camera for an indoor localization, we first transformed the coordinates from the camera into a NED coordinate system. Then, we used this coordinate to estimate the position and maintain the stability of the UAV during hovering.

3. Materials and Methods

This section will describe the materials and methods used in this work in two parts: the coordinate transformation and the system architecture. The coordinate transformation will explain how to convert the coordinate from a rigid body perspective to the camera coordinate system. Then, the system architecture will explain the architecture of the drone to achieve an indoor localization, which consists of the stabilization control mode and the pose estimation system.

3.1. The Coordinate Transformation

In this work, we utilized the ZED 2 camera to help our quadcopter to maintain its position in a particular coordinate. One of the features of the ZED 2 camera is positional tracking, consisting of an IMU sensor. The positional tracking coordinate provided by the ZED 2 camera has the opposite direction as the quadcopter. We have to transform the camera coordinate into the NED system of the quadcopter. To convert this coordinate, first we need to derive it from the rigid body of the quadcopter. The rigid body rotation illustrated in Figure 1 represents the distance between two points, the indicate point and the total point the quadcopter has, as presented in Equation (1); to obtain the displacement orientation point, we set this output of this equation with a constant number. Additionally, we can assume the position in the 3D coordinate (X, Y, Z) as presented in Figure 2.

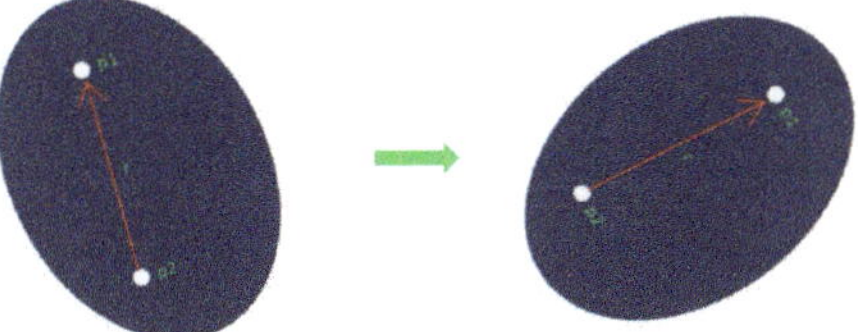

Figure 1. The rigid body position of r when it rotated to a certain angle.

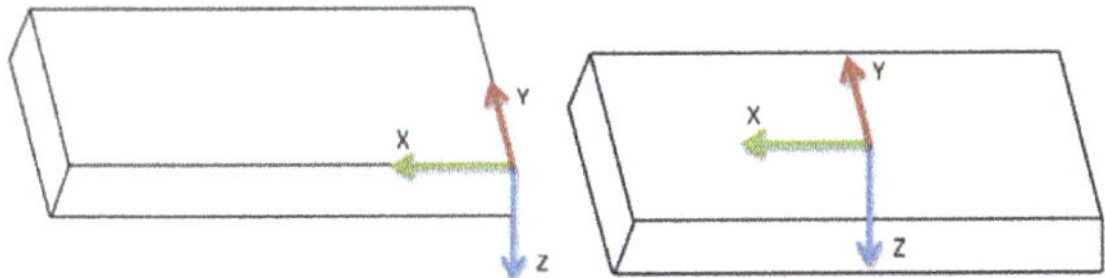

Figure 2. The illustration of the 3D coordinate system in the rigid body.

In order to determine the position in 3D coordinates, we set the frame of the reference as {O} and the body frame was represented as {B}. Because the body frame of the drone will change during the flight, the position can be obtained by comparing the body frame to the reference frame, as illustrated in Figure 3. In developing the quadcopter, we need to understand its orientation due to the rotation movement of the drone. We can describe the direction of a 3D rigid body by using the rotation matrix and Euler angle. The rotation matrix of the drone derives from Equations (2)–(4), and if we transpose the matrix using the left-hand rule, the equation becomes Equation (5).

$$\left| r_{p[i]} \right| = r_{p[i]} = constant \tag{1}$$

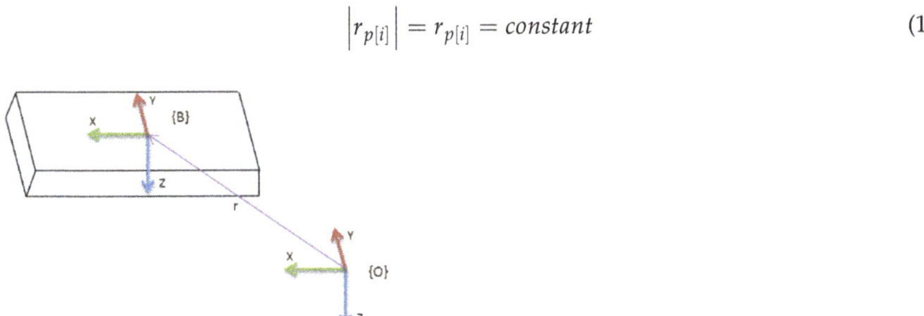

Figure 3. The coordinate position of the body frame and reference frame during displacement.

Another, in developing the quadcopter, we need to understand its orientation due to the rotation movement of the drone. We can describe a 3D rigid body's orientation using the rotation matrix and Euler angle. The rotation matrix of the drone is derived from Equations (2)–(4), and if we transpose the matrix using the left-hand rule, the equation is represented as Equation (5).

$$R_x(\theta) = \begin{bmatrix} 1 & 0 & 0 \\ 0 & \cos\theta & -\sin\theta \\ 0 & \sin\theta & \cos\theta \end{bmatrix} \tag{2}$$

$$R_y(\theta) = \begin{bmatrix} \cos\theta & 0 & \sin\theta \\ 0 & 1 & 0 \\ -\sin\theta & 0 & \cos\theta \end{bmatrix} \tag{3}$$

$$R_z(\theta) = \begin{bmatrix} \cos\theta & -\sin a\theta & 0 \\ \sin\theta & \cos\theta & 0 \\ 0 & 0 & 1 \end{bmatrix} \tag{4}$$

$$R_x(\theta)^T = \begin{bmatrix} 1 & 0 & 0 \\ 0 & \cos\theta & \sin\theta \\ 0 & -\sin\theta & \cos\theta \end{bmatrix} \tag{5}$$

From Equations (2)–(4), we can represent the orientation using the Euler angle rotation matrix. The Z, Y, and X transformed into α, β, and γ, derived from the ZYX Euler angle in Equation (6); therefore, the rotation matrix in each axis is represented in Equations (8)–(10), where Equation (7) denotes the identity matrix.

$$R(\alpha, \beta, \gamma) = I\,\mathrm{Rot}\left(\vec{z},\,\alpha\right)\cdot\mathrm{Rot}\left(\vec{y},\beta\right)\cdot\mathrm{Rot}\left(\vec{x},\,\gamma\right) \tag{6}$$

$$I = \begin{bmatrix} 1 & 0 & 0 \\ 0 & 1 & 0 \\ 0 & 0 & 1 \end{bmatrix} \tag{7}$$

$$\mathrm{Rot}\left(\vec{z},\,\alpha\right) = \begin{bmatrix} \cos a & -\sin a & 0 \\ \sin a & \cos a & 0 \\ 0 & 0 & 1 \end{bmatrix} \tag{8}$$

$$\mathrm{Rot}\left(\vec{y},\,\beta\right) = \begin{bmatrix} \cos\beta & 0 & \sin\beta \\ 0 & 1 & 0 \\ -\sin\beta & 0 & \cos\beta \end{bmatrix} \tag{9}$$

$$\mathrm{Rot}\left(\vec{x},\,\gamma\right) = \begin{bmatrix} 1 & 0 & 0 \\ 0 & \cos\gamma & -\sin\gamma \\ 0 & \sin\gamma & \cos\gamma \end{bmatrix} \tag{10}$$

This section aims to obtain the coordinate transformation from the camera view to the NED system, illustrated in Figure 4. In this work, we transform the coordinate using the homogeneous transformation, represented in Equations (11)–(13).

$$H = \begin{bmatrix} R & d \\ 0 & 1 \end{bmatrix} \tag{11}$$

$$R = \begin{bmatrix} i_x & j_y & k_z \\ i_x & j_y & k_z \\ i_x & j_z & k_z \end{bmatrix} \tag{12}$$

$$d = \begin{bmatrix} d_x, & d_y, & d_z \end{bmatrix}^T \tag{13}$$

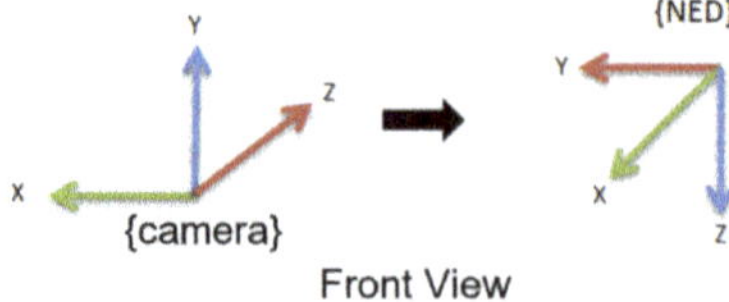

Figure 4. The illustration of transforming the camera coordinate to the NED system.

3.2. The System Architecture

The system architecture of our proposed method is illustrated in Figure 5. In this work, we utilized the ZED 2 stereo camera, which is equipped with the positional tracking feature. We also employed the Jetson Nano onboard computer to process the coordinate transformation, detect the object, and send the coordinate estimation to the flight controller.

The Pixhawk Cube Blank has been chosen as the flight controller to control the quadcopter orientation. This flight controller provided the ground control system software to witness the data transmission from the quadcopter movement and the camera input.

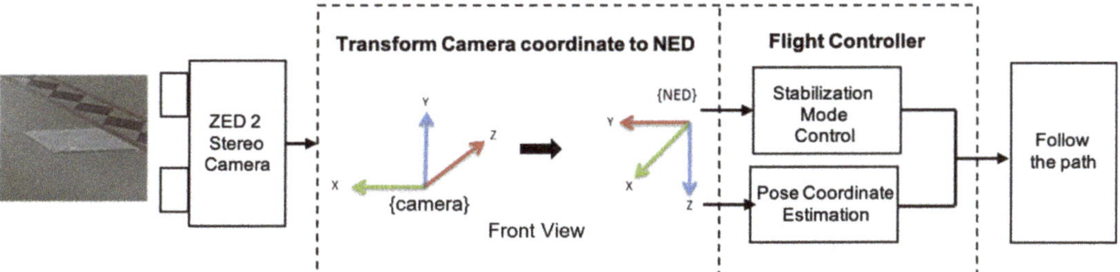

Figure 5. The system architecture of indoor localization.

In order to estimate the pose and let the quadcopter achieve a self-localization indoors, the camera will first detect the white square paper on the floor. Then, we collect the position, orientation, and confidence level from this sample image. Then, all this information collected from the camera will be sent to the Jeston Nano to process the coordinate transformation. The orientation and position data from the camera sent consists of 6 degrees of freedom: x, y, and z for the position vector and w, x, y, and z for the orientation. The process of transforming the camera coordinate into the NED frame is illustrated in Figure 6. This process was carried out on the onboard computer, where the number (2) from Figure 6 denoted the pose and confidence level data which is generated by the camera. Numbers (3) and (1) represented the offset configuration of the camera towards the flight controller to estimate the distance. To convert the camera coordinate into the NED frame, we utilized the homogenous transformation matrix, as presented in equation (11), by multiplying the value of offset camera configuration (1) and (3) with the pose and confidence level data.

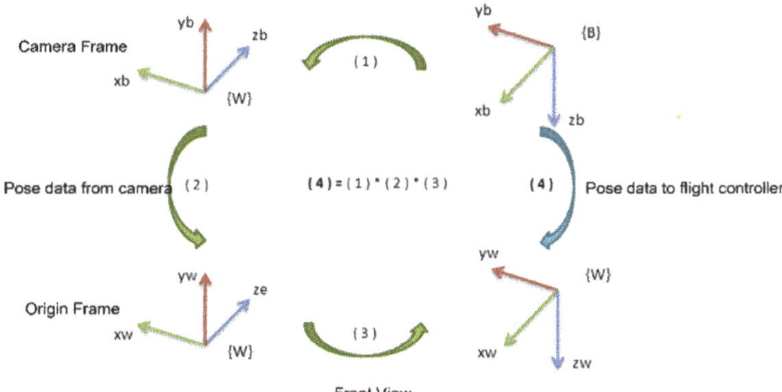

Figure 6. The illustration of the coordinate transformation process.

The results of this calculation are the rotation and translation of the quadcopter, which is represented in number (4).

When the coordinate transformation is ready, the data will be sent to the flight controller to activate the stabilization mode control. Because the flight controller is already firm regarding the attitude, we only send the camera's position and orientation. The altitude will be held at a particular position, which depends on the position data. In addition to the mode control, this data has also been used in the pose estimation. To make the drone follow the white square paper path on the floor, we combined the mode control stabilization and the pose estimation on the flight controller to ensure that the drone could move according to the path given.

4. Results

The experiment in this work was carried out in a real-time application. At first, we did experiments to understand how stable our system was when we pushed the drone using a rope. Then, we performed the indoor localization to follow the square and triangular paths. All the coordinates given in this experiment were measured in meters and were plotted in the cartesian mode.

4.1. The Stability Experiments

In the scheme for verifying the stability performance, we commanded the drone to remain hovering in some coordinates and to land at the same place after the remote control was released. In this experiment, the drone should stay at coordinate $(0, -0.5)$—this experiment's illustration is presented in Figure 7. As shown in Figure 7, first, we flew the drone using a remote control and commanded the drone to stay hovering at a specific position. After we released the remote control, the drone moved to the left and landed at the end of the track. This movement indicates that the drone cannot hover in the same place for some time.

Figure 7. The drone movement without activating the stability mode control.

The result of this experiment is represented in Figure 8. As seen in Figure 8, the graph was divided into three parts. The first graph was the drone position while hovering, printed in cartesian view, where the green line represented the track of the drone while hovering, the blue dot represented the take-off position, and the red dot denoted the landing position. The second one was the average error of X and Y generated by the drone movement, the red line represented the error on the X dan Y position, and the orange one indicated the average error of both positions. Additionally, the last was the time response of the X dan Y during the flight in a certain position, which has a blue line for the time response of the X and Y movement, and the orange represents the setpoint.

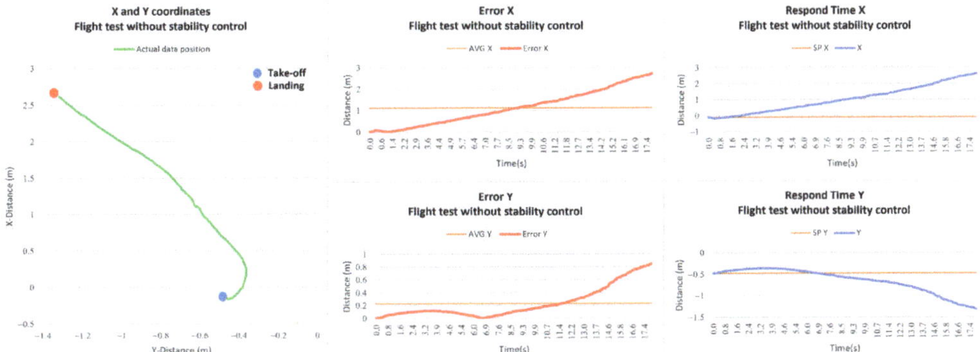

Figure 8. The response results from drone movement without activating the stability mode control.

From Figure 8, we can see that the drone was first taken off at the coordinate 0 m on the X-axis and about −0.5 m on the Y-axis. However, it dodged from the original position to the coordinates of 2.5 m on the X-axis and −1.4 m on the Y-axis. Additionally, then, the drone landed at these coordinates. From this movement, the average error produced is about 1 m on the X-axis and 0.2 m on the Y-axis. For the response system, the drone can only stay at coordinate 0 on the X-axis for about 2.4 s and the Y-axis for around 7 s. In this experiment, the drone failed to maintain its position if we turn off the control position mode.

Another stability experiment was conducted when the stability mode control was activated. In this experiment, we did the same scheme as presented in Figure 7, only in this experiment the coordinate position is about (0.08, −0,2) and lets the drone hover in this position. The result of this experiment is illustrated in Figure 9, where the drone was able to maintain its position, depicted on the first graph in Figure 9. The average error generated by the drone's movement is about 0.05 m for both the X- and Y-axes. The results of the time response position on the X and Y coordinate can be seen on the last graph, where the response results show that the drone was able to maintain its position with an acceptable error.

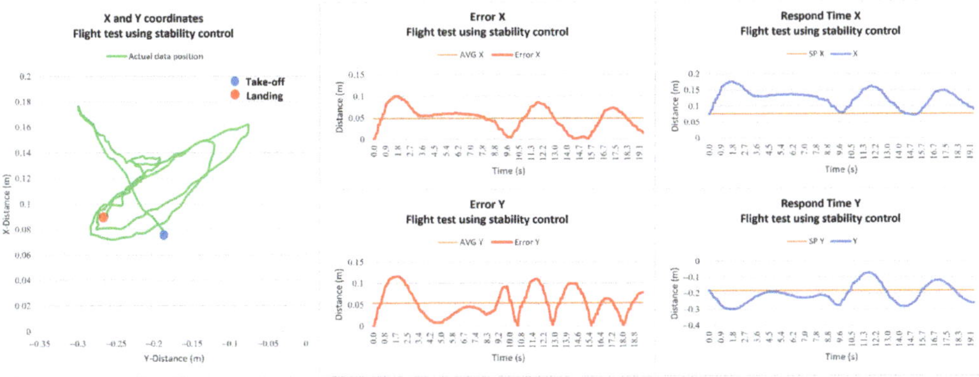

Figure 9. The response results from drone movement when activating the stability mode control.

To ensure our proposed position control worked well, we also compared the response results when we pulled the drone with the rope, as illustrated in Figure 10. In this experiment, we first flew the drone at some height in a particular coordinate. After the drone hovered, we pulled the rope to the right, as presented with the blue arrow, and the drone followed the rope path afterward; however, it landed away from its original position to the left. As for the response result represented in Figure 11, the original coordinate before the drone was pulled out is (0,0), yet after pulling out, the drone moved away from its original position and performed the angular path before landing at coordinate (0.5, −1).

Figure 10. Adding disturbance by pulling the rope during hovering without activating the stability mode control.

Figure 11. The response result from adding disturbance during hovering without stability mode control.

In comparison, we activated the stability mode control while pulling the drone during hovering, as presented in Figure 12. As we can see in Figure 12, the drone first flew at a certain height using the remote control. Then, after the remote was released in a few seconds, the drone was pulled on the Y-axis until the body tilted to the right; the drone condition after pulling the rope is depicted in the figure by the blue arrow. After a few seconds on the tilt condition, the drone succeeded in fixing its position to the original position. The graph response results are represented in Figure 13, where from the graph result, the drone was able to maintain its position after the disturbance given to it and survived the disturbance given.

Figure 12. Adding disturbance by pulling the rope during hovering by activating the stability mode control.

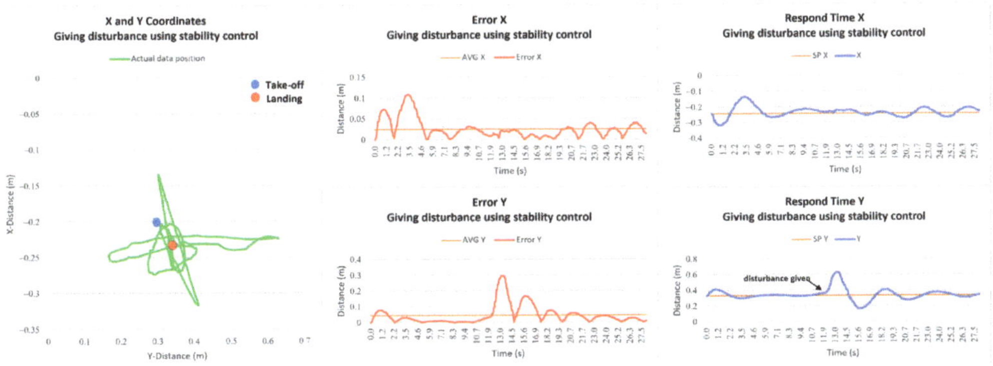

Figure 13. The response result from adding disturbance during hovering using stability mode control.

4.2. The Indoor Localization Experiments

Before conducting the indoor localization experiments, we first commanded the drone to move in certain positions and stay hovering several times by activating the stability mode control or without any activated stability mode control. Without activating the stability mode control, we commanded the drone to move from coordinate (0, 0.5) to coordinate (5, 0) and let the drone hover several times. The result of this experiment is represented in Figure 14, where we can see from the graph that the drone could move from the original point to the destination; however, it failed to stay hovering in the destination coordinate. The drone flew away from the end point coordinate and landed at the coordinate (8, −2.5); the average error generated at the X-axis is about 1.5 m and at Y-axis is approximately 1.8 m.

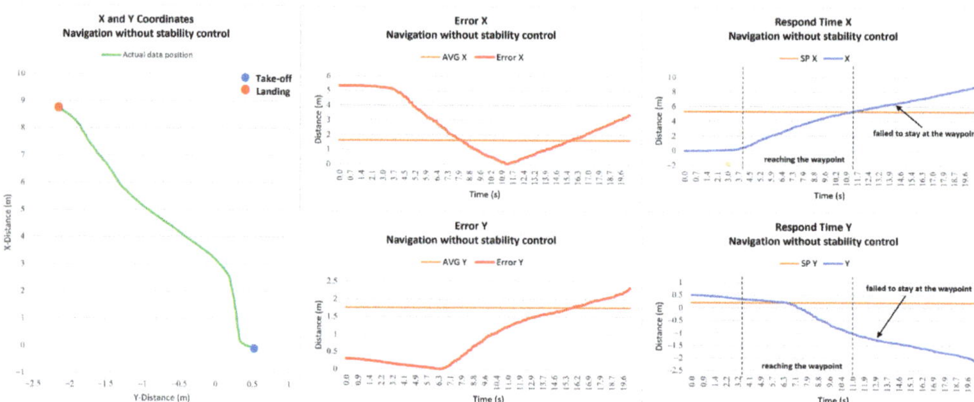

Figure 14. The response result of the simple indoor navigation without activating the stability mode control.

After activating the stability mode control, the drone can move from its original position to the destination as the system commands. On this occasion, we commanded the drone to move from coordinate (0, 1) to (5, 0.2). As seen in Figure 15, at first, the drone took off from its original position at coordinate (0, −1), then it tried to move to the destination at coordinate (5, 0.2) and stayed hovering in this end-point coordinate. Judging by the result given in Figure 15, it can be concluded that the drone was able to maintain its position due to the results of the average error at the X- and Y-axes at about 0 m and the drone remained hovering in this coordinate stably at a particular time. After ensuring that the drone can maintain its stability during hovering, the indoor localization practical can be verified. In this experiment, we ordered the drone to follow the rectangle and triangular path in a real-time application. To ensure that the drone moved accordingly to the path, we took the white rectangular paper with dimensions of about 40 cm × 40 cm and scattered it on the floor to resemble the triangular and rectangular path.

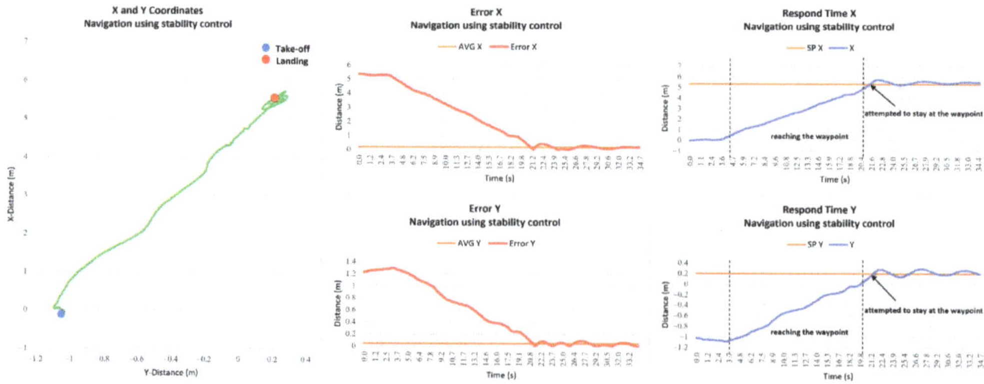

Figure 15. The response result of the simple indoor navigation by activating the stability mode control.

The rectangular path experiment is presented in Figure 16; in this experiment, we set the rectangular path as 5 m × 5 m using white paper indoors and let the drone be stopped and continued to hover in a particular coordinate, in this case, coordinates (0, 0), (5, 0), (5, 5), and (0, 5). As seen in Figure 16, a light spot above each white rectangular indicated that the ZED camera recognized the coordinate and commanded the drone to hover above the white rectangular. The response of this movement is represented in Figure 17; where the blue line denotes the path generated by the drone, the orange lines show the response system, and the red one presents the error during the maneuver. As for the response towards time for the X- and Y-axis, at first, the drone will remain hovering at coordinate (0, 0), then move to coordinate (5, 0), then coordinate (0, 5), and land at coordinate (0, 0) as the original position. Besides following the rectangular path, we resembled the triangular path using the same paper, which can be seen in Figure 18. The coordinates for the triangular path are (0, 0), (5, −4), (5, 4), and (0, 0) as the original position. The response system for this navigation is represented in Figure 19, where the graph produced errors which were almost the same as those of the rectangular path.

Figure 16. The rectangular path is to follow by the drone indoor.

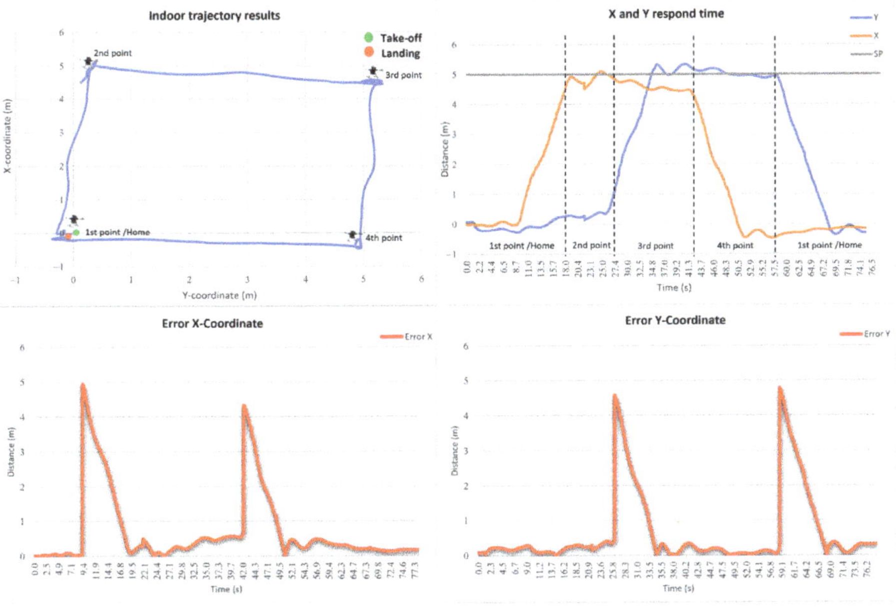

Figure 17. The response result of tracking the rectangular path indoors.

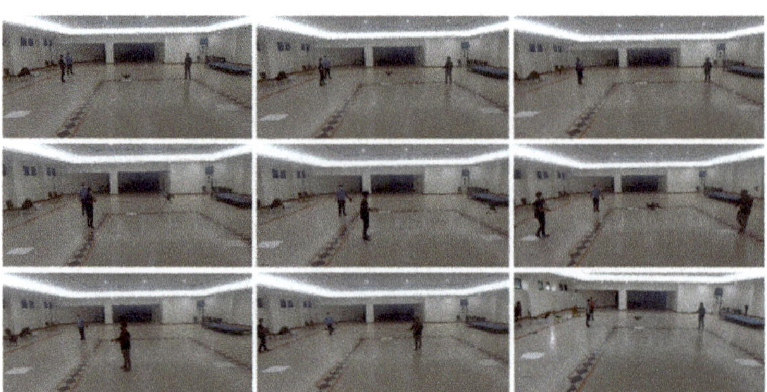

Figure 18. Tracking the triangular path indoor.

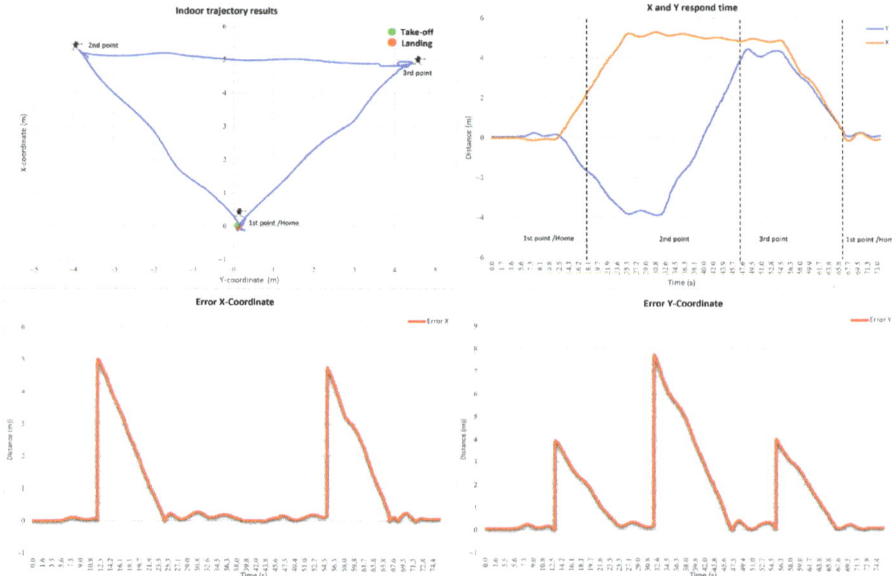

Figure 19. The response result of tracking the triangular path indoor.

5. Conclusions

This paper discussed alternative solutions to an indoor localization by using the features provided by the stereo camera. At first, we transformed the input camera coordinate into the NED system before performing the indoor localization, which aimed to control the drone's stability before being commanded to complete an indoor localization. All the experiments in this work have been done in a real-time application. The first experiment described that the drone could maintain its stability even if it added some disturbance. Moreover, for the next experiment, the drone succeeded in following both the rectangular and triangular path. However, following the path generated the error or oscillated it in particular coordinates. Therefore, in the future, we will focus on eliminating the oscillation which occurred during hovering to minimize the error.

Author Contributions: Conceptualization, A.R.F. and R.A.; methodology, A.R.F. and R.A.; validation, A.R.F. and R.A.; formal analysis, A.R.F. and R.A.; investigation, A.R.F. and A.H.; resources, A.H., A.S. and R.A.; data curation, R.A.; writing—original draft, R.A and A.R.F.; writing—review and editing, R.A and A.R.F.; visualization, R.A.; supervision, A.R.F. and R.A.; project administration, A.H. and A.S.; funding acquisition, A.R.F. All authors have read and agreed to the published version of the manuscript.

Funding: This research was fully funded by Politeknik Negeri Batam.

Data Availability Statement: Experimental video results can be found in #mdpi localization video.

Conflicts of Interest: The authors declare no conflict of interest.

References

1. Ostojić, G.; Stankovski, S.; Tejić, B.; Đukić, N.; Tegeltija, S. Design, control and application of quadcopter. *Int. J. Ind. Eng. Manag.* **2015**, *6*, 43–48.
2. Radiansyah, S.; Kusrini, M.D.; Prasetyo, L.B. Quadcopter applications for wildlife monitoring. *IOP Conf. Ser. Earth Environ. Sci.* **2017**, *54*, 012066. [CrossRef]
3. Mogili, U.R.; Deepak, B.B.V.L. Review on Application of Drone Systems in Precision Agriculture. *Procedia Comput. Sci.* **2018**, *133*, 502–509. [CrossRef]
4. Carrillo, L.R.G.; Rondon, E.; Dzul, A.; Sanchez, A.; Lozano, R. Hovering quad-Rotor Control: A comparison of nonlinear controllers using visual feedback. *IEEE Trans. Aerosp. Electron. Syst.* **2012**, *48*, 3159–3170. [CrossRef]
5. Hung, L.N.; Bon, L.S. A quadcopter-based auto cameraman system. In Proceedings of the 2016 IEEE Virtual Conference on Applications of Commercial Sensors (VCACS), Piscataway, NJ, USA, 15 June 2016–15 January 2017; pp. 1–8. [CrossRef]
6. Idrissi, M.; Salami, M.; Annaz, F. A Review of Quadrotor Unmanned Aerial Vehicles: Applications, Architectural Design and Control Algorithms. *J. Intell. Robot. Syst.* **2022**, *104*, 22. [CrossRef]
7. Analia, R.; Susanto; Song, K.-T. Fuzzy+PID attitude control of a co-axial octocopter. In Proceedings of the 2016 IEEE International Conference on Industrial Technology (ICIT), Taipei, Taiwan, 14–17 March 2016; pp. 1494–1499. [CrossRef]
8. Firdaus, A.R.; Tokhi, M.O. Robust sliding mode—Based interval Type-2 fuzzy logic observer for quadcopter UAVs. In Proceedings of the 2015 19th International Conference on System Theory, Control and Computing (ICSTCC), Cheile Gradistei, Romania, 14–16 October 2015; pp. 559–564. [CrossRef]
9. Everett, M.F. *LQR with Integral Feedback on a Parrot Minidrone*; Massachusetts Institute of Technology: Cambridge, MA, USA, 2015. Available online: http://mfe.scripts.mit.edu/portfolio/img/portfolio/16.31/16.31longreport.pdf (accessed on 20 August 2022).
10. Fedor, P.; Perduková, D. Use of Fuzzy Logic for Design and Control of Nonlinear MIMO Systems. In *Modern Fuzzy Control Systems and Its Applications*; IntechOpen: London, UK, 2017; p. 377. [CrossRef]
11. Jafar, A.; Fasih-UR-Rehman, S.; Fazal-UR-Rehman, S.; Ahmed, N.; Shehzad, M. A robust H∞ control for unmanned aerial vehicle against atmospheric turbulence. In Proceedings of the 2016 2nd International Conference on Robotics and Artificial Intelligence, Rawalpindi, Pakistan, 1–2 November 2016.
12. Pérez, R.; Galvan, G.; Vázquez, A.; Melo, S.; Alabazares, D. Attitude Control of a Quadcopter Using Adaptive Control Technique. In *Adaptive Robust Control Systems*; IntechOpen: London, UK, 2017.
13. Qiao, J.; Liu, Z.; Zhang, Y. Gain scheduling PID control of the quad-rotor helicopter. In Proceedings of the 2017 IEEE International Conference on Unmanned Systems (ICUS), Beijing, China, 27–29 October 2017; pp. 1594–1601.
14. Madani, T.; Benallegue, A. Backstepping control for a quadrotor helicopter. In Proceedings of the IEEE/RSJ International Conference on Intelligent Robots and Systems, Beijing, China, 9–15 October 2006; pp. 3255–3260.
15. Jiao, Q.; Liu, J.; Zhang, Y.; Lian, W. Analysis and design the controller for quadrotors based on PID control method. In Proceedings of the 33rd Youth Academic Annual Conference of Chinese Association of Automation (YAC), Nanjing, China, 18–20 May 2018.
16. Limbu, N.; Ahuja, I.; Sonar, H.; Solanki, S.; Jain, S.; Chung, H.; Chakraborty, D.; Narendra, L. Outdoor co-operative control of multiple quadcopters using decentralized GPS localisation. In Proceedings of the 2015 10th International Workshop on Robot Motion and Control (RoMoCo), Poznan, Poland, 6–8 July 2015; pp. 1–6. [CrossRef]
17. Gee, T.; James, J.; Van Der Mark, W.; Delmas, P.; Gimel'Farb, G. Lidar guided stereo simultaneous localization and mapping (SLAM) for UAV outdoor 3-D scene reconstruction. In Proceedings of the 2016 International Conference on Image and Vision Computing New Zealand (IVCNZ), Palmerston North, New Zealand, 21–22 November 2016; pp. 1–6. [CrossRef]
18. Wong, C.-C.; Vong, C.-M.; Jiang, X.; Zhou, Y. Feature-Based Direct Tracking and Mapping for Real-Time Noise-Robust Outdoor 3D Reconstruction Using Quadcopters. *IEEE Trans. Intell. Transp. Syst.* **2022**, *23*, 20489–20505. [CrossRef]
19. Tiemann, J.; Wietfeld, C. Scalable and Precise Multi-UAV Indoor Navigation using TDOA-based UWB Localization. In Proceedings of the International Conference on Indoor Positioning and Indoor Navigation (IPIN), Sapporo, Japan, 18–21 September 2017.
20. Fu, C.; Carrio, A.; Campoy, P. Efficient visual odometry and mapping for Unmanned Aerial Vehicle using ARM-based stereo vision pre-processing system. In Proceedings of the International Conference on Unmanned Aircraft Systems (ICUAS), Denver, CO, USA, 9–12 June 2015; pp. 957–962. [CrossRef]

21. Paredes, J.A.; Álvarez, F.J.; Aguilera, T.; Villadangos, J.M. 3D Indoor positioning of UAVs with spread spectrum ultra-sound and Time-of-Flight Cameras. *Sensors* **2018**, *18*, 89. [CrossRef] [PubMed]

22. Available online: https://www.stereolabs.com/ (accessed on 20 July 2022).

23. Lee, B.H.-Y.; Morrison, J.R.; Sharma, R. Multi-UAV control testbed for persistent UAV presence: ROS GPS waypoint tracking package and centralized task allocation capability. In Proceedings of the 2017 International Conference on Unmanned Aircraft Systems (ICUAS), Miami, FL, USA, 13–16 June 2017; pp. 1742–1750. [CrossRef]

24. Wang, R. IMM-EKF Based GPS/INS Combined Positioning Method for Drone. In Proceedings of the 2020 International Conference on Computing and Data Science (CDS), Stanford, CA, USA, 1–2 August 2020; pp. 160–164. [CrossRef]

25. Patrik, A.; Utama, G.; Gunawan, A.A.S.; Chowanda, A.; Suroso, J.S.; Shofiyanti, R.; Budiharto, W. GNSS-based navigation systems of autonomous drone for delivering items. *J. Big Data* **2019**, *6*, 53. [CrossRef]

26. Chen, M.; Xiong, Z.; Xiong, J.; Wang, R. A hybrid cooperative navigation method for UAV swarm based on factor graph and Kalman filter. *Int. J. Distrib. Sens. Netw.* **2022**, *18*, 1–14. [CrossRef]

27. Li, Y.; Scanavino, M.; Capello, E.; Dabbene, F.; Guglieri, G.; Vilardi, A. A novel distributed architecture for UAV indoor navigation. *Transp. Res. Procedia* **2018**, *35*, 13–22. [CrossRef]

28. Rahmaniar, W.; Wang, W.-J.; Caesarendra, W.; Glowacz, A.; Oprzędkiewicz, K.; Sułowicz, M.; Irfan, M. Distance Measurement of Unmanned Aerial Vehicles Using Vision-Based Systems in Unknown Environments. *Electronics* **2021**, *10*, 1647. [CrossRef]

29. Orgeira-Crespo, P.; Ulloa, C.; Rey-Gonzalez, G.; García, J.P. Methodology for Indoor Positioning and Landing of an Unmanned Aerial Vehicle in a Smart Manufacturing Plant for Light Part Delivery. *Electronics* **2020**, *9*, 1680. [CrossRef]

30. Ho, J.C.; Phang, S.K.; Mun, H.K. 2-D UAV navigation solution with LIDAR sensor under GPS-denied environment. *J. Physics Conf. Ser.* **2021**, *2120*, 012026. [CrossRef]

31. Zhou, Q.; Zou, D.; Liu, P. Hybrid obstacle avoidance system with vision and ultrasonic sensors for mul-ti-rotor MAVs. *Ind. Robot. Int. J.* **2018**, *45*, 227–236. [CrossRef]

 *electronics*

Article

Use of UAVs and Deep Learning for Beach Litter Monitoring

Roland Pfeiffer [1], Gianluca Valentino [1], Sebastiano D'Amico [2], Luca Piroddi [2,*], Luciano Galone [2], Stefano Calleja [1], Reuben A. Farrugia [1] and Emanuele Colica [2]

[1] Department of Communications & Computer Engineering, Faculty of Information and Communication Technology, University of Malta, MSD 2080 Msida, Malta
[2] Department of Geosciences, Faculty of Science, University of Malta, MSD 2080 Msida, Malta
* Correspondence: lucapiroddi@yahoo.it or luca.piroddi@um.edu.mt

Abstract: Stranded beach litter is a ubiquitous issue. Manual monitoring and retrieval can be cost and labour intensive. Therefore, automatic litter monitoring and retrieval is an essential mitigation strategy. In this paper, we present important foundational blocks that can be expanded into an autonomous monitoring-and-retrieval pipeline based on drone surveys and object detection using deep learning. Drone footage collected on the islands of Malta and Gozo in Sicily (Italy) and the Red Sea coast was combined with publicly available litter datasets and used to train an object detection algorithm (YOLOv5) to detect litter objects in footage recorded during drone surveys. Across all classes of litter objects, the 50%–95% mean average precision (mAP50-95) was 0.252, with the performance on single well-represented classes reaching up to 0.674. We also present an approach to geolocate objects detected by the algorithm, assigning latitude and longitude coordinates to each detection. In combination with beach morphology information derived from digital elevation models (DEMs) for path finding and identifying inaccessible areas for an autonomous litter retrieval robot, this research provides important building blocks for an automated monitoring-and-retrieval pipeline.

Keywords: beach litter; object detection; drone surveys; unmanned aerial vehicles (UAVs); deep learning; yolov5; geolocation; litter monitoring; beach cleaning; digital elevation models; unmanned aircraft systems

Citation: Pfeiffer, R.; Valentino, G.; D'Amico, S.; Piroddi, L.; Galone, L.; Calleja, S.; Farrugia, R.A.; Colica, E. Use of UAVs and Deep Learning for Beach Litter Monitoring. *Electronics* **2023**, *12*, 198. https://doi.org/10.3390/electronics12010198

Academic Editors: Umberto Papa, Marcello Rosario Napolitano, Giuseppe Del Core and Salvatore Ponte

Received: 1 December 2022
Revised: 23 December 2022
Accepted: 28 December 2022
Published: 31 December 2022

1. Introduction

Marine litter has been identified as a ubiquitous issue [1–4] with ecological [5] and socioeconomic impacts [6–9]. Marine litter is increasing in amount and results in a number of negative effects on marine flora and fauna [5], and research suggests more than 250,000 tons of plastic litter can be found in the world's oceans [1]. It follows various pathways, with one final sink for litter being the seafloor [10,11], where litter might accumulate either in its original form or in smaller pieces, which may result in the creation of microplastics due to fragmentation (or when already at a small initial size as primary microplastics) [12]. Alternatively, if the buoyancy of the litter remains high enough, it can accumulate along beaches and coastlines [13].

Monitoring and cleaning large areas repeatedly requires a substantial availability of personnel and a large number of person hours [14,15] and might not be possible in hard-to-reach areas. Therefore, airborne monitoring and the automatic retrieval of litter are important steps to streamline detection and mitigation efforts, reduce personnel costs and cover different types of terrain. This research was conducted within the scope of the BIOBLU project ("Robotic BIOremediation for coastal debris in BLUE Flag beach and in a Maritime Protected Area"), one part of which consisted of research establishing and evaluating the necessary components for a pipeline that automates these steps, while focusing on the aspects of litter detection using artificial intelligence and the geolocation of the detected items. This paper presents three essential components of this approach: drone surveys, object detection and geolocation of detected litter objects.

Drones or "unmanned aerial vehicles" (UAVs) can record footage at a much higher resolution than what can be achieved using aeroplane- or satellite-based surveys, as flights are conducted at a low altitude with high-resolution RGB cameras (20–48 MP). Therefore, UAV surveys can capture smaller objects that are usually not detected by aeroplane- or satellite-based surveys. In addition, UAV-based surveys can drastically cut costs when compared with these methods, but they come at the cost of lower coverage [16].

2. Materials and Methods

2.1. Artificial Intelligence for Object Detection

Object detection is the task of detecting the type as well as location (and maximum extent) of objects on an image [17]. Object detection algorithms (or models) can be binary (only detecting one type of object, e.g., "car") or multiclass (detecting multiple object categories, e.g., "person", "car", "traffic light").

One type of artificial intelligence algorithm commonly used for object detection tasks is the Convolutional Neural Network (CNN) [17]. A CNN consists of a series of convolutional and max pooling layers, which can extract features from images. These features are then typically passed on to a fully connected network. In order to train the object detection algorithm for the task of automatically detecting the objects of interest, a labelled dataset is required—i.e., a set of images in which the objects of interest have been labelled with rectangular boxes ("bounding boxes") corresponding to the object category.

In order to evaluate the performance of the trained model, this dataset is usually split into three parts: a training set, validation set and test set. The training set is used to train the model, and the performance of the trained model is then evaluated against the validation set. Depending on the training setup, the validation set may be used multiple times, and therefore the test set serves as a reference to evaluate the algorithm performance against data that it has never encountered before during training or validation. This gives an overview of how well the algorithm is able to "generalise"—i.e., to apply the behaviour learned during the training phase to new, unseen instances. If the performance on the test set is lower than that on the validation set, the algorithm suffers from "overfitting", which occurs when the model has been optimised to cater too closely to the characteristics of the validation data, and therefore struggles with new data that does not exhibit these same characteristics.

2.2. Object Detection Training Dataset Creation

The dataset used in the course of this research contained images from a variety of sources. Footage from drone surveys conducted on beaches in Malta and Gozo, Sicily, as well as along the Red Sea coast were used (see Figure 1 and Table 1) in addition to versions of existing datasets, including the TACO dataset [18] with manually adjusted classes and manually labelled versions of Kaggle datasets [19,20], as well as litter objects manually photographed by the authors using a Nokia X10 mobile phone with a 48 MP camera [21].

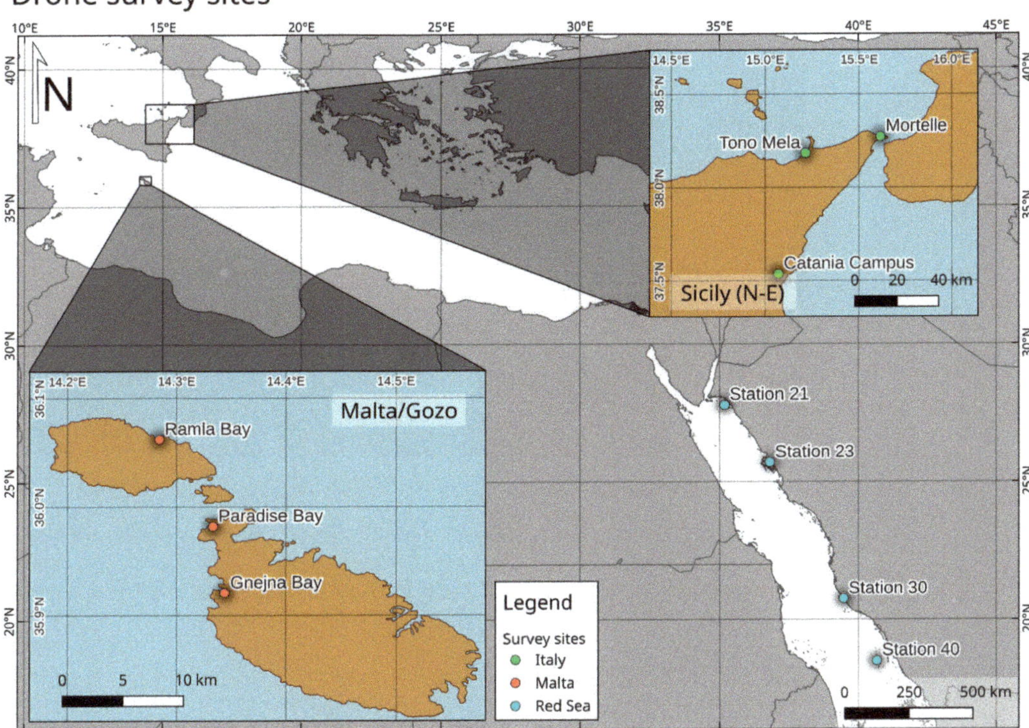

Figure 1. Main drone survey sites on Malta/Gozo (red dots) and Sicily (green dots), as well as survey locations of additional drone footage used for algorithm training that was recorded by [22] along the Red Sea coast (Coordinate Reference System: WGS 84 (EPSG:4326), background shapefile provided by [23]).

Table 1. Locations of survey sites (coordinates in WGS 84 (EPSG:4326)).

Location	Latitude	Longitude	Region	Reference/Source
Paradise Bay	35.981757	14.33372	Malta	Survey
Gnejna Bay	35.920815	14.344291	Malta	Survey
Ramla Bay	36.061839	14.284407	Malta	Survey
Tono Mela	38.185146	15.211505	Italy	Survey
Mortelle	38.273681	15.613148	Italy	Survey
Catania Campus	37.5369902	15.0698772	Italy	Survey
Station 21	27.785	35.1792	Red Sea	Martin et al. [22]
Station 23	25.7008	36.8118	Red Sea	Martin et al. [22]
Station 30	20.7501	39.4539	Red Sea	Martin et al. [22]
Station 40	18.5069	40.663	Red Sea	Martin et al. [22]

Surveys in Malta and Gozo (see Figure 2) were conducted using a DJI Phantom 4 Pro 2.0 (P4P2) drone equipped with a 20 MP RGB camera. Surveys in Italy were conducted using a DJI Mavic 2 Enterprise Advanced (M2EA) drone, equipped with a 48 MP RGB camera, and surveys at the Red Sea coast were conducted using a DJI Phantom 4 Pro [22]. Surveys were flown at a 10 m altitude, and footage was recorded in the form of still images (Malta, Gozo, Red Sea) or video (Italy). A 3D model of the survey site at Ramla Bay (Gozo) can be found at [24].

Figure 2. Setting up the DJI Phantom 4 Pro 2.0 (P4P2) drone for surveys in Ramla Bay (Gozo).

Since YOLOv5 requires images for training, still frames were extracted from drone survey videos at an interval of 1 image per second. In cases where multiple images showing the same location with little or no difference (e.g., when the drone was travelling at a slow speed), only one of those images was used in training in order to avoid duplicates.

Image Processing and Labelling

As using the drone survey images at full resolution for training initially exceeded the memory limits of the graphics processing unit (GPU) used for training, the images obtained from the drone surveys were cut into six square or near-square tiles (two rows, three columns of tiles) so that the resolution did not need to be reduced in the training process. The tile aspect ratio depends on the aspect ratio of the original image, and the maximum tile edge length was 1824 pixels.

Images were manually screened, and litter objects were labelled using the labelme software [25]. In addition, objects that were not litter but still prominent in the images were labelled as well to reduce ambiguity during training. In total, 67 classes were used for categorising labels. A table containing all 67 classes can be found in the Supplementary Materials in Table S1. For simplification and visualisation purposes, metaclasses were assigned based on their material and common waste separation schemes, and categories not aligning with these groups were classified as "Other". Instances that occurred in less than 10 images in the total dataset were assigned the metaclass "N img < 10". Grouped instance counts of different litter types can be seen in Figure 3. The number of annotations and images per class can be found in the Supplementary Materials in Table S1.

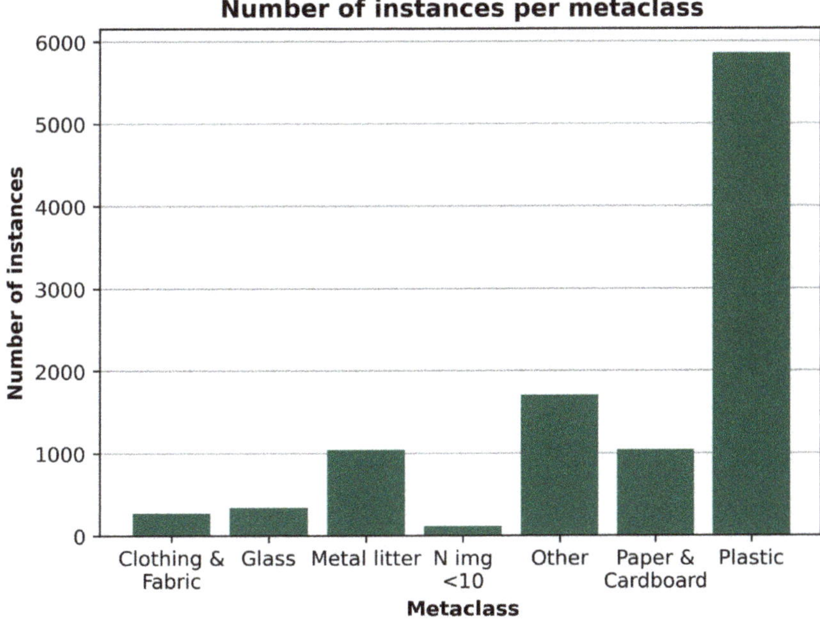

Figure 3. Number of annotations (i.e., instances) per material group. Bars depict meta groups that the 67 label classes were assigned to based on litter materials. Classes that were not litter objects were categorised as "Other", and small classes that were present on fewer than 10 images were labelled "N img < 10".

After image selection and labelling, the dataset consisted of a total of 4126 images and 10,611 annotations, with 1154 images showing no litter objects ("background images"). The dataset was split into a training, validation and test portion of the proportions of 0.6, 0.2 and 0.2, respectively. For the number of images and annotations in each set, see Table 2.

Table 2. Annotation and image counts for the training, validation and test set.

Set	Images	Annotations	Background Images
Training	2476	6124	701
Validation	825	2190	229
Test	825	2297	224
TOTAL	4126	10611	1154

2.3. Object Detection Algorithm Training

Common algorithms for object detection tasks are Convolutional Neural Networks (CNNs). In this paper, the YOLOv5 [26] architecture was used (derived from the original YOLO algorithm developed in 2016 [27]), as its single-stage architecture allows for faster detection speeds than other commonly used two-stage detectors (e.g., Faster R-CNN) [28]. Two-stage detectors first produce region proposals indicating regions of interest, and then they conduct object detection on those regions in a second step, while single-stage detectors perform both tasks in one neural net. The YOLO algorithm—instead of using region proposals—handles the full image, covers it with a grid and lets each cell handle those predictions whose BBOX centres fall within that cell [27]. The YOLOv5 network uses a CSP-Darknet53 as the backbone network, i.e., a Darknet53 Convolutional Neural Network following a Cross Stage Partial (CSP) Network strategy [26]. This backbone part of the algorithm is mainly used for extracting features from the input image. The next stage, the neck of the YOLOv5 algorithm, aggregates the features and allows the algorithm to generalise well across different scales, and uses fast Spatial Pyramid Pooling (SPPF) and CSP-PAN, i.e., a Cross Stage Partial Path Aggregation Network (PAN) with BottleneckCSP [26]. The last part of the YOLOv5 network—the head—uses a YOLOv3 head and is responsible for producing the final output of the predictions: the predicted classes, the corresponding bounding boxes, and the confidence per prediction [26]. For an overview of the general YOLOv5 architecture, see Figure 4.

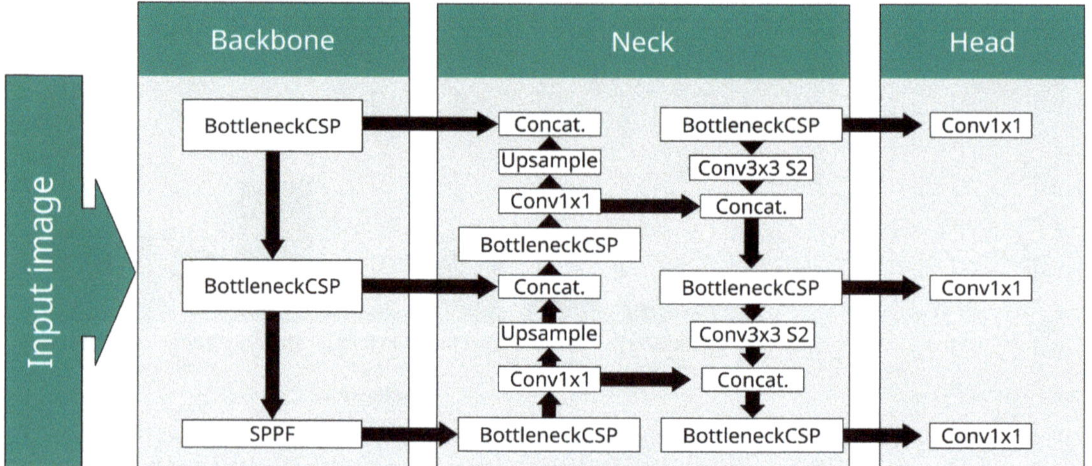

Figure 4. General architecture of the YOLOv5l network. The backbone consists of a Cross Stage Partial Darknet53 Convolutional Neural Network (CSP-Darknet53), responsible mainly for feature extraction. Spatial Pyramid Pooling (SPPF) provides feature pyramids that are used in the neck by a Cross Stage Partial Path Aggregation Network for feature aggregation. The head consists of a YOLOv3 head and provides the final output of the detector: the prediction classes, bounding boxes and confidence values.

Training was conducted on a cluster node running Ubuntu 20.04 LTS, CUDA version 11.4, NVIDIA driver version 470.141.03, utilising a NVIDIA A100 GPU with 80 GB GPU memory. For training, the yolov5l6.pt pretrained weights were used. The maximum number of epochs was set to 5000, the batch size was set to 16 and the image size was set to 1856 pixels, which allowed for the efficient use of the available GPU memory. All other settings were left at default values (specified in the YOLOv5 file "hyp.scratch-low.yaml" [26]).

The performance of the detection was measured in terms of precision (P), recall (R) and mean average precision (mAP). P is a measure of the likelihood of a detected object to have been detected correctly (i.e., is a measure of the accuracy of the predictions). R describes the proportion of objects that have been detected out of all objects that should have been detected (i.e., gives an estimate of the coverage of the algorithm).

P and R are calculated using ratios of True Positive (TP), False Negative (FN) as well as False Positive (FP) values (see Equations (1) and (2)). TP describes the proportion of correctly detected litter objects, FN describes the proportion of objects that were erroneously classified as the background (i.e., "missed" objects) and FP describes irrelevant background features that were erroneously detected by the algorithm.

$$P = TP/(TP + FP) \tag{1}$$

$$R = TP/(TP + FN) \tag{2}$$

Usually, P decreases with an increasing R [29]. If P is plotted as a function of R (a so-called precision–recall curve), the area under the curve (AUC) is a useful metric of assessing P over increasing R. In order to assess the performance of multiple classes, the average AUC across classes is calculated, resulting in the mAP metric. mAP values depend on the threshold of the overlap between the prediction and the ground truth box. This overlap is calculated using the Intersection-over-Union (*IoU*) ratio. *IoU* is calculated as depicted in Equation (3), where A and B are the ground-truth box and the prediction box, respectively.

$$IoU(A, B) = A \cap B / A \cup B \tag{3}$$

Predictions with bounding boxes that have an *IoU* value above a set threshold are regarded as correctly capturing the object, while boxes with an *IoU* below the threshold are considered *FP*. In this paper, we used the mAP metric of mAP@50-95 (average of mAPs of thresholds from 50% to 95%, in steps of 5%) to compare the performance of different classes.

A single metric that combines both P and R values is the *F1* score, which is the harmonic mean between P and R and is calculated using Equation (4) [17]. As such, the *F1* score is also a measure of whether P and R are both similarly high and penalises high differences between P and R [17].

$$F1 = 2/((1/P) + (1/R)) \tag{4}$$

2.4. Geolocation of Object Detections

In order to deliver useful information to a robot for debris retrieval, predictions made by the YOLOv5 algorithm need to be geolocated so that their coordinates can be communicated to the robot. In order to be able to retrieve coordinates for the predictions, the original footage needs to come with the GPS coordinates of the drone at the time of recording either embedded in the metadata (in case of image footage) or contained in a metadata-subtitle file (.srt, in the case of video footage).

After running the predictions, the results can then, in combination with the GPS information from the image metadata or video subtitle file, be georeferenced so that every prediction comes with a corresponding GPS coordinate. The geolocation of prediction boxes incorporates the following steps:

1. Calculation of pixel size of the footage.
2. Calculation of the distance (in m) between Meridians and Parallels at the latitude of recording.
3. Calculation of the horizontal and vertical distance of the prediction box centre from the image centre.
4. Transformation of the prediction distance to the real-world distance and the calculation of the prediction coordinates.

2.4.1. Pixel Size

A calculation of the real-world pixel size of the footage, also named the ground sampling distance (*GSD*), was conducted using Equation (5) provided by [30], which incorporates the drone camera's sensor width in millimetres (*Sw*), the flight altitude in metres (*a*), the focal length of the camera in millimetres (*f*, real focal length, not 35 mm equivalent) as well as the width of the recorded image in pixels (*imW*):

$$GSD = (Sw \cdot a \cdot 100) (f \cdot imW) \tag{5}$$

2.4.2. Distance between Meridians and Parallels

The distance between the Meridians and Parallels depends on the latitude, due to the spheroid shape of the Earth. Calculations for the geolocation are based on the WGS84 spheroid [31] with a semimajor axis of 6,378,137.0 metres (*a*) and first eccentricity of $8.1819190842622 \times 10^{-02}$ (*e*). From these values, the radii of curvature in metres along the Meridians (M) and Parallels (N) at latitude ϕ can be calculated using Equations (6) and (7), respectively, provided by [32]:

$$M = (a(1 - e^2))/(1 - e^2 \cdot sin^2\ \Phi)(3/2) \tag{6}$$

$$N = a/(1 - e^2 \cdot sin^2\ \Phi)(1/2) \tag{7}$$

From these, the distance in metres between Meridians (d_{lon}) and Parallels (d_{lat}) can be calculated using Equations (8) and (9), respectively:

$$d_{lon} = (N \cdot \pi)/180 \tag{8}$$

$$d_{lat} = (M \cdot \pi)/180 \tag{9}$$

2.4.3. Latitude and Longitude of Prediction Box

The horizontal and vertical offset in pixels of the prediction box centre (which is provided with the prediction from the algorithm) from the image centre along the latitude and longitude directions can be calculated using trigonometry, Euclidean distance calculations as well as d_{lat}, d_{lon} and GSD values (see Figure 5). For a visualisation of the workflow outlined above, see Figure 6.

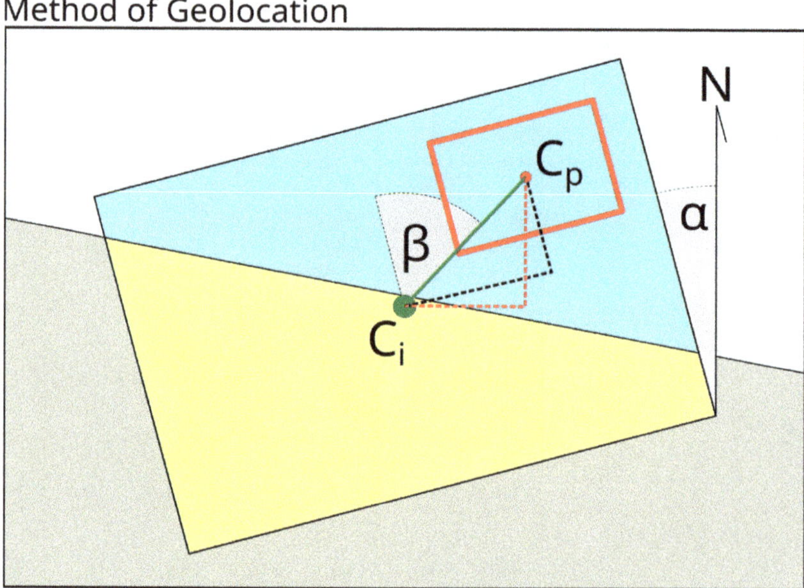

Figure 5. Schematic description of calculation of the location of the centre (Cp, red dot) of the prediction BBOX (red rectangle). The coordinates of the image centre (Ci, green dot) are known from image metadata. Horizontal and vertical offset of the prediction box centre (black dashed line) within the image are calculated from Cp and the image dimensions. Real-world offset of the BBOX centre along latitude and longitude (red dashed lines) is calculated using the Euclidean distance between Cp and Ci, the angle (β) of the Euclidean distance combined with the angle of drone yaw (α) and the previously calculated GSD-, d_{lat}- and d_{lon} values.

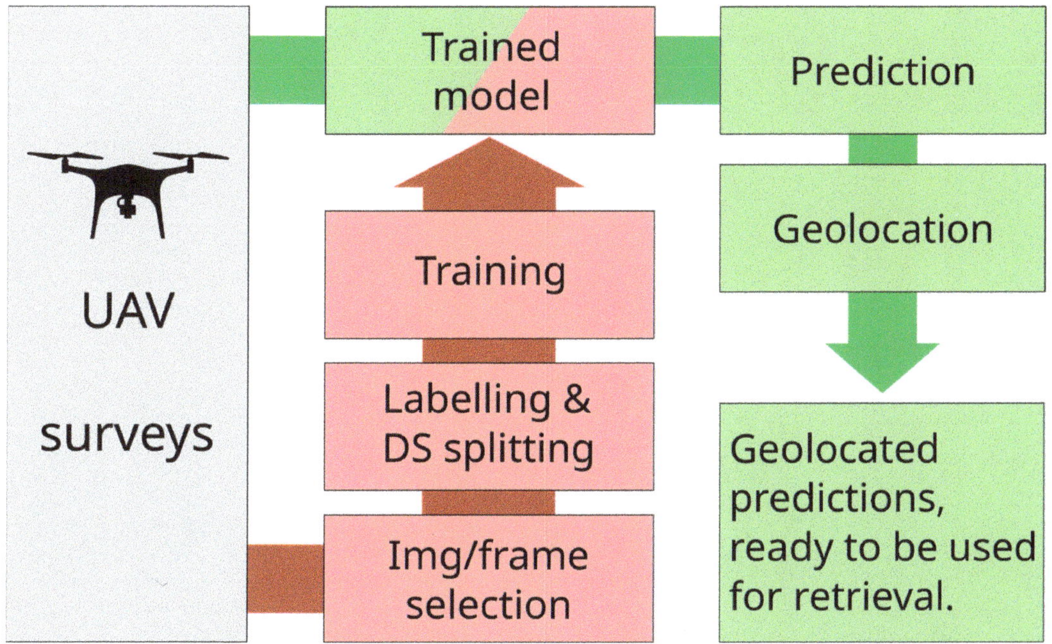

Figure 6. Visualisation of the general workflow with the two main pipelines of training (**red**) and prediction (**green**). For training, image footage from unmanned aerial vehicle (UAV) surveys was used (either as separate images or as frames from video file), and a subselection of image files was used to avoid objects appearing on multiple images. Images were subsequently labelled manually, and the algorithm was trained on the dataset, leading to a trained model. For prediction (**green**), this trained model could then be used on UAV survey footage (images or video footage) to predict litter objects. The predictions were then geolocated, leading to a set of geolocated predictions that are ready to be used in automated retrieval operations.

3. Results

Regarding the validation of the trained YOLOv5 algorithm against the test set of the labelled dataset, the performance metrics of the trained network across all classes were precision = 0.695, recall = 0.288, mAP50 = 0.314 and mAP50-95 = 0.252. Figure 7 provides a breakdown of the results, grouped by common waste separation categories. The overall *F1* score of the trained algorithm was 0.32 at a confidence of 0.235.

Considering that small classes are prone to overfitting [17] and the fact that we therefore focused on more abundant classes that appeared in more than 100 images, the top ten classes on which the algorithm performed most reliably were the categories "plastic bottle" (mAP50-95: 0.674), "metal can" ("mAP50-95: 0.516), "plastic bottlecap" (mAP50-95: 0.483), "plastic container" (mAP50-95: 0.447), "shoe" (mAP50-95: 0.43), "cardboard" (mAP50-95: 0.369), "pop tab" (mAP50-95: 0.366), "rope & string" (mAP50-95: 0.337), "wood" (mAP50-95: 0.324) and "glass bottle" (mAP50-95: 0.317). For the precision, recall and mAP50 values for these as well as the other classes, see Table S1 in the Supplementary Materials.

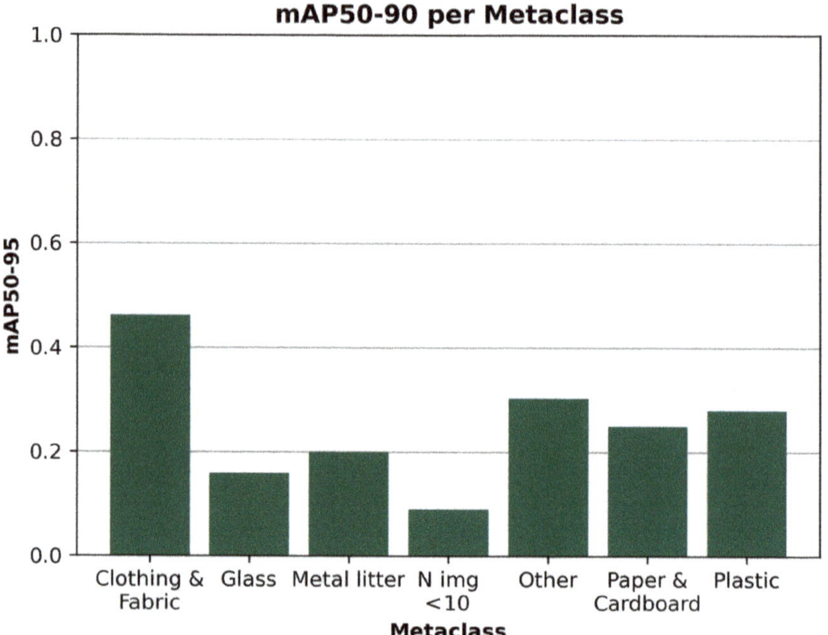

Figure 7. Mean average precision across intersect-over-union thresholds between 50% and 95% (mAP50-95) scores for grouped litter classes. Original classes were assigned metaclasses based on common waste separation protocols. Classes that were not common litter classes were grouped into "Other". Classes that occurred on fewer than 10 images in the dataset were assigned the metaclass "N img < 10".

For comprehensive, per-class mAP50-95 values, see Figure 8. Additionally, a table with the corresponding values can be found in the Supplementary Materials in Table S1. The per-class confusion matrix from the validation against the test set confirmed the above mentioned results and showed that the majority of the mislabelled objects were found in the "background" category, indicating False Negatives during the prediction (see Figure S1).

After the detections were geolocated, each detection was associated with latitude and longitude values. For an example of geolocated predictions from an image set recorded at Paradise Bay, see Figure 9.

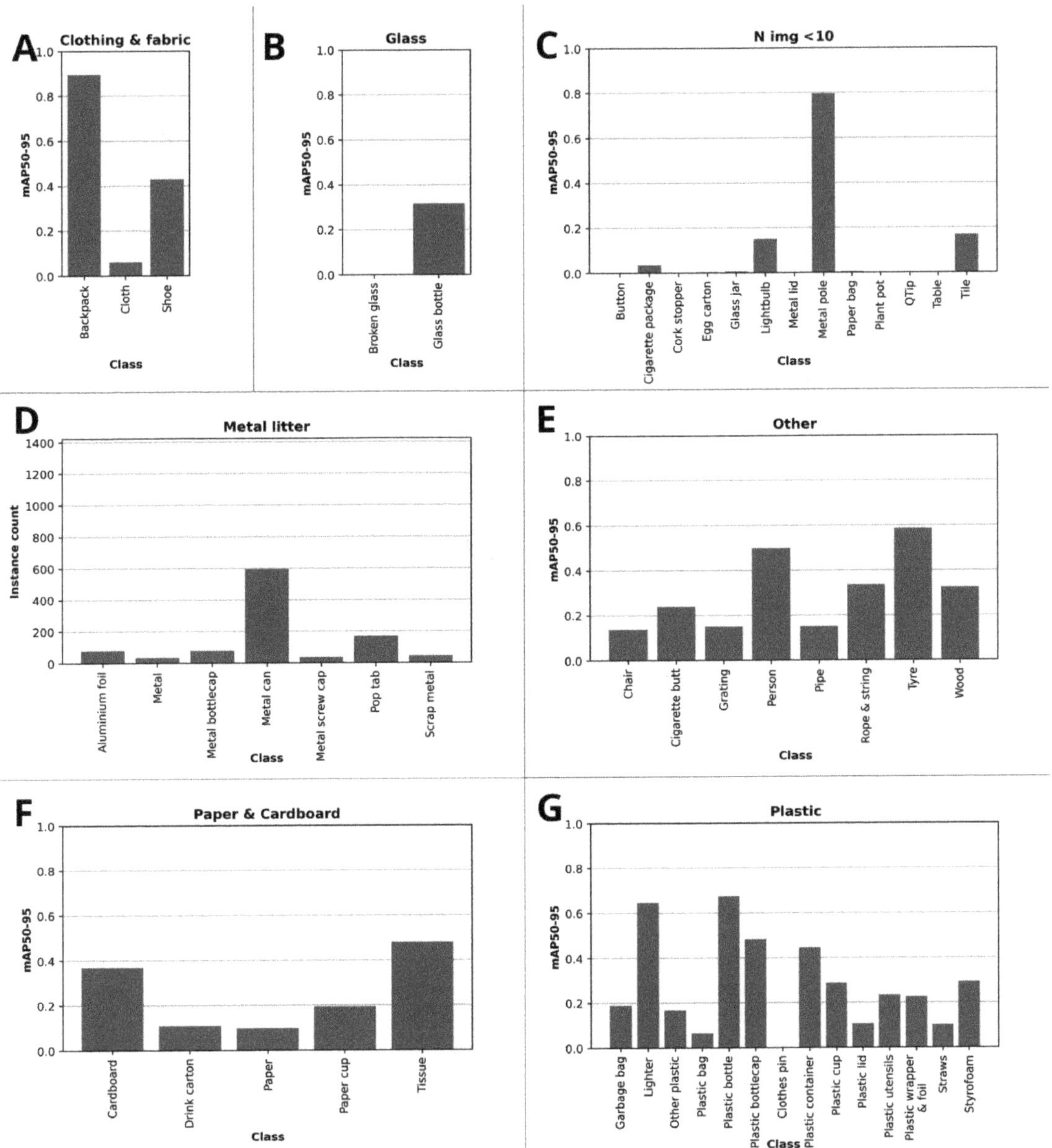

Figure 8. Per-class performance in subplots for each metaclass (**A–G**). X axes depict class name, y axes depict mean average precision across intersect-over-union thresholds between 50% and 95% (mAP50-95). Classes that were not litter objects were categorised as "Other", and small classes that were present on fewer than 10 images were labelled "N img < 10".

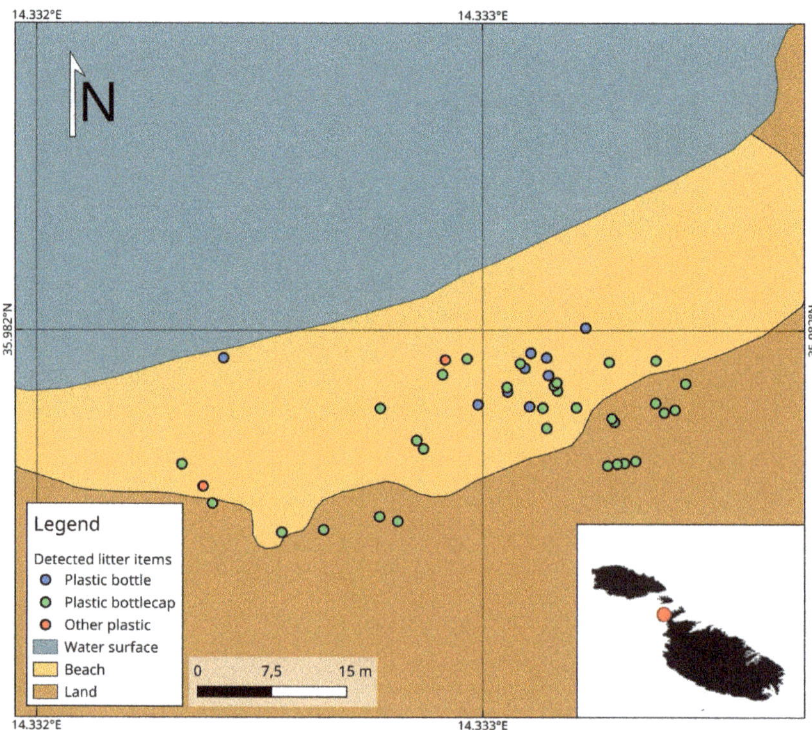

Figure 9. Geolocated object detections, colour-coded by their corresponding class, on Paradise Bay, at the northwestern coast of Malta. Coordinate Reference System: WGS 84 (EPSG:4326), overview shapefile provided by [23].

4. Discussion

The object detection and geolocation approach outlined in the sections above provides essential information that can be used for automatic retrieval. High-performance metrics were observed for classes that were frequent but also for some that were infrequent in the dataset. Since deep learning algorithms require large amounts of training data in order to be reliable and able to generalise [17], a recommendation would be to increase the amount of training data to bring classes that have a small number of images and/or instances in the training dataset to an even level compared to the larger classes. This would also make a comparison between the (previously imbalanced) classes more informative.

While drone surveys are mostly unaffected by ground morphology (except when encountering cliffs, for example, or flying at very low altitudes), automatic retrieval via robots requires detailed information about beach morphology, and in order to conduct ground surveys in a safe way, it is essential to identify those areas which are accessible to the robot and to distinguish them from inaccessible ones. This includes features such as boulders or rocky terrain, runoff channels or break-off edges due to erosion. Since these features can be subject to change over time, due to seasonality, weather events or long-term topography changes [33,34], it is important to collect up-to-date beach morphology information before deploying a robot for automatic retrieval missions.

UAVs are a useful platform for collecting beach morphology information. Common UAV-based methods include Lidar [35], a laser-based technique that has been shown to perform well for beach morphology monitoring and constructing digital elevation models (DEMs) [36,37], as well as Structure-from-Motion (SfM) photogrammetry, which utilises RGB images to recreate the 3D structure of the surveyed area and is commonly used for beach landform analyses and has been shown to perform well in comparison to aeroplane-based Lidar surveys by building high-resolution digital elevation models (DEMs) [38]. Furthermore, photogrammetry allows researchers to generate a high-resolution ortho-mosaic alongside the DEM, which facilitates the correct interpretation of the latter (see Figure 10).

Orthomosaic and Digital Elevation Model (DEM)

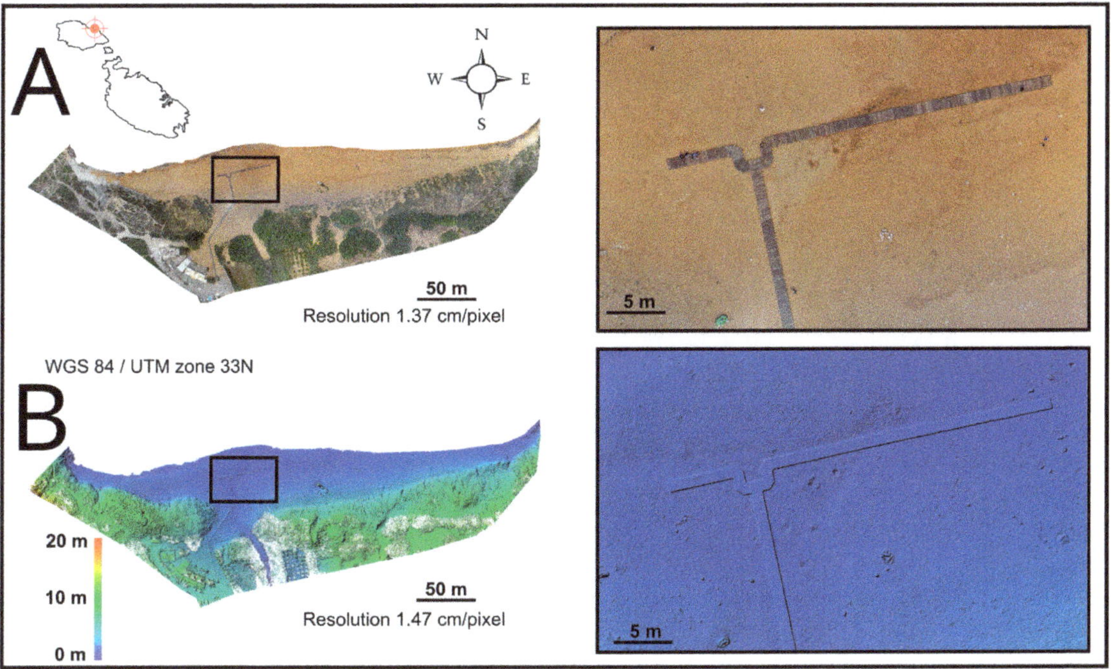

Figure 10. Examples of orthomosaics (**A**) and digital elevation models (**B**). Complete overviews on the left and zoomed-in sections on the right. Footage was collected in Ramla Bay Beach, Gozo.

The correct interpretation of the DEM is crucial when deploying autonomous robots for litter retrieval. The output derived from SfM photogrammetry can be used to detect inaccessible and/or impassable areas and optimise their path [36,37,39]. In addition, when surveys are repeated over time, DEMs and orthomosaics can be used to analyse beach morphology changes over time and assess local trends such as net losses or gains.

One example of using SfM photogrammetry for beach monitoring is a study by Colica et al. [38]. They used a DJI Phantom 4 Pro drone, equipped with a camera with a resolution of 20 Megapixels and 1″ Exmor R CMOS image sensor, to create a high-resolution DEM of Ramla Bay beach (Gozo). The acquisition interval of the images and the flight plan were programmed through the DJI Ground Station app, and the set parameters were an 85% forward and 70% side overlap of the images with a pixel resolution of 5472 × 3648 acquired at a flight altitude of about 60 m above sea level. The dataset includes 1021 nadiral images that were processed using the commercial software Agisoft Metashape [40], which allows one to set different parameters to control the photogrammetric reconstruction process including the accuracy parameter during image alignment, which controls the size and resolution of the images on which the software will detect the key points that are useful to the image alignment. In this phase, with the accuracy parameters set to the highest, a "sparse" point cloud containing approximately 722,000 points was produced. Subsequently, the 56 Ground Control Points (GCPs) measured with the Topcon HiPer HR

DGNSS receivers [41] in a Base + Rover configuration showed a horizontal accuracy of 3 mm ± 0.1 part per million and a vertical accuracy of 3.5 mm ± 0.4 part per million. Subsequently, the dense point cloud (about 94 million points) and depth maps with ultrahigh quality and mild filtering mode were calculated. From these, the DEM was then generated with a resolution of 1.47 cm/px and the orthomosaic with a resolution of 1.37 cm/px.

From the DEM, accessibility to a ground robot can be derived, e.g., by conducting a traversability analysis and providing a 2D costmap [39]. The costmap can be fed into a path-planning algorithm such as a D* algorithm [42] and then provide a series of waypoints to cover the accessible areas of interest as efficiently as possible (at a minimum cost, based on the cost map), as demonstrated by [39]. In addition, when surveys are repeated over time, DEMs should be constructed repeatedly as well in order to account for beach topology changes and identify newly inaccessible areas, for example. These repeatedly collected DEMs can also be used to analyse beach morphology changes over time and assess local trends such as net losses or gains.

Geolocating single frames from video footage or overlapping images from an image set will result in multiple detections of the same objects, as YOLOv5 does not natively allow for the tracking of objects. One possible approach to remedy these "double detections" is to cluster the geolocated points, e.g., by using clustering algorithms such as OPTICS or DBSCAN clustering [43,44], but clustering might not be possible on beaches where a high density of litter objects is present. In this case, using an algorithm that is capable of tracking objects across different images or video frames might be a worthwhile approach.

Clustering should also be considered when running object detection on drone video footage where the GPS recording frequency is not matched to the framerate of the recorded footage, as a mismatch of recording frequencies between the GPS and camera can distort GPS positioning along the UAVs flight direction.

5. Conclusions and Outlook

The object detection algorithm trained in the context of this paper performed well on recognising common beach litter categories such as plastic bottles, metal cans and plastic bottle caps, and the geolocation provided necessary information for later automatic retrieval when merged with additional information about accessibility derived from DEMs. For further improvement of the performance and reliability of the algorithm, more instances should be added to the classes that were underrepresented in the current dataset to reach a more balanced number of annotations and images per class. In order to reduce the number of double detections on overlapping footage, clustering and/or object tracking algorithms should be explored.

Supplementary Materials: The following supporting information can be downloaded at: https://www.mdpi.com/article/10.3390/electronics12010198/s1, Table S1: Name, number of annotations per class, number of images in which each class occurs, precision, recall, mean average precision at 50% intersect-over-union (IoU) threshold (mAP50) and at 50% to 95% IoU (in 5% steps, mAP50-95) values per class. Empty fields indicate classes that did not appear in the test set due to the low number of images available in the training set and the metaclass; Figure S1: Confusion matrix from the validation against the test set. The x-axis represents the actual categories of objects, while the y-axis represents the categories of the detections as predicted by the algorithm. Values in the matrix represent the proportion of predictions for detections of each category. Perfect predictions with no misclassifications will lead to a single line of 1.0 values running across the matrix from the top left to the bottom right. The fact that many misclassifications have been predicted as "background" and are therefore accumulating at the bottom of the matrix indicates False Negative (FN) predictions (i.e., missed objects that should have been detected).

Author Contributions: Conceptualization, G.V. and S.D.; methodology, G.V. and R.P.; software, R.P.; validation, R.P. and G.V.; formal analysis, R.P. and G.V.; investigation, R.P., G.V. and E.C.; resources, S.D., G.V. and E.C.; data curation, R.P. and S.C.; writing—original draft preparation, R.P.; writing—review and editing, R.P., S.D., G.V., L.P., L.G. and E.C.; visualization, R.P., E.C. and L.G.; supervision, G.V., S.D. and R.A.F.; project Administration, G.V.; funding acquisition, S.D., G.V. and R.A.F. All authors have read and agreed to the published version of the manuscript.

Funding: This research was financially supported by the BIOBLU project—Robotic BIOremediation for coastal debris in BLUE Flag beach and in a Maritime Protected Area (Code Area C2). Grant Number: J49C20000060007. Program INTERREG V A Italia-Malta 2014 2020 (Interreg V-A Cross Border Cooperation Italia-Malta projects). This work was partially supported by the projects "Coastal Satellite-Assisted Governance (tools, technique, models) for Erosion" (Coastal SAGE, grant agreement number SRF-2020-1S1) and "Satellite Investigation to study POcket BEach Dynamics" (SIPOBED, grant agreement number SRF-2021-2S1) financed by the Malta Council for Science & Technology (MCST, https://mcst.gov.mt/ (accessed on 30 November 2022)), for and on behalf of the Foundation for Science and Technology, through the Space Research Fund.

Data Availability Statement: The python scripts used in this study are publicly available under https://dsrg-ict.research.um.edu.mt/gianluca/bioblu (accessed on 30 November 2022). Trained weights for the object detection algorithm, as well as the labelled dataset and digital elevation models can be provided upon request but are not publicly available due to large file sizes.

Acknowledgments: The authors would like to thank Mario Vitti, Dario Guastella, Giovanni Cicceri, Giuseppe Sutera and Francesco Cancelliere for conducting drone surveys and collecting footage in Sicily, and for the valuable exchange of experience.

Conflicts of Interest: The authors declare no conflict of interest.

References

1. Eriksen, M.; Lebreton, L.C.M.; Carson, H.S.; Thiel, M.; Moore, C.J.; Borerro, J.C.; Galgani, F.; Ryan, P.G.; Reisser, J. Plastic Pollution in the World's Oceans: More than 5 Trillion Plastic Pieces Weighing over 250,000 Tons Afloat at Sea. *PLoS ONE* **2014**, *9*, e111913. [CrossRef] [PubMed]
2. Tekman, M.B.; Krumpen, T.; Bergmann, M. Marine Litter on Deep Arctic Seafloor Continues to Increase and Spreads to the North at the HAUSGARTEN Observatory. *Deep Sea Res. Part I Oceanogr. Res. Pap.* **2017**, *120*, 88–99. [CrossRef]
3. Bergmann, M.; Lutz, B.; Tekman, M.B.; Gutow, L. Citizen Scientists Reveal: Marine Litter Pollutes Arctic Beaches and Affects Wild Life. *Mar. Pollut. Bull.* **2017**, *125*, 535–540. [CrossRef] [PubMed]
4. Eriksson, C.; Burton, H.; Fitch, S.; Schulz, M.; van den Hoff, J. Daily Accumulation Rates of Marine Debris on Sub-Antarctic Island Beaches. *Mar. Pollut. Bull.* **2013**, *66*, 199–208. [CrossRef] [PubMed]
5. Kühn, S.; Bravo Rebolledo, E.L.; van Franeker, J.A. Deleterious Effects of Litter on Marine Life. In *Marine Anthropogenic Litter*; Bergmann, M., Gutow, L., Klages, M., Eds.; Springer International Publishing: Cham, Switzerland, 2015; pp. 75–116. ISBN 978-3-319-16510-3.
6. Mouat, J.; Lopez Lozano, R.; Bateson, H. *Economic Impacts of Marine Litter*; Kommunernes Internationale Miljøorganisation (KIMO): Lerwick, UK, 2010; 117p.
7. Ballancea, A.; Ryanb, P.G.; Turpieb, J.K. How Much Is a Clean Beach Worth? The Impact of Litter on Beach Users in the Cape Peninsula, South Africa. *S. Afr. J. Sci.* **2000**, *96*, 210–213.
8. Botero, C.M.; Anfuso, G.; Milanes, C.; Cabrera, A.; Casas, G.; Pranzini, E.; Williams, A.T. Litter Assessment on 99 Cuban Beaches: A Baseline to Identify Sources of Pollution and Impacts for Tourism and Recreation. *Mar. Pollut. Bull.* **2017**, *118*, 437–441. [CrossRef]
9. Williams, A.; Rangel-Buitrago, N.; Anfuso, G.; Cervantes, O.; Botero, C.-M. Litter Impacts on Scenery and Tourism on the Colombian North Caribbean Coast. *Tour. Manag.* **2016**, *55*, 209–224. [CrossRef]
10. García-Rivera, S.; Lizaso, J.L.S.; Millán, J.M.B. Composition, Spatial Distribution and Sources of Macro-Marine Litter on the Gulf of Alicante Seafloor (Spanish Mediterranean). *Mar. Pollut. Bull.* **2017**, *121*, 249–259. [CrossRef]
11. Ioakeimidis, C.; Zeri, C.; Kaberi, H.; Galatchi, M.; Antoniadis, K.; Streftaris, N.; Galgani, F.; Papathanassiou, E.; Papatheodorou, G. A Comparative Study of Marine Litter on the Seafloor of Coastal Areas in the Eastern Mediterranean and Black Seas. *Mar. Pollut. Bull.* **2014**, *89*, 296–304. [CrossRef]
12. Woodall, L.C.; Sanchez-Vidal, A.; Canals, M.; Paterson, G.L.J.; Coppock, R.; Sleight, V.; Calafat, A.; Rogers, A.D.; Narayanaswamy, B.E.; Thompson, R.C. The Deep Sea Is a Major Sink for Microplastic Debris. *R. Soc. Open Sci.* **2014**, *1*, 140317. [CrossRef]
13. Kusui, T.; Noda, M. International Survey on the Distribution of Stranded and Buried Litter on Beaches along the Sea of Japan. *Mar. Pollut. Bull.* **2003**, *47*, 175–179. [CrossRef]

14. Cruz, C.J.; Muñoz-Perez, J.J.; Carrasco-Braganza, M.I.; Poullet, P.; Lopez-Garcia, P.; Contreras, A.; Silva, R. Beach Cleaning Costs. *Ocean Coast. Manag.* **2020**, *188*, 105118. [CrossRef]
15. Martin, C.; Parkes, S.; Zhang, Q.; Zhang, X.; McCabe, M.F.; Duarte, C.M. Use of Unmanned Aerial Vehicles for Efficient Beach Litter Monitoring. *Mar. Pollut. Bull.* **2018**, *131*, 662–673. [CrossRef]
16. Muchiri, N.; Kimathi, S. A Review of Applications and Potential Applications of UAV. In Proceedings of the 2016 Annual Conference on Sustainable Research and Innovation, Juja, Kenya, 4 May 2016; p. 4.
17. Géron, A. *Hands-On Machine Learning with Scikit-Learn, Keras, and TensorFlow*, 2nd ed.; O'Reilly Media: Sebastopol, ON, Canada, 2019; ISBN 978-1-4920-3264-9.
18. Proença, P.F.; Simões, P. TACO: Trash Annotations in Context for Litter Detection. *arXiv* **2020**, arXiv:2003.06975.
19. Abid, M. Bottles and Cans Images. Available online: https://www.kaggle.com/datasets/moezabid/bottles-and-cans (accessed on 25 October 2022).
20. Abla, M. Garbage Classification (12 Classes). Available online: https://www.kaggle.com/datasets/126ab2c7f7e22add276bc29e4 4b97f635e3f6a04368afb20130a83518a9056b9 (accessed on 25 October 2022).
21. Nokia Nokia X10 Mobile. Available online: https://www.nokia.com/phones/en_int/nokia-x-10 (accessed on 25 October 2022).
22. Martin, C.; Zhang, Q.; Zhai, D.; Zhang, X.; Duarte, C.M. Enabling a Large-Scale Assessment of Litter along Saudi Arabian Red Sea Shores by Combining Drones and Machine Learning. *Environ. Pollut.* **2021**, *277*, 116730. [CrossRef]
23. US National Geospatial-Intelligence Agency. Administrative Boundaries World 1995. Available online: https://earthworks. stanford.edu/catalog/tufts-worldboundaries95 (accessed on 21 October 2022).
24. UM_GeoLab Ramla Bay May 2019—3D Model by UM_GeoLab (@UM_Geo_Lab). Available online: https://sketchfab.com/ models/f0d48f607b634fe4a2c8ab16d66c86ea/embed?autostart=1 (accessed on 26 November 2022).
25. Wada, K. Wkentaro/Labelme. Available online: https://github.com/wkentaro/labelme (accessed on 3 June 2022).
26. Ultralytics YOLOv5. Available online: https://github.com/ultralytics/yolov5 (accessed on 3 June 2022).
27. Redmon, J.; Divvala, S.; Girshick, R.; Farhadi, A. You Only Look Once: Unified, Real-Time Object Detection. *arXiv* **2016**, arXiv:1506.02640.
28. Ren, S.; He, K.; Girshick, R.; Sun, J. Faster R-CNN: Towards Real-Time Object Detection with Region Proposal Networks. *arXiv* **2016**, arXiv:1506.01497. [CrossRef]
29. Padilla, R.; Netto, S.L.; da Silva, E.A.B. A Survey on Performance Metrics for Object-Detection Algorithms. In Proceedings of the 2020 International Conference on Systems, Signals and Image Processing (IWSSIP), Niterói, Brazil, 1–3 July 2020; pp. 237–242.
30. Pix4D Support. TOOLS—GSD Calculator. Available online: http://support.pix4d.com/hc/en-us/articles/202560249-TOOLS-GSD-calculator (accessed on 16 June 2022).
31. US National Geospatial-Intelligence Agency. World Geodetic System 1984, Its Definition and Relationships with Local Geodetic Systems. Version 1.0.0; National Geospatial-Intelligence Agency (NGA) Standardization Document; 2014. Available online: https://earth-info.nga.mil/php/download.php?file=coord-wgs84 (accessed on 16 June 2022).
32. Bugayevskiy, L.M.; Snyder, J. *Map Projections: A Reference Manual*, 1st ed.; Taylor & Francis: London, UK, 1995; ISBN 978-0-429-15984-8.
33. Taveira-Pinto, F.; Silva, R.; Pais-Barbosa, J. Coastal Erosion Along the Portuguese Northwest Coast Due to Changing Sediment Discharges from Rivers and Climate Change. In *Global Change and Baltic Coastal Zones*; Schernewski, G., Hofstede, J., Neumann, T., Eds.; Coastal Research Library; Springer: Dordrecht, The Netherlands, 2011; pp. 135–151. ISBN 978-94-007-0400-8.
34. Bird, E.; Lewis, N. Causes of Beach Erosion. In *Beach Renourishment*; Bird, E., Lewis, N., Eds.; SpringerBriefs in Earth Sciences; Springer International Publishing: Cham, Switzerland, 2015; pp. 7–28. ISBN 978-3-319-09728-2.
35. National Oceanic and Atmospheric Administration What Is LIDAR. Available online: https://oceanservice.noaa.gov/facts/lidar. html (accessed on 8 November 2022).
36. Stockdonf, H.F.; Sallenger, A.H., Jr.; List, J.H.; Holman, R.A. Estimation of Shoreline Position and Change Using Airborne Topographic Lidar Data. *J. Coast. Res.* **2002**, *18*, 502–513.
37. Sallenger, A.H.; Krabill, W.B.; Swift, R.N.; Brock, J.; List, J.; Hansen, M.; Holman, R.A.; Manizade, S.; Sontag, J.; Meredith, A.; et al. Evaluation of Airborne Topographic Lidar for Quantifying Beach Changes. *J. Coast. Res.* **2003**, *19*, 125–133.
38. Colica, E.; Micallef, A.; D'Amico, S.; Cassar, L.F.; Galdies, C. Investigating the Use of UAV Systems for Photogrammetric Applications: A Case Study of Ramla Bay (Gozo, Malta). *Xjenza Online* **2017**, *5*, 125–131. [CrossRef]
39. Guastella, D.C.; Cantelli, L.; Melita, C.D.; Muscato, G. A Global Path Planning Strategy for a UGV from Aerial Elevation Maps for Disaster Response. In Proceedings of the 9th International Conference on Agents and Artificial Intelligence, Porto, Portugal, 24–26 February 2017; SCITEPRESS—Science and Technology Publications: Porto, Portugal, 2017; pp. 335–342.
40. Agisoft LLC Agisoft Metashape Professional Edition 2021. Available online: https://www.agisoft.com/features/professional-edition/ (accessed on 25 November 2022).
41. Topcon HiPer HR. Available online: https://www.topconpositioning.com/na/gnss-and-network-solutions/integrated-gnss-receivers/hiper-hr (accessed on 26 November 2022).
42. Stentz, A. *The D*Algorithm for Real-Time Planning of Optimal Traverses*; The Robotics Institute, Carnegie Mellon University: Pittsburgh, PA, USA, 1994; p. 34.

43. Ankerst, M.; Breunig, M.M.; Kriegel, H.-P.; Sander, J. OPTICS: Ordering Points to Identify the Clustering Structure. *SIGMOD Rec.* **1999**, *28*, 49–60. [CrossRef]
44. Ester, M.; Kriegel, H.-P.; Sander, J.; Xu, X. A Density-Based Algorithm for Discovering Clusters in Large Spatial Databases with Noise. *Knowl. Discov. Data Min.* **1996**, *96*, 1996.

Article

Post-Flood UAV-Based Free Space Optics Recovery Communications with Spatial Mode Diversity

Angela Amphawan [1,2,*], Norhana Arsad [3], Tse-Kian Neo [4], Muhammed Basheer Jasser [1,2] and Athirah Mohd Ramly [1,2]

[1] Photonics Laboratory, School of Engineering and Technology, Sunway University, Petaling Jaya 47500, Selangor, Malaysia; basheerj@sunway.edu.my (M.B.J.); athirahr@sunway.edu.my (A.M.R.)
[2] Elasticities Research Cluster, Sunway University, Petaling Jaya 47500, Selangor, Malaysia
[3] Faculty of Engineering & Built Environment, Universiti Kebangsaan Malaysia, Bangi 43600, Selangor, Malaysia; noa@ukm.edu.my
[4] Institute for Digital Education and Learning, Multimedia University, Cyberjaya 63100, Selangor, Malaysia; tkneo@mmu.edu.my
* Correspondence: angelaa@sunway.edu.my

Abstract: The deployment of unmanned aerial vehicles (UAVs) for free space optical communications is an attractive solution for forwarding the vital health information of victims from a flood-stricken area to neighboring ground base stations during rescue operations. A critical challenge to this is maintaining an acceptable signal quality between the ground base station and UAV-based free space optics relay. This is largely unattainable due to rapid UAV propeller and body movements, which result in fluctuations in the beam alignment and frequent link failures. To address this issue, linearly polarized Laguerre–Gaussian modes were leveraged for spatial mode diversity to prevent link failures over a 400 m link. Spatial mode diversity successfully improved the bit error rate by 38% to 55%. This was due to a 10% to 19% increase in the predominant mode power from spatial mode diversity. The time-varying channel matrix indicated the presence of nonlinear deterministic chaos. This opens up new possibilities for research on state-space reconstruction of the channel matrix.

Keywords: free-space optical communication; spatial mode diversity; unmanned aerial vehicle; emergency recovery communications; floods

Citation: Amphawan, A.; Arsad, N.; Neo, T.-K.; Jasser, M.B.; Mohd Ramly, A. Post-Flood UAV-Based Free Space Optics Recovery Communications with Spatial Mode Diversity. *Electronics* **2022**, *11*, 2257. https://doi.org/10.3390/electronics11142257

Academic Editor: Jiankang Zhang

Received: 7 March 2022
Accepted: 2 May 2022
Published: 19 July 2022

1. Introduction

The risk of floods is rising worldwide, due to an increase in the intensity and frequency of rainfalls as a consequence of global warming [1]. Floods cause fatalities, damage to buildings, deterioration of health conditions, severe economic losses, losses of livelihood, and disruption of global trade [2]. Globally, it is estimated that floods have directly affected 2.3 billion people and caused USD 662 billion in damages between 1995 and 2015 [3]. In Malaysia, destructive floods occur frequently in the three eastern states of the peninsular during the seasonal monsoon between October and March, affecting more than 4.8 million people annually [4]. In the recent December 2021 massive nationwide flood, as many as eight states were struck by a 1-in-a-100-year heavy rainfall spanning two weeks, which displaced thousands of residents and strained emergency services [5–7]. The aftermath of the floods also witnessed unscheduled water cuts and disruptions to the electricity supply [8,9]. In addition, telecommunication base stations were severely damaged, leading to the disruption of communication services [10,11]. This prevented the exchange of situational awareness and hampered the coordination of search and rescue operations.

To connect emergency responders to flood rescue centers and to victims, several recovery communications technologies are being explored, as shown in Figure 1. To compensate for flood-impaired ground radio base stations, satellites have been considered due its large capacity and wide coverage. This requires the design of satellite constellations comprising multiple satellites across several countries for reliable coverage in the flood-stricken area, and to backhaul unflooded cells via satellite [12–14]. High-altitude platforms (HAPs) such as airships, aerostats, and balloons have been used to ferry antennas in the affected area, for relaying data to neighboring ground base stations to a large number of users [15]. HAPs are located in the stratosphere, enabling a wide coverage and accommodating a large number of users [16]. Compared to satellites, HAPs provide a higher area throughput and resource utilization [15]. HAPs can also distribute the recording of the orbital paths of satellites and monitor the probability of a collision between satellites [17]. Data transmission from a HAP involves the control of the HAP flight trajectory and the transmit power of the HAP antenna [18–20].

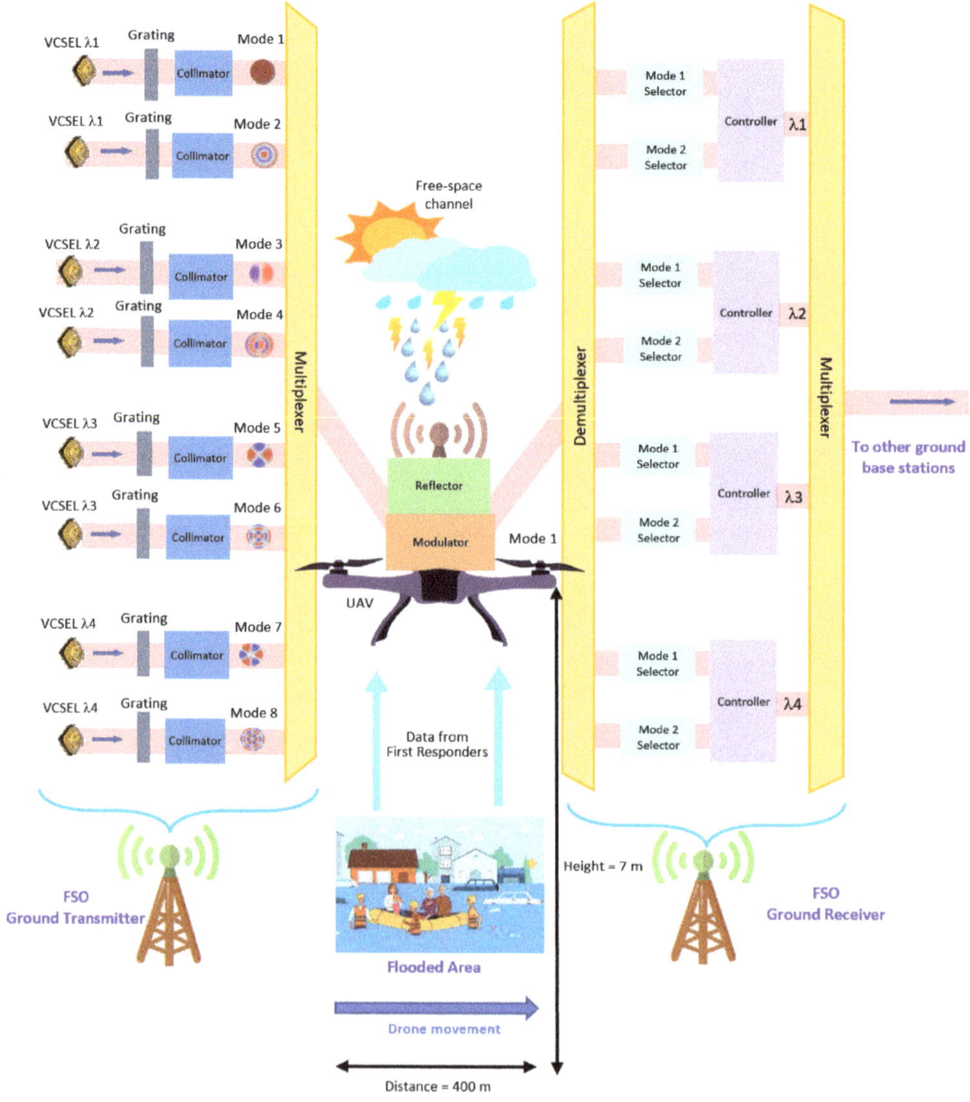

Figure 1. UAV-based spatial mode diversity free space optics (FSO) system for post-flood recovery communications.

Recently, the deployment of unmanned aerial vehicles (UAVs) during natural disasters has become increasingly prevalent for forwarding information from the affected area to undamaged ground base stations during rescue operations [21]. UAVs generally have limited onboard energy [22]. Typically, several UAVs are connected in topologies such as a mesh, star, or cluster. UAVs are linked to each other in order to reach the undamaged ground base station, in conjunction with routing and positioning algorithms [23–26].

The remainder of the paper organized as follows. Section 2 provides an overview of free space optics (FSO) for post-flood recovery communications and related work on spatial mode diversity using FSO. Section 3 elucidates the novelty and contributions of the paper. Section 4 describes the design of a UAV-based spatial mode diversity FSO system for post-flood recovery communications. Section 5 reports on the performance analysis of the proposed system.

2. Free Space Optics for Post-Flood Recovery Communications

Free space optics is an attractive solution during floods when conventional base stations spanning several cities have been damaged, providing connectivity to the flood-stricken area and ensuring minimal disruption in unaffected areas. Free space optics provides a high data bandwidth and rapid deployment, without spectrum licensing costs [27,28]. In addition, the light signals can penetrate easily through water droplets [29].

Although satellites and HAPs cover a larger area, the deployment time is longer [30]. Motivated by the rapid deployment and agility of UAVs for recovery communications [22,30], a UAV-based FSO is proposed as a wireless access technology for rapid recovery communications during floods.

Several UAV-based FSO strategies have previously been developed for recovery communications between base stations for recovery communications, focusing on downlink scheduling [31], and the placement of FSO transceivers [32,33], divergence angle [34], or hybrid radio frequency (RF)/FSO links [35,36]. An unveiled challenge to UAV-based FSO systems is to maintain a high SNR between the base station and UAV-based FSO relay. This is largely unattainable due to minute but rapid UAV propeller and body movements, which result in swift fluctuations in the beam alignment and water scattering from flood waters. Consequently, this would lead to frequent link failures. To address this issue, spatial mode diversity is proposed for improving the resilience of the system from UAV movements.

Harnessing spatial modes as independent information carriers using space division multiplexing (SDM) techniques has recently gained traction for tackling the impending data capacity crunch [37]. Spatial modes provide an additional degree of freedom in wireless communications. Characteristically, in a SDM system, independent data-carrying beams are structured on distinct spatial modes that can be multiplexed at the transmitter aperture for co-transmission and demultiplexed at the receiver aperture, with minimal interference [37].

To increase the resilience of an FSO system, multiplexing in the amplitude, frequency, polarization, and time domains have been employed. Recently, a new degree of freedom based on spatial modes is being explored. The simultaneous transmission of several spatial modes can be realized without the separation between apertures, thus reducing the device footprint. Individual spatial modes encounter different refractive index perturbations from atmospheric turbulence, despite propagating in the same path [38,39]. Spatial modes may be leveraged for transmission of independent data streams to improve link reliability and prevent network interruption.

Several approaches have been demonstrated for spatial mode diversity in FSO systems. In [40], spatial mode diversity was employed in conjunction with multiple transmitter-receiver aperture pairs to improve FSO link reliability under atmospheric turbulence. Three aperture pairs were used and each aperture pair utilized a Gaussian beam and an OAM beam in the uplink and downlink direction simultaneously for carrying the same data stream. In [41], spatial mode diversity from a three-mode photonic lantern coupling FSO receiver was designed in conjunction with a digital maximal ratio combining to enhance the worst signal-to-noise ratio (SNR) by more than 10 dB and to mitigate the interruption probability under atmospheric turbulence. In [42], the three-mode diversity reception in free space optical links from low-Earth orbital satellites to a ground terminal was demonstrated by mode coupling to a few-mode fiber for tracking power variations under different elevation angles. In [43], to reduce the outage probability, aperture diversity was realized through several transmitter apertures transmitting the same data stream on fundamental Gaussian beams whilst mode diversity was realized through multiple receiver apertures that decompose the incoming beam to several orbital angular momentum (OAM) modes, in conjunction with digital signal processing. In [44], OAM modes were used for spatial mode diversity through a 2 km FSO link to improve the link resilience and system capacity. The overall channel capacity was maximized by selecting optimal OAM mode numbers at each value of the SNR and turbulence strength. Spatial mode diversity has also been demonstrated using photonic crystal fibers [45–48]. Another approach for spatial mode diversity was demonstrated using Hermite-Gaussian (HG) and Laguerre-Gaussian (LG) modes of identical size [39]. This was shown to improve the bit error rate by up to 54% without an increase in the total transmit power or radius of the receive aperture. Unmanned aerial vehicle (UAV)-based FSO strategies have been explored for improving coverage between base stations for emergency communications during natural disasters, focusing on downlink scheduling [31] and the placement of FSO transceivers [32,33] and hybrid RF/FSO links [35,36].

3. Contributions

The disastrous aftermath of floods has reinforced the need for complementary access networks for critical recovery communications from the flood-stricken area to unaffected areas [21]. In the event where many conventional base stations spanning several cities are damaged, unlicensed, high-bandwidth wireless access technology is required for complementary post-flood recovery communications, to expedite communications to the flood-stricken area. Connecting first responders and healthcare workers to flood victims and hospitals is an integral part of recovery communications. Several UAV-based FSO strategies have previously been developed for recovery communications between base stations for recovery communications, focusing on downlink scheduling [31], the placement of FSO transceivers [32,33], divergence angle [34], or hybrid RF/FSO links [35,36]. An unveiled challenge to UAV-based FSO systems is to maintain a high SNR between the base station and UAV-based FSO relay. This is largely unattainable due to minute but rapid UAV propeller and body movements, which result in swift fluctuations in the beam alignment and water scattering from flood waters. Consequently, this would lead to frequent link failures. To address this issue, spatial mode diversity is proposed using linearly polarized modes for improving the resilience of the system from UAV movements.

4. Methods

The proposed UAV-based spatial mode diversity FSO system for post-flood recovery communications is illustrated in Figure 1. Four vertical cavity surface-emitting lasers (VCSELs) generating continuous-wave optical signals on the fundamental modes at wavelengths of 850 nm, 880 nm, 910 nm, and 940 nm were structured into eight distinct linearly polarized Laguerre–Gaussian beams, $LP\ lm$, where $l = 0, 1, 2, 3, 4$ and $m = 1, 3$. The LG beams were intensity-modulated at 20 Gbps by pre-processed binary data collected from individual sensors on their respective channels. The LP modes and wavelengths used for the respective channels are shown in Table 1. The ground transmitter was enclosed in a case for protection from rain.

Table 1. Channel Characteristics.

Channel/Mode, i	Mode, LP lm	Signal	Type	Wavelength
1	LP 01	Sensor 1	Primary	850 nm
2	LP 03	Sensor 1	Backup	850 nm
3	LP 11	Sensor 2	Primary	880 nm
4	LP 13	Sensor 2	Backup	880 nm
5	LP 21	Sensor 3	Primary	910 nm
6	LP 23	Sensor 3	Backup	910 nm
7	LP 31	Sensor 4	Primary	940 nm
8	LP 33	Sensor 4	Backup	940 nm

LP lm modes were generated from the first diffraction order in the Fourier plane using the binarized electric field displayed on individual liquid crystal gratings, as follows. It is known that the Fourier transform of a linear translation is a complex phase shift [49]. Thus, to produce a translation in the Fourier plane, the transverse modal field is first multiplied by a complex phase shift:

$$\exp[j(\tau_x x + \tau_y y)] \tag{1}$$

where τ_x and τ_y are linear tilt constants in the horizontal and vertical directions, respectively.

The complex tilted field of the *LP* mode, $a + jb$, is binarized such that, in regions represented by

$$\left(b \pm \frac{1}{2}\right)^2 + a^2 \geq \frac{1}{4}, \quad a \leq 0 \tag{2}$$

an expanded laser beam is transmitted through a liquid crystal display, and in other regions, no light passes through. The binarized beam may be expressed as a Fourier series expansion [45–48,50]:

$$g(x_1, y_1) = m_o + \frac{4}{\pi} \sum_{n=1}^{\infty} m_n \ \cos\{n \ [\xi(x_1, y_1) + \tau_x x_1 + \tau_y y_1]\} \tag{3}$$

where m_o is the constant term and n is the *n-th* diffraction order. In the Fourier plane, the diffraction orders are translated linearly and separated:

$$G(x_2, y_2) = M_o(x_2, y_2) \ + \ \sum_{n=1}^{\infty} [M_n(x_2 + n\tau_x, \ y_2 + n\tau_y) + M_n^*(n\tau_x - x_2, n\tau_y - y_2)] \tag{4}$$

where x_2 and y_2 are spatial coordinates in the Fourier plane, $*$ is the complex conjugate, and $M_n(x_2, y_2)$ is the n-th diffraction order in the Fourier plane.

Each binarized field was then Fourier transformed by a 400 cm focal-length achromatic convex lens, and the linearly polarized Laguerre–Gaussian mode, *LP lm*, was extracted from the first diffraction order in the Fourier plane, M1.

The transverse modal field distribution of a linearly polarized *LP lm* mode is expressed as [51]:

$$E = R^l L_{m-1}^l (VR^2) \exp\left(-VR^2/2\right) \cos \phi \tag{5}$$

where R is the normalized radius, ϕ is the azimuthal angle of the transmitted transverse modal field, V is the normalized frequency, L_{m-1}^l is the generalized Laguerre polynomial whereby l is the azimuthal mode number, and m is the radial mode number.

For channel diversity, each sensor transmitted data on two *LP* modes so that they experience distinct refractive index fluctuations and modal power coupling. For any data stream, the first mode was used as the primary channel, and a second mode was used as the backup channel during adverse weather conditions, such as rain. Different combinations of the azimuthal mode number and radial mode numbers were used for various primary and backup channels, as shown in Table 1. The backup channels operated on different *LP* modes from the primary channels, thus experiencing distinct refractive index fluctuations and mode coupling.

The optical signals from all channels were then amplified, multiplexed, aligned, and transmitted to the UAV, which was flown at a height of 7 m above ground, in a straight line toward the ground receiver, located 40 m away from the transmitter. The horizontal distance travelled and height of the flying drone is shown in Figure 1. Following this, the optical beams are reflected from the UAV to the ground receiver. System parameters are provided in Table 2. The beam waist is the location along the propagation direction where the beam radius is minimum. The Rayleigh length is the distance from the beam waist, in the direction of beam propagation, where the beam radius is increased by a factor of the square root of 2. The beam divergence is an angular measure of the increase in beam diameter or radius with distance from the grating. The photodetector responsivity is a measure of optical-to-electrical conversion efficiency of a photodetector.

Table 2. Systems Parameters.

Parameter	Value
Beam waist	0.025 m
Operating wavelengths	850 nm, 880 nm, 910 nm and 940 nm
Rayleigh range	0.5 m
Propagation distance	400 m
Transmitted beam diameter	3 mm
Beam divergence	0.5 mrad
Photodetector responsivity	0.6 A/W
Focal lengths of transmitter lens	40 cm

The transverse electric field of the set of eight *LP* modes used for the transmission may be modeled as:

$$X = [X_1 \ X_2 \ X_3 \dots \ X_8]^T \tag{6}$$

where X_i is the transverse electric field of the transmitted i-th mode and T is the transpose of the matrix. The estimated received electric field after propagating through the FSO channel is

$$Y = [Y_1 \ Y_2 \ Y_3 \dots \ Y_8]^T, \tag{7}$$

where Y_i is estimated electric field of the i-th received mode and T is the transpose of the matrix. Y is related to the channel matrix, H by [52]

$$Y = HX + N, \tag{8}$$

where the channel matrix H contains the power coupling coefficients,

$$h_{i,j}, \quad i,j \in [1,N] \tag{9}$$

A charge-coupled device was used to collect the intensity distribution of the received optical signals. The power coupling coefficients were computed for all channels using a modal decomposition method [53].

Multiplicative slow fading and inter-symbol interference from adjacent modes $i \neq j$ were assumed. N is additive noise from the surroundings.

The received electric field of the i-th mode may be estimated as the summation of the individual components of the transmitted electric field, X_j [54]:

$$Y_i = \sum_j h_{i,j} X_j + N \tag{10}$$

The power coupling coefficient $h_{i,j}$ is the overlap integral between the received transverse electric field, Y_i, and transmitted transverse electric field, X_j [55]:

$$h_{i,j} = \frac{\left| \int\limits_A Y_i(x,y) \cdot X_j^*(x,y)\, dA \right|^2}{\int\limits_A |Y_i(x,y)|^2 dA \ \int\limits_A |X_j(x,y)|^2 dA} \tag{11}$$

where A is the plane of the transverse electric field. When the set of modes propagate through the atmosphere, it encounters spatially and temporally varying refractive indices, due to random pressure temperature variations. Spatial modes individually experience different refractive index perturbations, even when propagating in the same path [39]. This causes unique random wavefront aberrations and power spreading for each mode into adjacent modes, evident by the degradation of power coupled into the original mode $h_{i,j}$. Under atmospheric turbulence, these power fluctuations are slowly time-varying and vary significantly slower than the signal [56,57].

The Kolmogorov model for turbulent flow is the basis for many contemporary theories and models for emulating atmospheric turbulence for spatial modes [52,58–60]. However, these models of turbulence typically only provide statistical averages for the random variations of the atmosphere. Thus, they may be considered insufficient to represent temporal intensity fluctuations and modal crosstalk for a UAV-based system. Hence, in our work, in order to demonstrate the performance of mode diversity for maintaining the signal quality for composite channels, the time-varying optical signal-to-interference-and-noise ratio (SINR) was evaluated under various weather conditions over a time interval of 30 min and sampling interval of 1 s. The optical SINR for the i-th channel is given by:

$$\begin{aligned} SINR_i &= \frac{P_i}{I_i + N} \\ &= \frac{\left| h_{i,\, j=i} \right|^2}{\left| h_{i,\, j \neq i} \right|^2 + N} \end{aligned} \tag{12}$$

where P_i is the received power from the i-th desired mode ($j = i$), either in the primary channel or the backup channel. The measured interference power I_i arises from crosstalk from undesirable spatial modes ($j \neq i$). The measured additive noise power, N, constituting thermal noise, ambient light from sunlight, and surrounding objects is minimal compared to the interfering modal power.

For optimal signal detection, the ground receiver requires knowledge of the current channel matrix, H. The main challenge lies in the alignment between the ground transmitter to the UAV and the alignment between the UAV to the ground receiver, in order to prevent tip-tilt errors, despite a line-of-sight link between the ground transmitter and ground receiver. To address this issue at the ground transmitter, the conjugates of a known sequence of complex LP mode fields was transmitted toward the UAV and then reflected back to the ground transmitter to determine the optimum angle for pointing the optical beam from the ground transmitter toward the UAV. Adjustments to the pointing angle were made on the ground transmitter using a rotating shaft, translation stages, and goniometers until the exact LP mode was acquired on the reflected optical signal. Similarly, at the ground receiver, beam tracking of the signals was performed to avoid tip-tilt errors using a rotation shaft, translation stages, and goniometers. Using the conjugate of a known LP mode, the direction of arrival of the received optical beam was determined for maximizing spatial acquisition. After the optimal angles were determined, the pointing angle and angle of arrivals were fixed, as the UAV travelled in a straight line from the initial point. An alternative method for beam tracking is to measure the displacement offsets of the arriving at the detector, which translate into angles of arrival variations [61].

The SINR threshold $SINR\gamma$ was set at the minimum level for turbulence-free transmission under heavy rain, such that:

$$SINR_\gamma = \frac{\sum_{i=1}^{8} P_i}{\bar{r}_i} \tag{13}$$

where $\sum_{i=1}^{8} P_i$ refers to the sum of the transmitted power in all desired modes and $\bar{r}_i$ is the average receiver noise power per mode. The calculated $SINR\gamma$ value was 25 dB.

As shown in Table 1, two different LP modes were used, for the primary channel and backup channels, under the same weather condition. The time-varying primary and backup channels were used for obtaining the composite channel, with the aid of a micro-electromechanical systems (MEMS) optical switch for temporal switching between the two channels. The MEMS switch is controlled dynamically using an algorithm designed LabVIEW software by National Instruments (Austin, TX, USA), for automated switching between the two channels based on the comparison of the current SINR to the threshold SNR value, $SINR\gamma$. While the received SINR of the primary channel is equal to greater than $SINR\gamma$, the received data from the primary channel are forwarded to the composite channel by the MEMS switch. On the other hand, when the received SINR of the primary channel is less than $SINR\gamma$, the received data from the backup channel is forwarded to the composite channel by the MEMS switch.

A key parameter for indicating the strength of the atmospheric turbulence is the refractive index structure parameter, C_n^2, which was evaluated by transmitting the concerned mode and then calculating the scintillation index at the receiver using the intensity profile measurements from a charge-coupled-device camera. The scintillation index is given by the Rytov variance [62]:

$$\sigma^2 \approx \frac{\langle I^2 \rangle - \langle I \rangle^2}{\langle I \rangle^2} \tag{14}$$

where I is the on-axis beam intensity and the angle brackets <> denote the ensemble average. The approach for measuring the scintillation index is similar to that in [61]. The refractive index structure parameter C_n^2 is then evaluated by substituting the scintillation index into:

$$\sigma^2 = 1.23 C_n^2 k^{7/6} L^{11/6} \tag{15}$$

where wave number $k = 2\pi/\lambda$ and L is the link distance. Using Equation (13) to Equation (14), the C_n^2 values were computed for various weather conditions based on the data transmission experiments using the designed FSO transceiver. The turbulence strength and turbulence fluctuation are indicated by C_n^2 and the Rytov variance σ^2, respectively. The values of C_n^2 in the atmosphere are typically in the range from 10^{-17} m$^{-2/3}$ for weaker turbulence strength to 10^{-13} m$^{-2/3}$ for stronger turbulence strength. The value of σ^2 is typically related to the refractive index inhomogeneities C_n^2, optical wavelength λ, and the propagating distance L. A higher value of σ^2 generally represents stronger turbulence fluctuation [62]. Weak turbulence is associated with $\sigma^2 < 1$, and moderate fluctuation conditions are characterized by $\sigma^2 \approx 1$. Strong fluctuations are associated with $\sigma^2 > 1$ [62].

5. Results and Discussion

To evaluate the effectiveness of spatial mode diversity using linearly polarized modes for maintaining the signal quality of a UAV-based free space optical system, the time-varying SINR of the primary, backup, and composite channels were measured under various weather conditions: (i) light rain, (ii) medium rain, (iii) heavy rain, (iv) clear day. The rain level was classified as 'light', 'medium', and 'heavy' for precipitation rates of accumulation of about 0 to 2 mm per h, 2 to 10 mm per h, and 10 to 50 mm per h, respectively [63]. For each weather condition, the minimum and maximum C_n^2 values were computed to determine the range of turbulence strengths.

Although the signals were received at 20 Gbps, the acquisition and evaluation of the channel matrix and interfering modes at high data rates were computationally intensive and required time. Thus, the SINR was sampled at 1 Gbps. In addition, the hovering time of the UAV was 45 min, so a 30 min interval was selected to give the UAV sufficient time to return to the ground station.

For each case, the time-varying SINR is plotted for the primary channel, backup channel, and the composite channel. The composite channel comprises various temporal segments of data from the primary channel and the backup channel such that when the SINR is equal to or more than the threshold value of 25 dB, the received data from the primary channel are forwarded to the composite channel. On the other hand, when the SINR on the primary channel drops below 25 dB, the received data from backup channel are forwarded to the composite channel instead.

Figure 2 shows typical plots of the SINR versus time under light rain. The average SINR was 28.5 dB, with a standard deviation of 5.6 dB. Channel 1 was the primary channel (first graph) and Channel 2 was the backup channel (second graph), utilized when the signal in Channel 1 declined below 25 dB. The third graph in Figure 2 shows that the utilization of Channel 2 as a backup channel successfully prevented the SINR from dropping below 25 dB, especially between 1157 s and 1172 s.

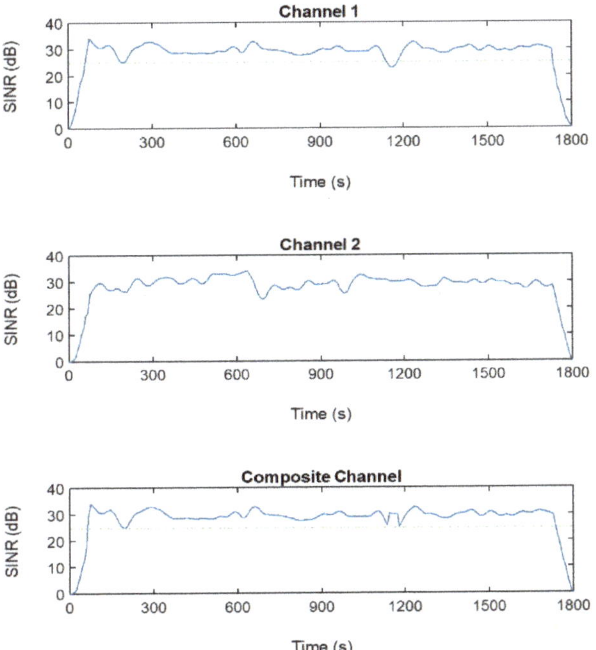

Figure 2. SINR versus time under light rain. Green line shows $SINR\gamma$ value of 25 dB.

Figure 3 shows a typical plot of the SINR versus time under medium rain. The average SINR was 27.5 dB, with a standard deviation of 6.0 dB, underlining deeper variations in the SINR as compared to those under light rain. Channel 3 was the primary channel (first graph) and Channel 4 was the backup channel (second graph), utilized when the SINR in Channel 3 decreased below 25 dB. The third graph in Figure 3 illustrates that employing Channel 4 as a backup channel successfully prevented the composite signal from depreciating below an SINR of 25 dB, preventing link failures to the cloud, especially between 670 s and 701 s, and 1352 s and 1396 s.

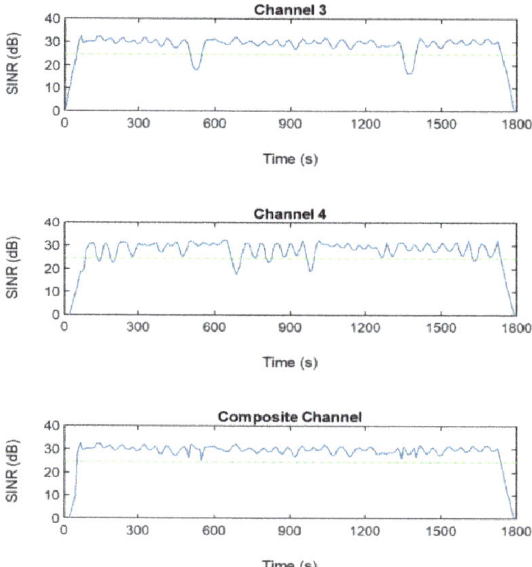

Figure 3. SINR versus time under medium rain. Green line shows $SINR\gamma$ value of 25 dB.

Figure 4 shows typical plots of the SINR versus time under heavy rain. The average SINR was 25.9 dB, with a standard deviation of 6.5 dB, showing more pronounced fluctuations in the SINR as compared to those under medium rain. Channel 7 was the primary channel (first graph) whilst Channel 8 was the backup channel (second graph). The number of regions with compromised SINR in heavy rain was higher than that under medium rain. The SINR for the composite channel, the third plot in Figure 4, shows that leveraging Channel 7 as a backup channel successfully alleviated a large portion of the composite signal from deteriorating below 25 dB, including during a predominant dip between 667 s and 710 s.

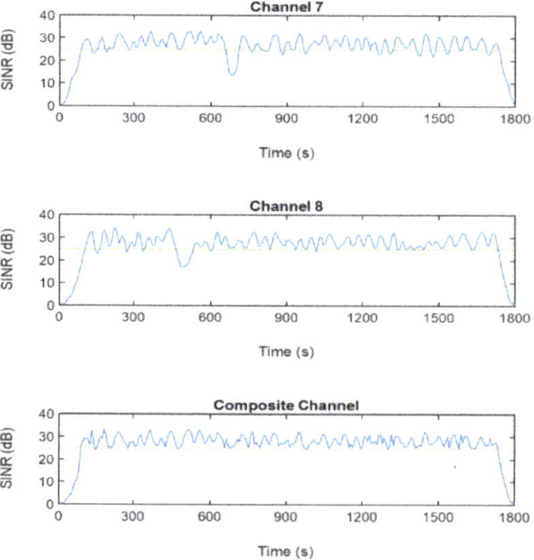

Figure 4. SINR versus time under heavy rain. Green line shows $SINR\gamma$ value of 25 dB.

On a clear day, the average SINR was 30.4 dB. Thus, the backup channel was not required, as the SINR remained above 25 dB, as shown in Figure 5.

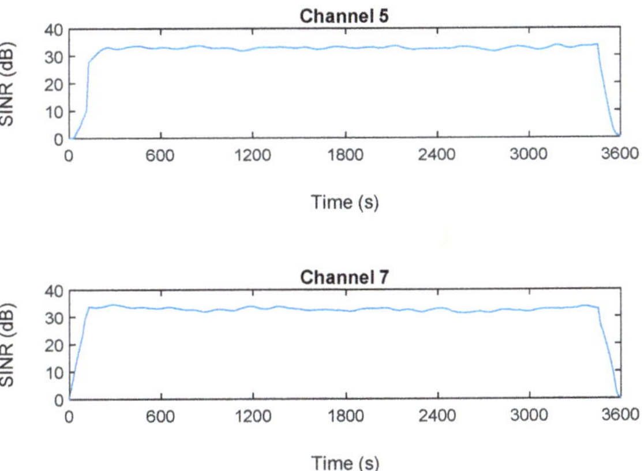

Figure 5. SINR versus time on a clear day.

Uplink and downlink transmissions demonstrated similar SINR characteristics. Overall, a significant improvement in the SINR and outage performance was observed due to spatial mode diversity.

At the receiver, the received optical beam for each channel is demultiplexed to separate spatial modes. It is well-established that refractive index fluctuations in clouds from pressure and temperature fluctuations cause rain [64]. This leads to aberration of the beam wavefront from random power and phase perturbations [65]. This results in the signal initially launched in a particular mode to spread to other modes [66]. The amount of spreading is influenced by the strength of the atmospheric turbulence.

On a clear day, C_n^2 values were found to be between 7×10^{-16} m$^{-2/3}$ and 4×10^{-15} m$^{-2/3}$, peaking in the afternoon. On a clear night, the C_n^2 values were found to be lower, between 8×10^{-17} m$^{-2/3}$ and 1×10^{-16} m$^{-2/3}$. For light rain, the C_n^2 values were in the range of 4×10^{-16} m$^{-2/3}$ to 7×10^{-16} m$^{-2/3}$. For moderate rain, the C_n^2 values were slightly higher, in the range of 6×10^{-16} m$^{-2/3}$ to 9×10^{-16} m$^{-2/3}$, and for heavy rain, the C_n^2 values were higher, in the range of 8×10^{-16} m$^{-2/3}$ to 2×10^{-15} m$^{-2/3}$. The C_n^2 values were found to vary more on a clear day than during rainfall.

The power coupling coefficients were computed for all channels using a noninterferometric modal decomposition method [53], under various weather conditions, at the minimum and maximum C_n^2 values. With knowledge of the power coupling coefficients, the modal crosstalk matrices were constructed, without and with spatial mode diversity. The power coupling coefficients were normalized to the total received optical power for the set of modes present.

Samples of inter-modal crosstalk matrices before mode diversity are shown in Figures 6 and 7, at the minimum and maximum C_n^2 values, respectively. Samples of inter-modal crosstalk matrices after mode diversity are shown in Figures 8 and 9. For clear weather conditions in the day and at night, the power in the dominant mode was more than 75%. Under rain, power from the transmitted mode leaked into adjacent modes. The heavier the rain, the more substantial the spreading of power into adjacent modes due to increased atmospheric turbulence.

Under light rain, the power in the dominant mode was in the range of 68% to 75% before channel diversity reception and 79% to 83% after mode diversity and power were successfully maintained in the dominant modes through the switch to the backup channels during SINR dips. Under medium rain, the power in the dominant mode was in the range of 64% to 68% prior to mode diversity and improved to 75% to 78% in the presence of mode diversity. Under heavy rain, the power in the dominant mode was approximately 56% to 63% without mode diversity reception and increased to 71% to 75% with mode diversity reception. Uplink and downlink data transmissions demonstrated similar power coupling coefficient characteristics, whereas uplink and downlink data transmissions demonstrated similar power coupling coefficient characteristics.

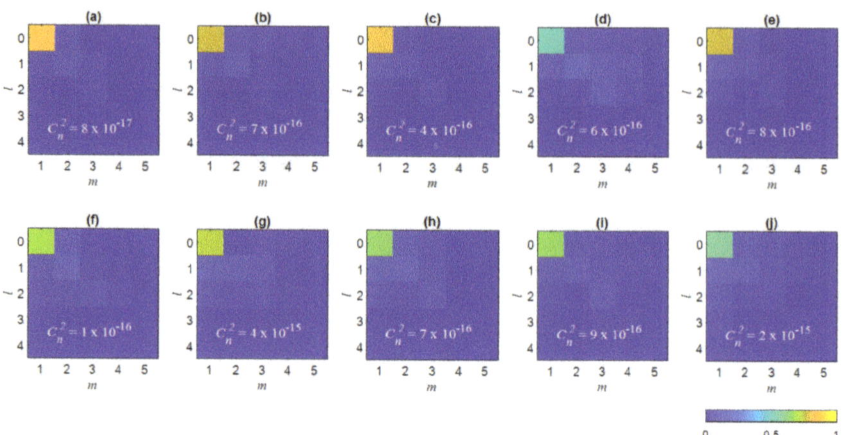

Figure 6. Variation in normalized power coupling coefficients for on *LP 01* mode before spatial mode diversity: (**a**,**f**) clear night, (**b**,**g**) clear day, (**c**,**h**) light rain, (**d**,**i**) medium rain, (**e**,**j**) heavy rain. Top panel (**a**–**e**) shows the power coupling coefficients at low-refractive-index structure values, and bottom panel shows (**f**–**j**) those at high-refractive-index structure values. Color legend of normalized power coupling coefficients for all plots is on the bottom right corner.

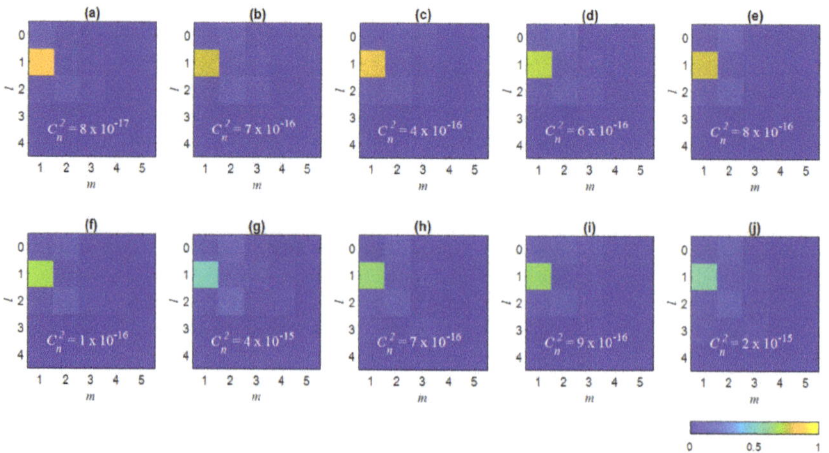

Figure 7. Variation in normalized power coupling coefficients on *LP11* mode before spatial mode diversity: (**a**,**f**) clear night, (**b**,**g**) clear day, (**c**,**h**) light rain, (**d**,**i**) medium rain, (**e**,**j**) heavy rain. Top panel (**a**–**e**) shows the power coupling coefficients at low-refractive-index structure values, and bottom panel shows (**f**–**j**) those at high-refractive-index structure values. Color legend of normalized power coupling coefficients for all plots is on the bottom right corner.

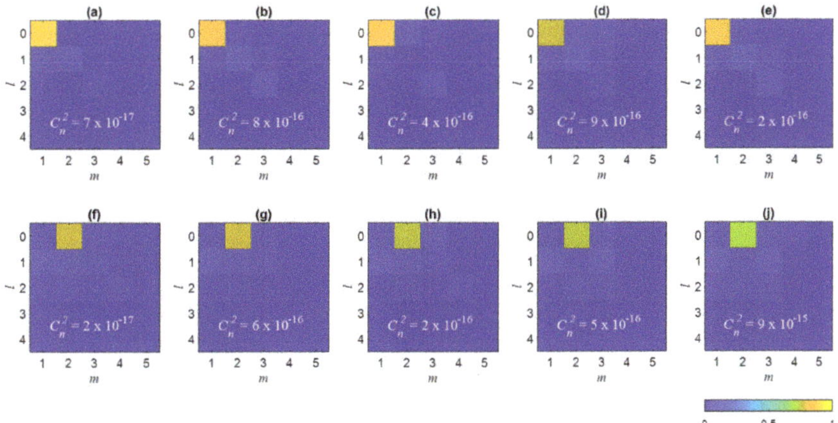

Figure 8. Variation in normalized power coupling coefficients on *LP01* mode after spatial mode diversity reception: (**a,f**) clear night, (**b,g**) clear day, (**c,h**) light rain, (**d,i**) medium rain, (**e,j**) heavy rain. Top panel (**a–e**) shows the power coupling coefficients for primary channel at low-refractive-index structure values, and the bottom panel shows (**f–j**) those for backup channel at high-refractive-index structure values. Color legend of normalized power coupling coefficients for all plots is on the bottom right corner.

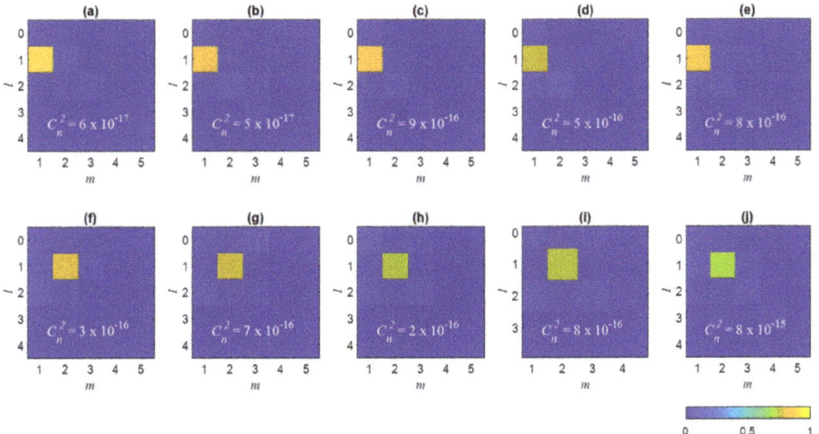

Figure 9. Variation in normalized power coupling coefficients on *LP11* mode after spatial mode diversity: (**a,f**) clear night, (**b,g**) clear day, (**c,h**) light rain, (**d,i**) medium rain, (**e,j**) heavy rain. Top panel (**a–e**) shows the power coupling coefficients for primary channel at low-refractive-index structure values, and the bottom panel shows (**f–j**) those for backup channel at high-refractive-index structure values. Color legend of normalized power coupling coefficients for all plots is on the bottom right corner.

Overall, 25% to 44% of power was leaked to other modes during rainfall, prior to mode diversity. The power leakage was more significant for adjacent modes to the dominant modes and gradually decreased for modes further away from the dominant modes. With increased turbulence strength, the crosstalk was more prevalent and the power distribution of the modes was more dispersed. After mode diversity reception, an increase of 10% to 19% in the dominant mode power was observed. A larger amount of power was successfully maintained in the dominant modes through the switch to the backup channels during SINR dips.

The inter-modal crosstalk matrices for the primary channel, secondary channel, and composite channel after mode diversity is shown Figure 10. The power in the dominant mode of each channel was improved under all weather conditions. Without the mode diversity scheme, there were interruptions in the data transmission from sensors. By incorporating mode diversity, uninterrupted data transmission from sensors was observed, even under heavy rain due to the backup channels on different LG modes experiencing different channel degradations from the primary channels.

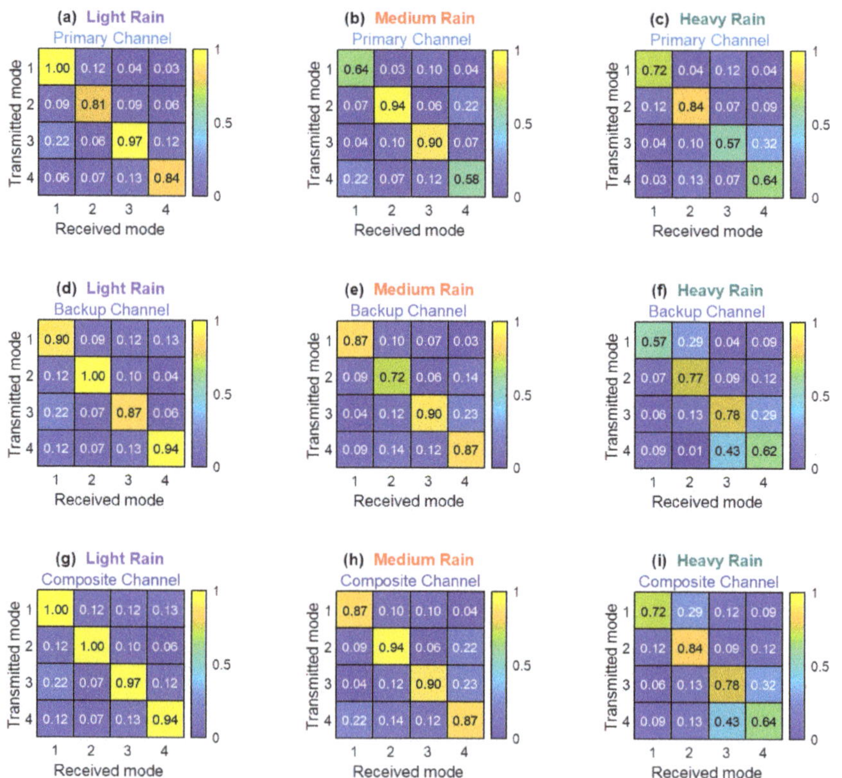

Figure 10. Inter-modal crosstalk matrices for primary channel and secondary channel before mode diversity, and composite channel after mode diversity. The transmitted mode number and received mode number, i, are given in Table 1. The color legend shows the normalized power coupling coefficient.

The average diversity gain from spatial mode diversity was 0.97 dB. The average bit error rate (BER) versus SINR for all channels under various weather conditions was measured and compared, without and with spatial mode diversity, as shown in Figure 11. It was observed that spatial mode diversity successfully enhanced the BER by 38% to 55% from the original BER. After spatial mode diversity, a satisfactory BER below the 7% forward error correction (FEC) limit of 3.8×10^{-3} was attained, at SINRs higher than 20 dB, under all levels of rain. An FEC of 7% was used to provide a 0.3 dB coding gain to better mitigate atmospheric turbulence.

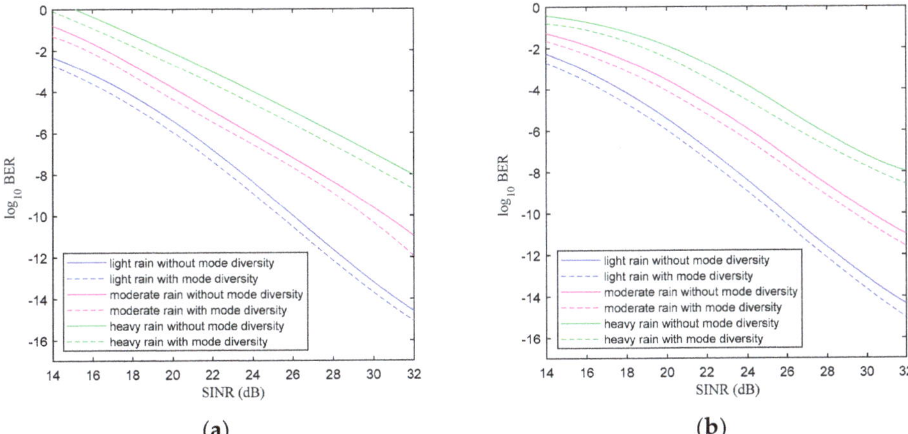

Figure 11. BER for all channels under varying rain levels. (**a**) Composite Channel 1,2. (**b**) Composite Channels 3,4.

Through UAV-FSO spatial diversity, despite mobility restrictions due to floods, vital health statistics from injured flood victims could be forwarded to a patient cloud-based health monitoring system accurately and timely, for hospitals to make critical preparations for relevant medical treatment upon the victims' arrival.

To verify the experimental measurements, the average values of C_n^2 were computed using [67]:

$$
\begin{aligned}
C_n^2 = {} & 3.8 \times 10^{-14}w + 2.0 \times 10^{-15}T - 2.8 \times 10^{-15}W \\
& + 2.9 \times 10^{-17}W^2 - 1.1 \times 10^{-19}W^3 - 2.5 \times 10^{-15}S \\
& + 1.2 \times 10^{-15}S^2 - 8.5 \times 10^{-17}S^3 - 5.3 \times 10^{-13}
\end{aligned}
\tag{16}
$$

where T is the average air temperature, W is the average relative humidity, S is the average wind speed from the local weather station, and w is a scaling coefficient. The average values of C_n^2 using [67] were 3×10^{-15} on a clear day, 1×10^{-16} on a clear night, 6×10^{-16} under light rain, 8×10^{-16} under moderate rain, and 2×10^{-15} under heavy rain. This concurred with earlier experimental values computed using the scintillation index within 1×10^{-15}. The refractive index structure parameter C_n^2 was in the range of 1×10^{-15} to 1×10^{-13}. Based on the turbulence strength classification in [62], this indicated weak to moderat turbulence.

A probability distribution function of the C_n^2 values from the experiment could be computed to determine if this adhered to relevant channel probability distribution functions for FSO spatial mode diversity and FSO aerial communications such as Malaga, log normal, Gamma-Gamma, Nagakami, and others [68–72]. Log normal and Gamma-Gamma channel models are considered as special cases of the Malaga channel model, for weak turbulence and moderate-strong turbulence, respectively [72].

Nevertheless, comparison of the time-varying SINR plots from the experiment with existing channel models do not provide accurate time-varying error analysis, as existing channel models are probabilistic [73]. Existing channel models only provide the statistical averages for the random variations and do not describe the evolution of the channel characteristics with time. In simulations modelling the FSO channel, the time-varying atmospheric turbulence is typically emulated by continuously upgrading the random phase patterns to model random irradiance fluctuations, based on a known channel probability distribution function. This is only valid when the atmospheric turbulence is stationery during the laser pulse width [74]. Expanding on the previous work, we performed the false-nearest-neighbors algorithm on the measured time-varying channel matrix, which revealed that the fraction of false nearest neighbors converged to 0. This indicated the presence of nonlinear deterministic chaos. This opens up new possibilities for research on state-space reconstruction and the dynamic attractor for modeling the nonlinear FSO-UAV channel matrix.

6. Conclusions

A 4 × 2-channel UAV-based spatial mode diversity FSO system was developed for post-flood recovery communications, achieving 2.4 Mbps for a distance of 400 m, under various weather conditions. Spatial mode diversity was realized by using distinct linearly polarized spatial modes in the primary and backup channels. Backup channels were used when the SINR in the primary channels declined below 25 dB. SINR plots from the composite channel showed that dips in the SINR from atmospheric fluctuations were successfully mitigated through spatial mode diversity. The power in the dominant modes increased by 10% to 19% through temporal switching to backup channels during SINR dips in the primary channels. Hence, this provided uninterrupted transmission, even under heavy rain. The spatial mode diversity scheme improved the BER by 38% to 55% compared to the original BER. In the future, the number of modes used for spatial mode diversity may be extended in conjunction with the maximal-ratio combination for better signal quality. The UAV-based spatial mode diversity FSO system is valuable for rapid recovery communications after natural disasters, especially when conventional ground base stations are damaged. Further analysis of nonlinear time-varying deterministic channel characteristics would be valuable for emulating atmospheric turbulence under various weather conditions, toward the implementation of more efficient turbulence compensation approaches.

Author Contributions: Conceptualization, A.A. and T.-K.N.; formal analysis, A.A., N.A., T.-K.N. and M.B.J.; funding acquisition, A.A.; investigation, A.A. and A.M.R.; resources, N.A.; software, M.B.J. and T.-K.N.; supervision, A.A. and T.-K.N.; validation, N.A. and M.B.J.; visualization, T.-K.N. and A.M.R.; writing—original draft and revisions A.A., T.-K.N. and M.B.J. All authors have read and agreed to the published version of the manuscript.

Funding: The authors are grateful for the financial support received from the Sunway University Individual Research Grant GRTIN-IGS-DCIS[S]-01-2022 and the Sunway University Elasticities Research Cluster Grant STR-RCGS-E-CITIES[S]-004-2021.

Acknowledgments: The authors wish to express their appreciation to Telekom Malaysia (TM) Berhad and TM Research & Development Sdn. Bhd. for collaborating with us on this research.

Conflicts of Interest: The authors declare no conflict of interest.

References

1. Tabari, H. Climate change impact on flood and extreme precipitation increases with water availability. *Sci. Rep.* **2020**, *10*, 13768. [CrossRef]
2. Veettil, P.C.; Raghu, P.T.; Ashok, A. Information quality, adoption of climate-smart varieties and their economic impact in flood-risk areas. *Environ. Dev. Econ.* **2020**, *26*, 45–68. [CrossRef]
3. Kundzewicz, Z.W.; Su, B.; Wang, Y.; Wang, G.; Wang, G.; Huang, J.; Jiang, T. Flood risk in a range of spatial perspectives—From global to local scales. *Nat. Hazards Earth Syst. Sci.* **2019**, *19*, 1319–1328. [CrossRef]
4. Baharuddin, K.A.; Wahab, S.F.A.; Ab Rahman, N.H.N.; Mohamad, N.A.N.; Kamauzaman, T.H.T.; Noh, A.Y.M.; Majod, M.R.A. The record-setting flood of 2014 in Kelantan: Challenges and recommendations from an emergency medicine perspective and why the medical campus stood dry. *Malays. J. Med. Sci.* **2015**, *22*, 1–7.
5. Reuters. Malaysia Floods Hit Seven States Forcing Thousands to Evacuate. 2022. Available online: https://edition.cnn.com/2022/01/02/asia/malaysia-floods-evacuation-intl-hnk/index.html (accessed on 23 April 2022).
6. Hassan, H. Peninsular Malaysia hit by '1-in-100-year' rainfall, government says amid severe flooding. *The Straits Time*, 20 December 2021.
7. Malaysia National Disaster Management Agency (NADMA). Malaysia-Floods and Landslides, Update. In European Union Civil Protection and Humanitarian Aid (ECHO) Daily Flash, European Civil Protection and Humanitarian Aid. 3 January 2022. Available online: https://reliefweb.int/report/malaysia/malaysia-floods-and-landslides-update-nadma-met-malaysia-media-echo-daily-flash-3 (accessed on 23 April 2022).
8. Chen, G. Floods: 472 areas in Petaling, KL and Hulu Langat hit by water cuts as treatment plants shut down. *The Star*, 19 December 2021.

9. TNB shuts down 333 power substations in six states hit by floods. *Malay Mail*, 19 December 2021.

10. Over 100 Celcom 2G/4G Network Sites impacted by Flood, Restoration in Progress. *Malaysian Wireless*, 23 December 2021.

11. MCMC: 276 Communication Tower Still Down, PJ Among Highest. *Business Today*, 25 December 2021.

12. Hu, J.; Huang, H.; Yang, L.; Zhu, Y. A multi-objective optimization framework of constellation design for emergency observation. *Adv. Space Res.* **2020**, *67*, 531–545. [CrossRef]

13. Wang, X.; Zhang, H.; Bai, S.; Yue, Y. Design of agile satellite constellation based on hybrid-resampling particle swarm optimization method. *Acta Astronaut.* **2021**, *178*, 595–605. [CrossRef]

14. Liu, Y.; Wan, Z.; Dai, Y.; Zhao, Y.; Liu, Q.; Ji, C. Emergency Constellation Design Based on Micro SAR Satellite. In Proceedings of the 2021 Global Reliability and Prognostics and Health Management, Nanjing, China, 15–17 October 2021; pp. 1–5.

15. Alam, S.; Kurt, G.K.; Yanikomeroglu, H.; Zhu, P.; Dao, N.D. High Altitude Platform Station Based Super Macro Base Station Constellations. *IEEE Commun. Mag.* **2021**, *59*, 103–109. [CrossRef]

16. Alexandre, L.C.; Linhares, A.; Neto, G.; Sodre, A.C. High-Altitude Platform Stations as IMT Base Stations: Connectivity from the Stratosphere. *IEEE Commun. Mag.* **2021**, *59*, 30–35. [CrossRef]

17. Kurt, G.K.; Khoshkholgh, M.G.; Alfattani, S.; Ibrahim, A.; Darwish, T.S.J.; Alam, S.; Yanikomeroglu, H.; Yongacoglu, A. A Vision and Framework for the High Altitude Platform Station (HAPS) Networks of the Future. *IEEE Commun. Surv. Tutorials* **2021**, *23*, 729–779. [CrossRef]

18. Deka, R.; Mishra, V.; Ahmed, I.; Anees, S.; Alam, M.S. Performance Analysis of HAPS Assisted Dual-Hop Hybrid RF/FSO System. In Proceedings of the 2021 IEEE 94th Vehicular Technology Conference (VTC2021-Fall), Virtual, 27–30 September 2021; pp. 1–6.

19. Na, Z.; Wang, Y.; Xiong, M. Joint trajectory and power optimization for NOMA-based high altitude platform relaying system. *Wirel. Net.* **2021**, 1–12. [CrossRef]

20. Ke, M.; Gao, Z.; Huang, Y.; Ding, G.; Ng, D.W.K.; Wu, Q.; Zhang, J. An Edge Computing Paradigm for Massive IoT Connectivity Over High-Altitude Platform Networks. *IEEE Wirel. Commun.* **2021**, *28*, 102–109. [CrossRef]

21. Hu, B.; Wang, L.; Chen, S.; Cui, J.; Chen, L. An Uplink Throughput Optimization Scheme for UAV-Enabled Urban Emergency Communications. *IEEE Internet Things J.* **2021**, *9*, 4291–4302. [CrossRef]

22. Alzahrani, B.; Oubbati, O.S.; Barnawi, A.; Atiquzzaman, M.; Alghazzawi, D. UAV assistance paradigm: State-of-the-art in applications and challenges. *J. Netw. Comput. Appl.* **2020**, *166*, 102706. [CrossRef]

23. Lin, Y.; Wang, T.; Wang, S. Trajectory Planning for Multi-UAV Assisted Wireless Networks in Post-Disaster Scenario. In Proceedings of the 2019 IEEE Global Communications Conference (GLOBECOM), Waikoloa, HI, USA, 9–13 December 2019; pp. 1–6.

24. Arafat, M.Y.; Moh, S. Localization and Clustering Based on Swarm Intelligence in UAV Networks for Emergency Communications. *IEEE Internet Things J.* **2019**, *6*, 8958–8976. [CrossRef]

25. Yang, Z.; Liu, H.; Chen, Y.; Zhu, X.; Ning, Y.; Zhu, W. UEE-RPL: A UAV-Based Energy Efficient Routing for Internet of Things. *IEEE Trans. Green Commun. Netw.* **2021**, *5*, 1333–1344. [CrossRef]

26. Lin, N.; Liu, Y.; Zhao, L.; Wu, D.O.; Wang, Y. An Adaptive UAV Deployment Scheme for Emergency Networking. *IEEE Trans. Wirel. Commun.* **2021**, *21*, 2383–2398. [CrossRef]

27. Dhasarathan, V.; Singh, M.; Malhotra, J. Development of high-speed FSO transmission link for the implementation of 5G and Internet of Things. *Wirel. Net.* **2019**, *26*, 2403–2412. [CrossRef]

28. Chowdhury, M.Z.; Shahjalal, M.; Hasan, M.; Jang, Y.M. The role of optical wireless communication technologies in 5G/6G and IoT solutions: Prospects, directions, and challenges. *Appl. Sci.* **2019**, *9*, 4367. [CrossRef]

29. Wang, F.; Liu, Y.; Jiang, F.; Chi, N. High speed underwater visible light communication system based on LED employing maximum ratio combination with multi-PIN reception. *Opt. Commun.* **2018**, *425*, 106–112. [CrossRef]

30. Alsabah, M.; Naser, M.A.; Mahmmod, B.M.; Abdulhussain, S.H.; Eissa, M.R.; Al-Baidhani, A.; Noordin, N.K.; Sait, S.M.; Al-Utaibi, K.A.; Hashim, F. 6G Wireless Communications Networks: A Comprehensive Survey. *IEEE Access* **2021**, *9*, 148191–148243. [CrossRef]

31. Sun, X.; Yu, L.; Zhang, T. Latency Aware Transmission Scheduling for Steerable Free Space Optics. In *IEEE Transactions on Mobile Computing*; IEEE: Piscataway, NJ, USA, 2021; p. 1. [CrossRef]

32. Wu, D.; Sun, X.; Ansari, N. An FSO-Based Drone Assisted Mobile Access Network for Emergency Communications. *IEEE Trans. Netw. Sci. Eng.* **2019**, *7*, 1597–1606. [CrossRef]

33. Ansari, N.; Wu, D.; Sun, X. FSO as backhaul and energizer for drone-assisted mobile access networks. *ICT Express* **2020**, *6*, 139–144. [CrossRef]

34. Zhang, T.; Sun, X.; Wang, C. On Optimizing the Divergence Angle of an FSO based Fronthaul Link in Drone Assisted Mobile Networks. *IEEE Internet Things J.* **2021**, *9*, 6914–6921. [CrossRef]

35. Swaminathan, R.; Sharma, S.; Vishwakarma, N.; Madhukumar, A.S. HAPS-Based Relaying for Integrated Space–Air–Ground Networks With Hybrid FSO/RF Communication: A Performance Analysis. *IEEE Trans. Aerosp. Electron. Syst.* **2021**, *57*, 1581–1599. [CrossRef]

36. Shah, S.; Siddharth, M.; Vishwakarma, N.; Swaminathan, R.; Madhukumar, A.S. Adaptive-Combining-Based Hybrid FSO/RF Satellite Communication With and Without HAPS. *IEEE Access* **2021**, *9*, 81492–81511. [CrossRef]

37. Trichili, A.; Park, K.H.; Zghal, M.; Ooi, B.S.; Alouini, M.S. Communicating using spatial mode multiplexing: Potentials, challenges, and perspectives. *IEEE Commun. Surv. Tutor.* **2019**, *21*, 3175–3203. [CrossRef]

38. Trichili, A.; Cox, M.A.; Ooi, B.S.; Alouini, M.-S. Roadmap to free space optics. *J. Opt. Soc. Am. B* **2020**, *37*, A184–A201. [CrossRef]
39. Cox, M.A.; Cheng, L.; Rosales-Guzmán, C.; Forbes, A. Modal diversity for robust free-space optical communications. *Phys. Rev. Appl.* **2018**, *10*, 024020. [CrossRef]
40. Li, L.; Song, H.; Zhang, R.; Zhao, Z.; Liu, C.; Pang, K.; Song, H.; Du, J.; Willner, A.N.; Almaiman, A.; et al. Increasing system tolerance to turbulence in a 100-Gbit/s QPSK free-space optical link using both mode and space diversity. *Opt. Commun.* **2020**, *480*, 126488. [CrossRef]
41. Zheng, D.; Li, Y.; Zhou, H.; Bian, Y.; Yang, C.; Li, W.; Qiu, J.; Guo, H.; Hong, X.; Zuo, Y.; et al. Performance enhancement of free-space optical communications under atmospheric turbulence using modes diversity coherent receipt. *Opt. Express* **2018**, *26*, 28879–28890. [CrossRef]
42. Arikawa, M.; Ito, T. Performance of mode diversity reception of a polarization-division-multiplexed signal for free-space optical communication under atmospheric turbulence. *Opt. Express* **2018**, *26*, 28263–28276. [CrossRef]
43. Song, H.; Li, L.; Pang, K.; Zhang, R.; Zou, K.; Zhao, Z.; Du, J.; Song, H.; Liu, C.; Cao, Y.; et al. Demonstration of using two aperture pairs combined with multiple-mode receivers and MIMO signal processing for enhanced tolerance to turbulence and misalignment in a 10 Gbit/s QPSK FSO link. *Opt. Lett.* **2020**, *45*, 3042–3045. [CrossRef]
44. Yousif, B.B.; Elsayed, E.E. Performance Enhancement of an Orbital-Angular-Momentum-Multiplexed Free-Space Optical Link Under Atmospheric Turbulence Effects Using Spatial-Mode Multiplexing and Hybrid Diversity Based on Adaptive MIMO Equalization. *IEEE Access* **2019**, *7*, 84401–84412. [CrossRef]
45. Amphawan, A.; Chaudhary, S.; Neo, T.-K.; Kakavand, M.; Dabbagh, M. Radio-over-free space optical space division multiplexing system using 3-core photonic crystal fiber mode group multiplexers. *Wirel. Net.* **2020**, *27*, 211–225. [CrossRef]
46. Amphawan, A.; Chaudhary, S.; Ghassemlooy, Z.; Neo, T.-K. 2×2-channel mode-wavelength division multiplexing in Ro-FSO system with PCF mode group demultiplexers and equalizers. *Opt. Commun.* **2020**, *467*, 125539. [CrossRef]
47. Chaudhary, S.; Amphawan, A. Solid core PCF-based mode selector for MDM-Ro-FSO transmission systems. *Photon-Netw. Commun.* **2018**, *36*, 263–271. [CrossRef]
48. Chaudhary, S.; Amphawan, A. Selective excitation of LG 00, LG 01, and LG 02 modes by a solid core PCF based mode selector in MDM-Ro-FSO transmission systems. *Laser Phys.* **2018**, *28*, 075106. [CrossRef]
49. Goodman, J.W. *Introduction to Fourier Optics*; Macmillan Learning: New York, NY, USA, 2017.
50. Chaudhary, S.; Amphawan, A. High-speed MDM-Ro-FSO system by incorporating spiral-phased Hermite Gaussian modes. *Photon-Netw. Commun.* **2018**, *35*, 374–380. [CrossRef]
51. Snyder, A.W.; Love, J.D. *Optical Waveguide Theory (Science Paperbacks)*; Chapman and Hall: London, UK, 1983.
52. Amhoud, E.-M.; Ooi, B.S.; Alouini, M.-S. A Unified Statistical Model for Atmospheric Turbulence-Induced Fading in Orbital Angular Momentum Multiplexed FSO Systems. *IEEE Trans. Wirel. Commun.* **2019**, *19*, 888–900. [CrossRef]
53. Amphawan, A.; O'Brien, D. Modal decomposition of output field for holographic mode field generation in a multimode fiber channel. In Proceedings of the International Conference on Photonics, Langkawi, Malaysia, 5–7 July 2010; pp. 1–5.
54. Rogel-Salazar, J.; Treviño, J.P.; Chávez-Cerda, S. Engineering structured light with optical vortices. *J. Opt. Soc. Am. B* **2014**, *31*, A46–A50. [CrossRef]
55. Amphawan, A.; Payne, F.; O'Brien, D.; Shah, N. Derivation of an analytical expression for the power coupling coefficient for offset launch into multimode fiber. *J. Lightwave Technol.* **2009**, *28*, 861–869. [CrossRef]
56. Andrews, L.C.; Phillips, R.L. *Laser Beam Propagation through Random Media*; SPIE Optical Engineering Press: Bellingham, WA, USA, 1998; pp. 47–50.
57. Ren, Y.; Huang, H.; Xie, G.; Ahmed, N.; Yan, Y.; Erkmen, B.I.; Chandrasekaran, N.; Lavery, M.; Steinhoff, N.K.; Tur, M.; et al. Atmospheric turbulence effects on the performance of a free space optical link employing orbital angular momentum multiplexing. *Opt. Lett.* **2013**, *38*, 4062–4065. [CrossRef] [PubMed]
58. Andrews, L.C. An Analytical Model for the Refractive Index Power Spectrum and Its Application to Optical Scintillations in the Atmosphere. *J. Mod. Opt.* **1992**, *39*, 1849–1853. [CrossRef]
59. Funes, G.; Vial, M.; Anguita, J.A. Orbital-angular-momentum crosstalk and temporal fading in a terrestrial laser link sing single-mode fiber coupling. *Opt. Express* **2015**, *23*, 23133–23142. [CrossRef]
60. Anguita, J.A.; Neifeld, M.A.; Vasic, B.V. Modeling channel interference in an orbital angular momentum-multiplexed laser link. In *Free-Space Laser Communications IX*; International Society for Optics and Photonics: San Diego, CA, USA, 2009; Volume 7464, p. 74640U.
61. Amphawan, A.; Mishra, V.; Nisar, K.; Nedniyom, B. Real-time holographic backlighting positioning sensor for enhanced power coupling efficiency into selective launches in multimode fiber. *J. Mod. Opt.* **2012**, *59*, 1745–1752. [CrossRef]
62. Maurer, T.; Driggers, R.G.; Vollmerhausen, R.H.; Friedman, M.H. NVTherm improvements. In *Infrared and Passive Millimeter-Wave Imaging Systems: Design, Analysis, Modeling, and Testing, Orlando, USA*; International Society for Optics and Photonics: Bellingham, WA, USA, 2002; Volume 4719, pp. 15–23.
63. U.M. Office. National Meteorological Library and Archive Fact Sheet 3—Water in the Atmosphere. 2012. Available online: https://www.metoffice.gov.uk/binaries/content/assets/metofficegovuk/pdf/research/library-and-archive/library/publications/factsheets/factsheet_3-water-in-the-atmosphere.pdf (accessed on 23 April 2022).
64. Falkovich, G.; Fouxon, A.; Stepanov, M.G. Acceleration of rain initiation by cloud turbulence. *Nature* **2002**, *419*, 151–154. [CrossRef]

65. Mahalov, A.; McDaniel, A. Long-range propagation through inhomogeneous turbulent atmosphere: Analysis beyond phase screens. *Phys. Scr.* **2018**, *94*, 034003. [CrossRef]
66. Zhao, L.; Liu, H.; Hao, Y.; Sun, H.; Wei, Z. Effects of Atmospheric Turbulence on OAM-POL-FDM Hybrid Multiplexing Communication System. *Appl. Sci.* **2019**, *9*, 5063. [CrossRef]
67. Sadot, D.; Kopeika, N.S. Forecasting optical turbulence strength on the basis of macroscale meteorology and aerosols: Models and validation. *Opt. Eng.* **1992**, *31*, 200–212. [CrossRef]
68. Sarker, N.A.; Badrudduza, A.S.M.; Islam, S.R.; Islam, S.H.; Kundu, M.K.; Ansari, I.S.; Kwak, K.S. On the Intercept Probability and Secure Outage Analysis of Mixed (α–κ–μ)-Shadowed and Málaga Turbulent Models. *IEEE Access* **2021**, *9*, 133849–133860. [CrossRef]
69. Singh, H. Performance analysis of FSO system with spatial diversity using two-point Padé approximation. *Opt. Quantum Electron.* **2021**, *53*, 1–25. [CrossRef]
70. Wenjing, G.; Ziyuan, S.; Yueying, Z.; Lei, Y. Channel Modeling for Ground-to-UAV Free-Space Optical Communication Systems. In Proceedings of the 26th Optoelectronics and Communications Conference, Hong Kong, China, 3–7 July 2021; Optica Publishing Group: Washington, DC, USA, 2021; p. S4B.6.
71. Khallaf, H.S.; Kato, K.; Mohamed, E.M.; Sait, S.M.; Yanikomeroglu, H.; Uysal, M. Composite Fading Model for Aerial MIMO FSO Links in the Presence of Atmospheric Turbulence and Pointing Errors. *IEEE Wirel. Commun. Lett.* **2021**, *10*, 1295–1299. [CrossRef]
72. El Saghir, B.M.; El Mashade, M.B.; Aboshosha, A.M. Performance analysis of modulating retro-reflector FSO communication systems over Málaga turbulence channels. *Opt. Commun.* **2020**, *474*, 126160. [CrossRef]
73. Cox, M.A.; Mphuthi, N.; Nape, I.; Mashaba, N.; Cheng, L.; Forbes, A. Structured light in turbulence. *IEEE J. Sel. Top. Quantum Electron.* **2020**, *27*, 1–21. [CrossRef]
74. Taylor, G.I. Statistical theory of turbulence-II. *Proc. R. Soc. London. Ser. A Math. Phys. Sci.* **1935**, *151*, 444–454. [CrossRef]

electronics

Article

Cross-Layer Optimization Spatial Multi-Channel Directional Neighbor Discovery with Random Reply in mmWave FANET

Yifei Song [1], Liang Zeng [1,*], Zeyu Liu [2], Zhe Song [3], Jie Zeng [1] and Jianping An [1]

1   School of Cyberspace Science and Technology, Beijing Institute of Technology, Beijing 100081, China; 3120215686@bit.edu.cn (Y.S.); zengjie@tsinghua.edu.cn (J.Z.); an@bit.edu.cn (J.A.)
2   Xuteli School, Beijing Institute of Technology, Beijing 100081, China; 1120181541@bit.edu.cn
3   School of Information and Electronics, Beijing Institute of Technology, Beijing 100081, China; 3220195085@bit.edu.cn
*   Correspondence: liang@bit.edu.cn

Abstract: MmWave FANETs play an increasingly important role in the development of UAVs technology. Fast neighbor discovery is a key bottleneck in mmWave FANETs. In this paper, we propose a two-way neighbor discovery algorithm based on a spatial multi-channel through cross-layer optimization. Firstly, we give two boundary conditions of the physical (PHY) layer and media access control (MAC) layer for successful link establishment of mmWave neighbor discovery and give the optimal pairing of antenna beamwidth in different stages and scenarios using cross-layer optimization. Then, a mmWave neighbor discovery algorithm based on a spatial multi-channel is proposed, which greatly reduces the convergence time by increasing the discovery probability of nodes in the network. Finally, a random reply algorithm is proposed based on dynamic reserved time slots. By adjusting the probability of reply and the number of reserved time slots, the neighbor discovery time can be further reduced when the number of nodes is larger. Simulations show that as the network scale is 100 to 500 nodes, the convergence time is 10 times higher than that of the single channel algorithm.

Keywords: flying ad-hoc network (FANET); millimeter-wave (mmWave); neighbor discovery; unmanned aerial vehicle (UAV)

Citation: Song, Y.; Zeng, L.; Liu, Z.; Song, Z.; Zeng, J.; An, J. Cross-Layer Optimization Spatial Multi-Channel Directional Neighbor Discovery with Random Reply in mmWave FANET. *Electronics* **2022**, *11*, 1566. https://doi.org/10.3390/electronics11101566

Academic Editor: Raed A. Abd-Alhameed

Received: 19 February 2022
Accepted: 2 May 2022
Published: 13 May 2022

Publisher's Note: MDPI stays neutral with regard to jurisdictional claims in published maps and institutional affiliations.

1. Introduction

Unmanned aerial vehicles (UAVs) are aircrafts that are controlled by a remote radio or an autonomous program without a human onboard [1]. In recent years, with the fast improvement of technologies such as artificial intelligence and the Internet of Things, UAVs have entered a rapid development stage, whose variety is more abundant, and their application scenarios have further expanded from the military field to major national application fields such as urban anti-terrorism and disaster relief [2]. Taking emergency disaster rescue as an example, its typical disaster rescue scenario is shown in Figure 1. In rescue operations, UAVs can provide a variety of functions through large-scale, long-term persistence, and multi-level deployment, including but not limited to, providing real-time information acquisition as the final sensor, expanding the search range as a network communication relay, and providing relief materials as an execution platform [3]. With the continuous development of UAV technology, the types of tasks involved in specific applications such as disaster relief will become more extensive, and the role it will play is expected to become more important [4].

A multi-UAV system organized in a meshed manner is referred to as a flying ad-hoc network (FANET), where multiple UAVs can collaboratively carry out complex missions [1]. FANETs have the advantages of no fixed facilities required, flexible self-organization, and a mobile platform, which is the information basis of a drone swarm to complete real-time sharing and collaborative work [5], and is one of the most important technologies of UAVs.

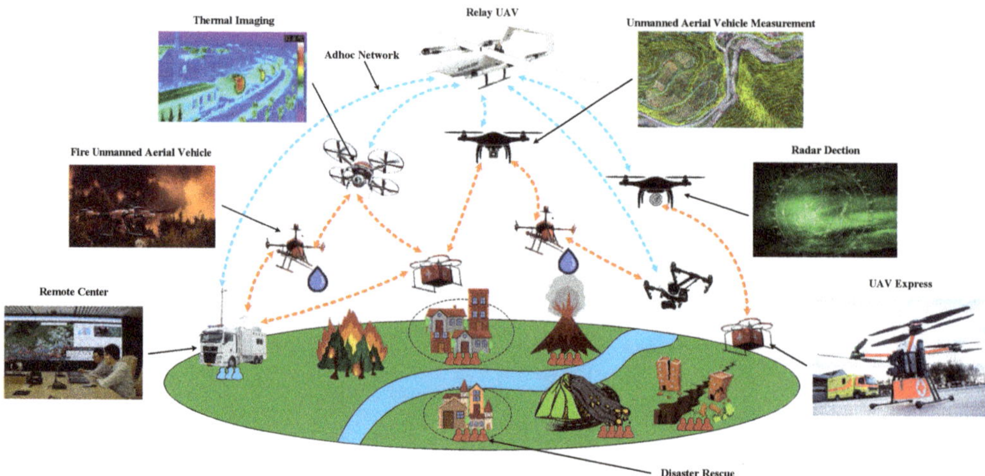

Figure 1. The important role of drones in disaster relief.

FANETs for drone swarm need to bear the various output mission-related data of different sensors [6]. As the resolution of onboard sensors becomes higher and the tasks become more strenuous and arduous, the data traffic of mission-related UAV is rapidly growing [7]. For the frequency bands sub-6 GHz which is occupied in a large amount by mobile communications and other services, the traditional wireless services have occupied a large percentage of the available spectrum, and have been unable to meet the rapidly developing communication needs of the UAV business. The mmWave frequency band has the advantages of a wide available bandwidth, an antenna aperture which is small and easy to install, beam flexibility among other advantages [8], which makes FANETs based on mmWave a research hotspot in recent years [9–11].

A recognized challenge in the research of FANETs based on mmWave is how to perform fast neighbor discovery [1]. This is due to the following reasons:

- In FANETs, neighbor discovery is the basis for self-organization, as well as a precondition for routing and resource allocation. In FANETs, each UAV node needs to discover its neighbors in a 3D space. This brings difficulties to discover neighbors in itself;
- In contrast to lower communication frequencies, the path loss in the mmWave band is much higher at the same distance. To ensure the connectivity of the links, the antenna major lobe beam with the mmWave band tends to be narrower for D2D (Device-to-Device) communication at the same communication distance;
- The two reasons above bring up that the neighbor discovery algorithm of FANETs based on mmWave needs to solve the problem of scanning a larger spatial range with a narrower beam. This makes the neighbor discovery process inevitably slower, especially in asynchronous FANETs.

As shown in Table 1, neighbor discovery methods are mostly IEEE 802.11-based with an omnidirectional antenna. The "hidden terminal" and "deafness" problems are more challenging in directional transmission [12,13]. Mostly directional neighbor discovery algorithms are proposed based on TDMA. Neighbor discovery with omnidirectional antenna assistance has been proposed in [13,14]. However, the different communication distances between omnidirectional antennas and directional antennas may result in different neighbor sets. To eliminate the reliance on omnidirectional antennas, ref. [15–18] assume time synchronization, which can ensure that all nodes are synchronized when switching antennas. However, this requires equipment, such as GPS, which will increase the overhead of node hardware. There are also some works that use asynchronous systems, such as [19,20]. The authors let the node enter the transmission mode with probability p and send the

HELLO packet, and then, enter the reception mode with probability $1 - p$ to monitor the channel. This probability-based neighbor discovery method is simple, but it has the following disadvantages. First, the probability-based neighbor discovery method cannot guarantee the time of neighbor discovery, and some neighbors may not be discovered. Using the scan-based method [17,21,22], the upper bound of the neighbor discovery time can be guaranteed. Second, some methods such as the 1-way neighbor discovery in [20], only receive HELLO messages but do not reply. This results in that only the receiving node knows the existence of the sending node, while the sending node still does not know the existence of the receiving node, that is, after the sole node discovery process, there is still no bidirectional communication link. Therefore, a 2-way handshakes mechanism is necessary [22–24].

Table 1. Related Works.

Algorithm	Feature	Years	Reference
BD-SBA	The link layer is based on IEEE 802.11. Carrier sense of the BD-SBA algorithm is performed in the first broadcast step which can reduce the collision of broadcasting the scanning request frames.	2019	[12]
MDND	Use the 2.4-GHz band with the omnidirectional antennas. The data transmissions are performed by using the 60-GHz band with directional antennas.	2015	[13]
Gossip-Based Algorithm	Nodes gossip about their neighbors' location information to accelerate the discovery. However, the synchronous algorithms need additional hardware or GPS support.	2005	[19]
1-way ND	Each device periodically transmits advertisement messages to announce its neighbors. When the neighbors receive the advertisement information, it determines that the neighbor discovery is successful.	2015	[20]
2-way ND	Once a device receives an advertisement message, it provides an active response to its neighbor. This allows neighbor nodes to discover each other.	2015	[20]
I-SBA	An extra mode named 'idle' is added in every time slot. When the node receives a new advisement information successfully, it will not always respond but with probability p.	2012	[25]
PRA	Send HELLO packets with probability p at the beginning of each time slot in a random direction, which makes all nodes transmit and receive in different directions that decreases the collision probability.	2008	[15]

This paper also tries to solve the problem of fast neighbor discovery in the mmWave FANETs network from our point of view. The main contributions of this paper are listed below:

- A 3D spatial scanning method combining the PHY layer and MAC layer is proposed. Under the condition of a closed physical layer link, the matching mode of mmWave antenna beamwidth is optimized, and the spatial scanning time is fully reduced through cross-layer optimization design.
- A space division multi-channel reply mechanism based on mmWave narrow beam is proposed. By increasing the number of reply channels existing at the same time in 3D space, the efficiency of the neighbor discovery algorithm is improved and the neighbor discovery time is greatly shortened.
- Furthermore, two different reply mechanisms of fixed reply and random reply are given, the average convergence time of neighbor discovery time under the two reply mechanisms is derived, and the optimization strategies under different conditions are given.

The rest of this paper is arranged as follows: In Section 2, we will give the PHY layer and MAC layer joint modeling method of the mmWave FANETs. In Section 3, we will introduce our neighbor discovery algorithm in detail, including two steps: optimizing paired spatial scanning and space division multi-channel reply. Section 4 will give the performance comparison and simulation of the algorithm. The simulation shows that our algorithm has great superiority in average time for neighbor discovery. Section 5 provides a conclusion.

2. System Model

2.1. Network and Node Model

The antenna 3D-model of a UAV is shown In Figure 2. The beam of the antenna can be viewed as a spherical cone. The width of the beam is the solid angle of the spherical cone, which is denoted by θ. The angle between the beam and the x-axis is denoted by γ. The angle between the beam and the z-axis is denoted by ϕ. The surface cap area A, on a unit radius sphere centered at the cone apex, of a spherical cone of angular width θ, is

$$A = \int_{cos(\frac{\theta}{2})}^{1} 2\pi dx = 2\pi(1 - cos(\frac{\theta}{2})) \tag{1}$$

and the area of a unit radius sphere is denoted as S. The number of the beams is denoted by K. To cover the whole sphere with a fixed K non-overlapping cones, the solid angle without overlapping denoted as $\theta_{no\_overlap}$ can be obtained from the following equation by considering only the numerical value of the area.

$$K = \frac{S}{A} = \frac{4\pi}{2\pi(1 - cos(\frac{\theta_{no\_overlap}}{2}))} = \frac{2}{1 - cos(\frac{\theta_{no\_overlap}}{2})} \tag{2}$$

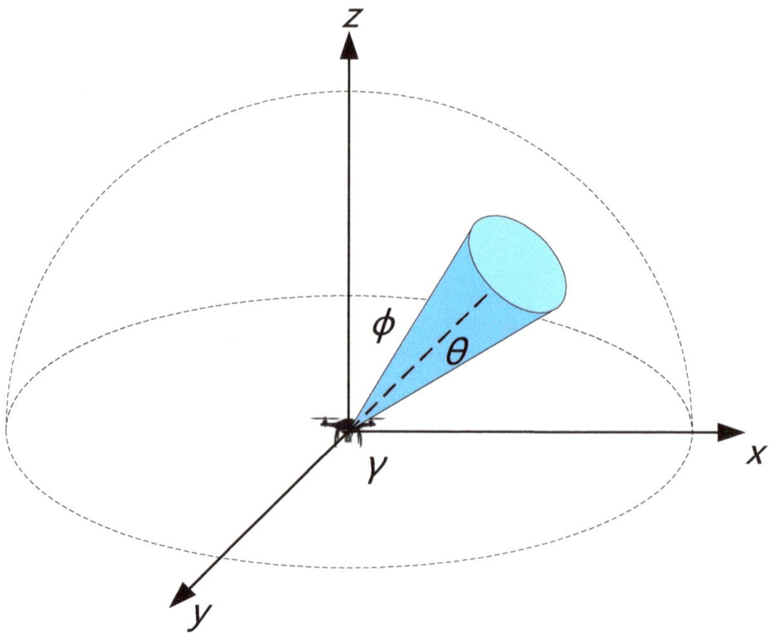

Figure 2. The 3D model of the direction antenna.

In fact, it is impossible to cover the whole sphere with K non-overlapping cones. Considering the overlapping, the whole sphere can be coverd with 12 pentagons simply when K is 12 [26,27]. The solid angle with overlapping denoted as $\theta_{overlap}$ is

$$\theta_{overlap} = \varepsilon \cdot \theta_{no\_overlap} \tag{3}$$

in which the ε is a coefficient greater than 1 because of the overlapping. The ε is not a fixed constant which varies with K. The relationship between K and θ_{overlap} is

$$K = \frac{2}{1 - cos(\frac{\varepsilon \cdot \theta_{\text{overlap}}}{2})} \tag{4}$$

While in the planar case the beamwidth increases linearly with the increase of the inverse of K, with 3D beams this increase is proportional to $(1 - cos(\frac{\theta}{2}))$.

We discuss the flying ad hoc with a cluster head as shown in Figure 3. Assuming that the number of nodes in the network is N, each node in the network has a unique ID. Let node 0 be the cluster head, and node 1 to node $N - 1$ are cluster members, which is denoted with n $\in [0, N - 1]$. It is assumed that all cluster members in the network are one-hop reachable to the cluster head.

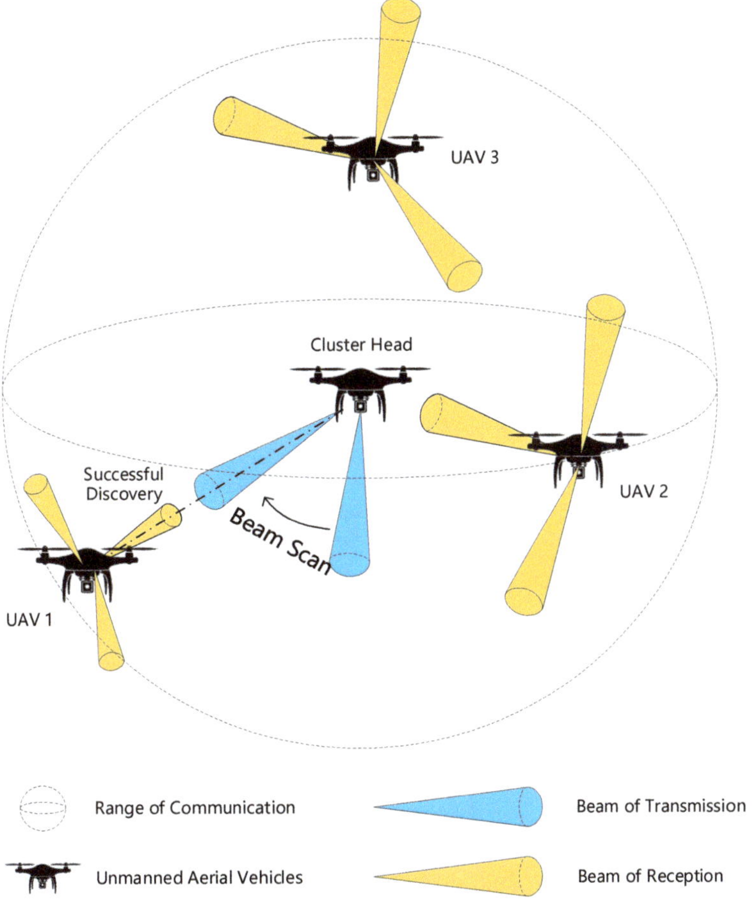

Figure 3. Neighbor discovery in sector model.

Both the cluster head node and the cluster members work in the mmWave frequency band, and the common frequency band is 24 GHz~32 GHz. As mentioned above, the free space loss is relatively large because of the mmWave frequency band. The node needs to be equipped with a mmWave directional antenna with variable beamwidth, and the need for long-distance wireless communication can be achieved through a very narrow antenna beam. This will cut the coverage of the node into sectors. In Figure 3, the cluster head and node 1 can discover each other because of the alignment of sectors. Node 2 points at the cluster head but the cluster head does not. They cannot discover each other. The cluster head and node 3 cannot discover their neighbor because of the misalignment.

We use a directional graph $G = (V, E)$ to express the network link relationship, where V represents a node, and E represents a directional link. For example, if there is a directed link between node i and node j, then $(i, j) \in E$.

Therefore, the condition for successful neighbor discovery in FANETs with a cluster head is that the cluster head node can successfully discover all neighbor nodes, that is, a two-way link establishment is completed between the cluster head node and all cluster members. The convergence time of the neighbor discovery algorithm is the interval from the start of the discovery process from the cluster head node to the time of successful completion of a two-way link establishment between the cluster head and each cluster member.

2.2. Bi-Directional Communication Conditions

Based on the above model, this section discusses the link establishment conditions of two nodes i and j. It is assumed that each node works in half-duplex mode, that is, the transmission and reception cannot be performed at the same time. P_t, G_t, G_r and P_{r_min} represent the transmit power, the transmit antenna gain, the receive antenna gain and the receiving sensitivity, respectively. K_t is the number of transmitting antennas and K_r is the number of receiving antennas. Each node in the network is equipped with multiple mmWave antennas with variable beamwidths or adjustable beamwidths, each antenna can be used for transmitting and receiving. $\theta_t, \theta_r \in [0, 2\pi]$ represent the beamwidth of the transmit antenna and the beamwidth of the receive antenna, respectively. For a specific θ_t and θ_r, K_t and K_r beam for transmission and reception are required to cover the entire space, respectively. According to the basic knowledge of the antenna, the transmit antenna gain $G_t \propto \frac{1}{\theta_t}$, the receive antenna gain $G_r \propto \frac{1}{\theta_r}$, $K_t \propto \frac{1}{\theta_t}$ and $K_r \propto \frac{1}{\theta_r}$. It can be further known that $G_t \propto K_t$, $G_r \propto K_r$. That is to say, the antenna gain is proportional to the number of sectors.

According to the relationship of K_t and K_r, there are three cases as follows:

- Wide transmission and narrow reception ($K_t < K_r$).
  The larger K_r is, the more time is needed for a node to scan the same area, in order to ensure that the receiving node has enough time to receive a complete beacon frame.
- Narrow transmission and wide reception ($K_t > K_r$).
  The larger K_t is, the more sectors are needed to scan when sending node broadcasts the packets.
- Symmetric transmission and reception ($K_t = K_r$).
  Although both K_t and K_r are smaller than the above, there is the problem that nodes need more time to point at each other in neighbor discovery.

From the above, the successful link establishment of a pair of nodes with IDs i and j requires two conditions of the PHY layer and the MAC layer to be satisfied at the same time:

* PHY layer: Considering the symmetric channel, the two links of the transmit node i and the receive node $j(i \rightarrow j)$, as well as the transmit node j and the receive node $i(j \rightarrow i)$. Taking $(i \rightarrow j)$ as an example, it needs to satisfy: $P_{r\_j} = \frac{P_{t\_j} * G_{t\_i} * G_{r\_i}}{L_p} \geq P_{r\_min}$, where L_p is the path loss between node i and node j. The formula above indicates that the physical layer link establishment condition of node i and node j is that the receive power must be greater than the receiving sensitivity requirement.
* MAC layer: If nodes i and j want to successfully establish a link between them in both directions $i \rightarrow j$ and $j \rightarrow i$, it is necessary for node i and node j to use the correct transmit and receive antennas pointing at each other, as shown in Figure 4a. Figure 4b shows that the MAC layer also needs to solve the multi-node collision problem within the same coverage area, because of the fact that all cluster members in the network are one-hop neighbor nodes after passing through the cluster head as described above.

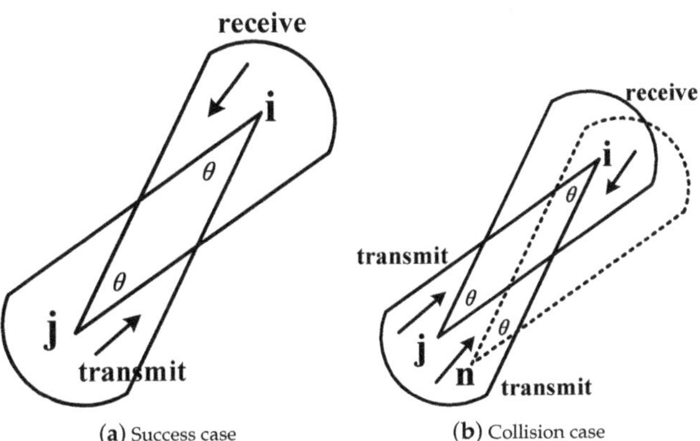

(a) Success case   (b) Collision case

Figure 4. Link established cases.

2.3. Cross-Layer Optimization Method

Considering the above two conditions comprehensively, any pair of nodes with IDs i and j can obtain different transmit and receive antenna gains by adjusting the beamwidth of the transmit and receive antennas to meet the link establishment conditions of the physical layer; at the same time, the adjustment of the beamwidth will also affect the difficulty of two nodes pointing at each other in the airspace and the probability of collision problem caused by multiple nodes. For qualitative analysis, the narrower the node beamwidth, the higher the antenna gain, and the easier it is to meet the physical layer link establishment conditions; but at the same time, the more beam positions that need to be scanned, the more difficult it is for two antennas to point at each other. Therefore, cross-layer optimization of the PHY layer and MAC layer is needed to optimize the neighbor discovery time of the whole network. We adopt the joint optimization method as shown in the figure below to carry out the study. The longest communication distance between cluster members and cluster heads of the whole network is obtained from the application layer and provided to the PHY layer, the number of network nodes is obtained from the application layer and provided to the MAC layer. The PHY layer calculates the required gains of the transmit and receive antennas according to the link establishment condition (1), and provides it to the MAC layer at the same time; the MAC layer, taking the shortest neighbor discovery time of the whole network as the optimization goal, calculates the transmit and receive beamwidth that the node needs to use in the scanning and ACK phases, respectively, and feeds it back to the PHY layer according to the number of nodes in the network and the gain requirements of the transmit and receive antennas. Subsequent sections of this paper will quantitatively analyze the influence of the transmit and receive antenna beamwidth on the neighbor discovery time.

On this basis, we can further reduce the neighbor discovery time to adapt to scenarios with very strict requirements on time, such as rescue operations. The concept of a multi-receive channel is further considered, that is, allowing a single node to have M receivers at the same time when it has K_r receiving beam positions, as shown in Figure 5. The quantitative analysis of the influence of different numbers of receivers on neighbor discovery time will be given in the following sections.

Note that in the process discussed in this paper, we consider that the nodes are not motive. Besides, all nodes have the following features:

- Time is divided into time slots, but it does not mean that all nodes have perfect time synchronization;
- Communication is in half-duplex mode, which means a node cannot transmit and receive simultaneously;

- The node does not rotate since the pointing relation of directional antennas of two nodes will be greatly changed when the nodes rotate.

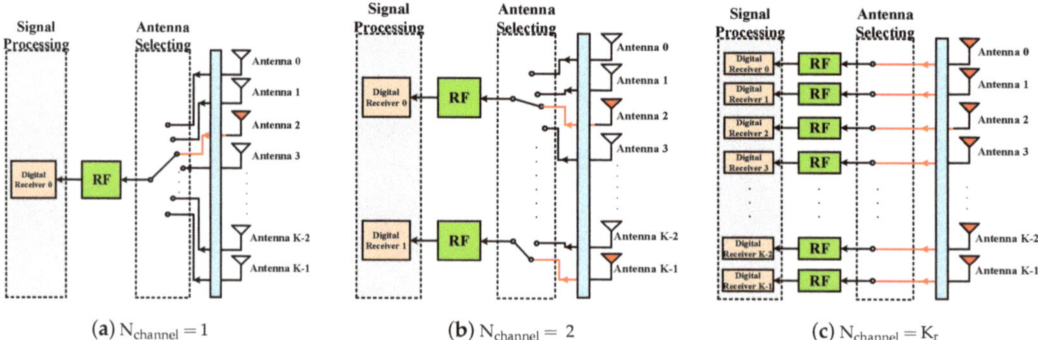

(a) $N_{channel} = 1$ (b) $N_{channel} = 2$ (c) $N_{channel} = K_r$

Figure 5. Multi-channel model.

3. Cross-Layer Based Spatial Multi-Channel Directional Neighbor Discovery with Random Reply (CSM-RR) Algorithm

Initially, the nodes boot up in a fast scan mode, in which they switch sectors in turn counterclockwise for reception, with each sector residing for one communication time slot. When the cluster head switches on, the neighbor discovery process begins, and the whole process can be divided into two steps:

- Beacon dispatch based on the scan.
- Random reply mechanism based on the dynamic reserved slot.

3.1. Beacon Dispatch Based on Scanning

Before the cluster head starts the neighbor discovery, the cluster members need to align their activated sectors to the cluster head and lock them first. In the initial phase, all nodes are in fast scan mode switching sectors and waiting for beacon frames sent by the cluster head. As shown in Figure 6, when the cluster head switches on, it stays in each sector for K_r time slots and sends beacon frames to ensure that all cluster members in this sector can receive the beacon frame after a fast scan. When a cluster member receives a beacon frame, it would stop the sector switching immediately and lock its own beam pointing. The cluster head will broadcast beacon frames several times in all sectors.

To describe more easily, the beam model is simplified to a 2D sector in the following example. Take the beacon dispatch for a single sector as an example in Figure 7. In the beginning, all cluster members in Figure 7a are in fast scan mode. In Figure 7b, the cluster head starts to send beacon frames in the first sector. After a beacon frame is received by cluster member 1, it locks the sector pointing to the cluster head immediately and enters the waiting state. Cluster member 2 is also in the sector where the beacon frame is sent, but the receiving sector which is activated by cluster member 2 does not point to the cluster head, hence no beacon frame is received. In Figure 7c,d, cluster member 2 is still in fast scan mode, switching its own receiving sector. The cluster head sends beacon frames continuously until cluster member 2 switches to the sector pointing to the cluster head. After a beacon frame is received by cluster member 2, it locks the sector and answers the waiting state, as shown in Figure 7d,f.

	Sector 1				Sector 2			⋯		Sector K_t		
Cluster Head	Beacon 1	Beacon 2	⋯	Beacon K_r	Beacon 1	Beacon 2	⋯	Beacon K_r	●●●		Beacon 1	Beacon 2	⋯	Beacon K_r

Member Node	Receiving	Receiving	⋯	Receiving	Receiving	Receiving	**Received!**	Waiting	
	S_1	S_2	⋯	S_K	S_1	S_2		S_i	

Figure 6. The sector switches in beacon dispatch.

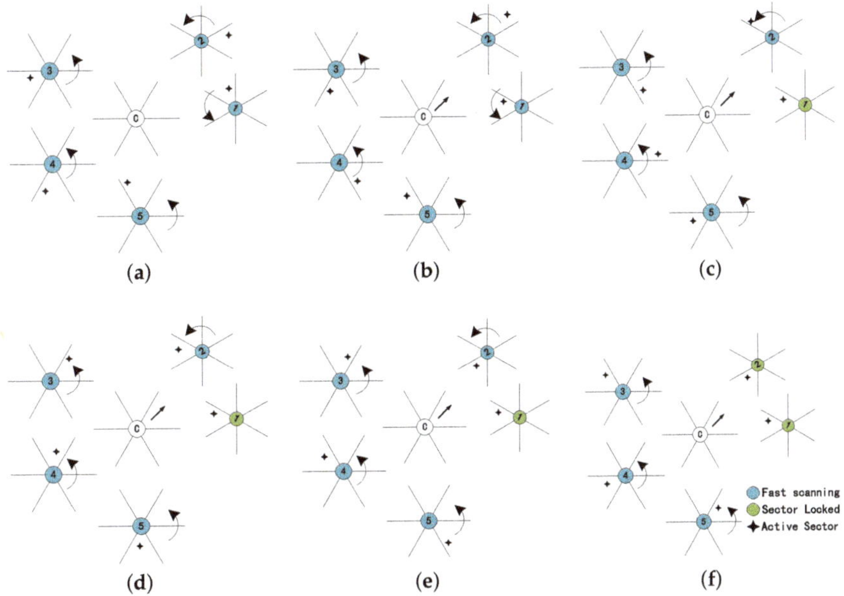

Figure 7. An example of beacon dispatch. (**a**) All cluster members are in fast scan mode. (**b**) The cluster head starts to send beacon frames in the first sector. (**c**) Cluster member 1 has locked the sector. (**d**) The cluster head sends beacon frames continuously until cluster member 2 switches to the sector pointing to the cluster head. (**e**) Cluster member 2 has pointed at the cluster head. (**f**) Cluster member 2 has locked the sector.

3.2. Random Reply Mechanism Based on Dynamic Reserved Slot

When beacon dispatch is over, it can be considered that all cluster members point at the cluster head and lock it. At this time, the cluster head starts to broadcast the HELLO packet once in each sector, and when the cluster members receive the HELLO packets, they reply according to the protocol agreement in the subsequent reply phase. When the cluster head receives all the replies, it is considered that all cluster members have been discovered.

Subsequently, an end packet of neighbor discovery is broadcasted in each sector, at which point the neighbor discovery process ends.

The HELLO packet contains three pieces of information, which are:

1. The start time of the reply phase by cluster members;
2. The number of time slots reserved for the reply phase of cluster members of the current round;
3. The number of cluster members that have been discovered.

Then the random reply mechanism based on the dynamic reserved slot is proposed here for neighbor discovery. The cluster head broadcasts HELLO packets on each sector after beacon dispatch and then starts waiting for replies from the cluster members. After receiving the HELLO packet, cluster members can choose whether to reply to the HELLO message in this round with probability P_{reply}, which can reduce the collision of reply packets. When choosing to reply to the HELLO packets in this round, the cluster member will select a time slot to reply randomly for reducing the collision of two reply packets. If two nodes choose the same time slot to send reply packets, a collision occurred and neither of their reply packets could be received correctly by the cluster head. Once the cluster head receives a reply message from a cluster member, it adds the cluster member to the neighbor list. When the number of discovered neighbors cannot reach the number of nodes ($N_{\text{accessed}} \neq N$), the cluster head starts a new round of neighbor discovery process until all cluster members are discovered.

When there are few nodes, the reply time slot of each node is very easy to plan. As shown in Figure 8, it is called the fixed reply mechanism when the number of the reserved slot (N_{reserved}) is set to a constant equal to $N - 1$ and the probability of reply P_{reply} is set to 1 After receiving the HELLO packet, the cluster member just selects the time slot with the same index as its own ID to reply in the reply phase. The replies from cluster members will not collide because the sufficient time slots are reserved.

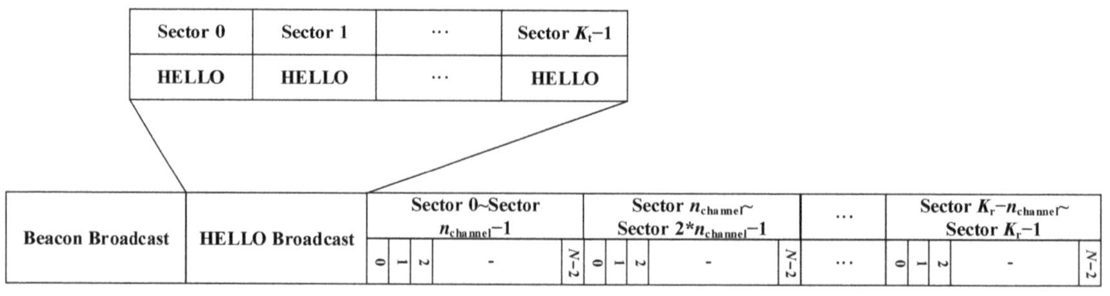

Figure 8. Fixed reply mechanism.

The analysis for neighbor discovery with fixed reply is as follows. Since the cluster head has the ability of multi-channel reception, it can receive data packets from N_{channel} aerials at the same time. When $N_{\text{channel}} = K_r$, the whole reply phase of cluster members only lasts for $N - 1$ time slots. The duration of the neighbor discovery process of the fixed reply mechanism, excluding the beaconing phase, can be expressed as:

$$t_{\text{discovery}} = 2K_t \cdot t_{\text{Hello}} + \frac{K_r}{N_{\text{channel}}} \cdot (N - 1) \cdot t_{\text{reply}} \tag{5}$$

in which $t_{\text{discovery}}$ is the time for replying, t_{Hello} is the duration of the HELLO packet, K_t is the number of transmitting sectors, K_r is the number of receiving sectors, t_{reply} is the duration of the REPLY packet.

The overall duration of the discovery process with the fixed reply mechanism, including the beaconing phase, can be expressed as:

$$t_{all} = t_{align} + t_{discovery} = K_t \cdot K_r \cdot t_{beacon} + 2K_t \cdot t_{Hello} + \frac{K_r}{N_{channel}} \cdot (N-1) \cdot t_{reply} \tag{6}$$

in which t_{all} is the time for cluster head finding all neighbors, t_{align} is the time for dispatch, t_{beacon} is the duration of the beacon frame.

As shown in Algorithms 1 and 2, the random reply mechanism is different from the fixed reply mechanism. The $n_{reserved}$ and p_{reply} are not fixed in each round of neighbor discovery. These two parameters are adjusted continuously with r, the number of rounds of the neighbor discovery. The duration of the neighbor discovery process of the random reply mechanism, excluding the beaconing phase, can be expressed as:

$$t_{discovery} = 2K_t \cdot t_{Hello} + \sum_{i=0}^{r} \left(\frac{K_r}{N_{channel}} \cdot n_{reserved}(i) \cdot t_{reply} \right) \tag{7}$$

The overall duration of the discovery process with the random reply mechanism, including the beaconing phase, can be expressed as:

$$t_{all} = t_{align} + t_{discovery} = K_t \cdot K_r \cdot t_{beacon} + 2K_t \cdot t_{Hello}$$
$$+ \sum_{i=0}^{r} \left(\frac{K_r}{N_{channel}} \cdot n_{reserved}(i) \cdot t_{reply} \right) \tag{8}$$

in which r is the number of the replying round.

Algorithm 1 Random reply algorithm in cluster head

During the dispatch stage:
$n = 0$
while $n < K_t$ **do**
  select the nth sector
  broadcast the beacon for K_r slots
  $n++$
end while
During the randomly reply stage:
$r = 0$
while $r < threshold$ **do**
  if $present - mode = transmission$ **then**
    $N_{reserved}(r) = \beta \times N_{reserved}(r-1)$
    $n = 0$
    while $n < K_t$ **do**
      select the nth sector
      encapsule the neighbor list into HELLO packet
      encapsule the $N_{reserved}(r)$ into HELLO packet
      broadcast the HELLO packet for 1 slot
      $n++$
    end while
  end if
  if $present - mode = reception$ **then**
    if received a REPLY packet **then**
      add the identity to the neighbor list
    end if
    $r++$
  end if
end while

Algorithm 2 Random reply algorithm in cluster member

During the dispatch stage:
while has not received the beacon **do**
    change the active sector
    listening
end while
point at the cluster head and lock the sector
During the randomly reply stage:
if has received the HELLO packet **then**
    if neighbor list has self-identity **then**
        return
    else
        $s = random(0, N_{reserved})$
        transmit the REPLY packet in s slot with probability $p_{reply}(r)$
        $p_{reply}(r) = \alpha \times p_{reply}(r-1)$
    end if
end if

Take the topology in Figure 7 as an example, and assume $n_{receiver} = K_r$. Figure 9 shows the neighbor discovery process for $N = 6$. In *round* $= 0$, the cluster head broadcasts HELLO packets and specifies the number of reserved time slots is $n_{reserved}(0)$ in this round. After the cluster members receive HELLO packets, node 1 and node 2 select time slot 2 to reply simultaneously, then collision occurs. Node 3 does not reply in this round according to the probability. Node 4 selects time slot 0 and node 5 selects time slot 1 to reply, the cluster head receives their REPLY packets successfully and adds node 4 and node 5 to the neighbor list. In *round* $= 1$, the cluster head broadcasts HELLO packets and the neighbor list including node 4 and node 5, and specifies the reserved time slot in this round is $n_{reserved}(1)$. Node 2 selects time slot 2 to reply and the cluster head receives REPLY packets successfully. Both node 1 and node 3 select time slot 0 to reply, and the reason why the cluster head also receives REPLY packets successfully is that the cluster head has the ability of multi-channel reception, and node 1 and node 3 are in different sectors of the cluster head, thus, the REPLY packets do not collide.

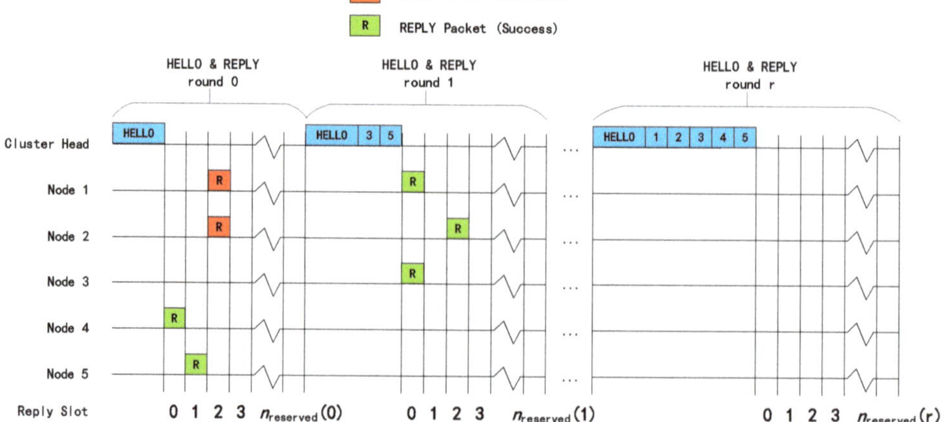

Figure 9. Neighbor discovery with random reply mechanism ($N = 6$).

4. Simulation Results and Discussion

Here, we conduct the performance comparisons of CSM-RR with other directional neighbor discovery algorithms under different numbers of nodes and sectors. Firstly, we compare the CSM-RR with the scanning-based algorithm. Then we compare the CSM-RR with the probabilistic algorithm. Finally, we show how the various control parameters (K, N_{channel}, α, β, and distribution of position) affect the performance of the CSM-RR protocol.

4.1. Comparison with Existing Methods

In order to verify the CSM-RR protocol, we implement it with the OMNeT++ simulation IDE. For comparative research, we also implemented other neighbor discovery algorithms based on the directional antenna, including the Pure Random Algorithm (PRA) proposed in the literature [15], the Improved Scan-based Algorithm (I-SBA) discussed in the literature [25], and the 1-way neighbor discovery algorithm (1-way) and 2-way neighbor discovery algorithm (2-way) proposed in the literature [20]. None of these directional neighbor discovery methods need the assistance of an omnidirectional antenna, and all of them divide the time into time slots. According to the way they respond, these algorithms can be divided into two categories: deterministic reply and probabilistic reply.

PRA and 1-way both send HELLO packets with probability p at the beginning of each time slot and enter the reception with probability $1 - p$. The difference is that after receiving the HELLO packet, PRA immediately replies to the REPLY packet, while 1-way does not reply to the REPLY packet.

Furthermore, 2-way sends the HELLO packets with probability p at the beginning of each time slot and enters receiving state with probability $1 - p$. At the beginning of the time slot, I-SBA not only selects transmission or reception, but also has the probability to enter the idle state. After receiving the HELLO packet, both methods reply to the REPLY packet with a probability, which is not certain.

In Figure 10, we compare the performance of the PRA, 1-way algorithm, and the fixed reply mechanism proposed in this paper under different numbers of nodes. The discovery time of a fixed mechanism, even with only the single channel, is much less than the PRA and 1-way algorithm. The fixed mechanism has almost no collisions because the reply slot of the cluster member is fixed. In addition, when adopting multi-channel, the time required for CSM-RR is much shorter.

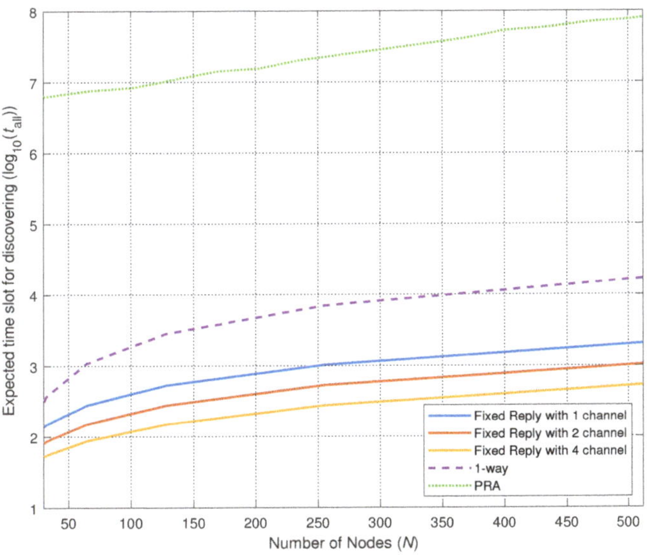

Figure 10. Comparison between neighbor discovery algorithm with fixed reply ($K = 4$).

In Figure 11, we compare the performance of the I-SBA, 2-way algorithm, and the random reply mechanism proposed in this paper under different numbers of nodes. The random mechanism takes less time to discover the same number of neighbors. When the number of nodes is large, the discovery time for I-SBA and 2-way algorithm increases to a large extent because of the collision. Moreover, the use of multi-channel neighbor discovery can further reduce the time of neighbor discovery.

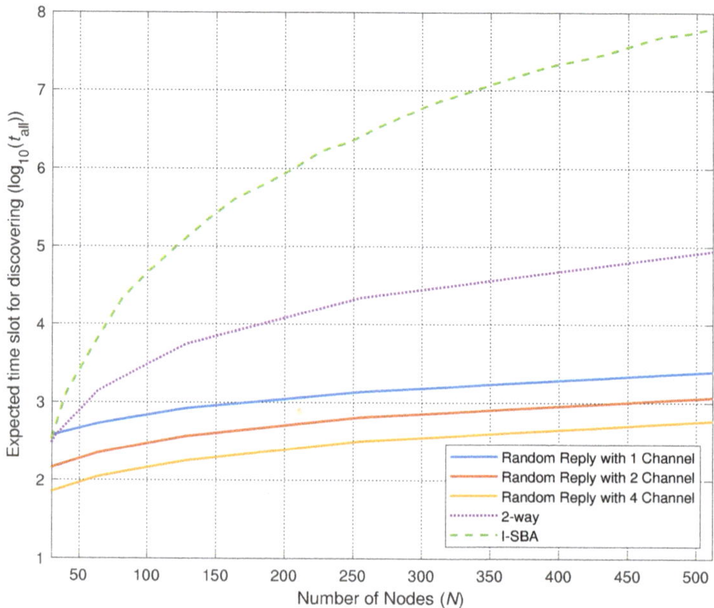

Figure 11. Comparison between neighbor discovery algorithm with random reply ($K = 4$).

In Figure 12, we compare the performance of 1-way, 2-way and CSM-RR protocols under different numbers of sectors. With the increase in the number of sectors, the neighbor discovery time of CSM-RR with the fixed reply mechanism and 1-way algorithm increases. Because the number of sectors increases, the number of transmission sectors that need to be scanned during transmission need to increase, which results in an increase in the neighbor discovery time. However, the neighbor discovery time of 2-way decreases gradually with the increase of the number of sectors. That is because, with the increase in the number of sectors, the number of neighbors in each sector will be reduced, which reduces the number of collisions when the nodes reply to the REPLY packet. In the CSM-RR algorithm with a random reply mechanism, when the number of sectors decreases, the neighbor discovery time decreases with the increase of the number of sectors. This is because the increase of sectors effectively reduces the collision of the REPLY packets when the number of nodes goes large, so that the probability of cluster members being discovered increases. When the number of sectors continues to increase, the time spent broadcasting the HELLO packet by scanning all sectors becomes the main factor affecting the neighbor discovery time, resulting in a longer neighbor discovery time.

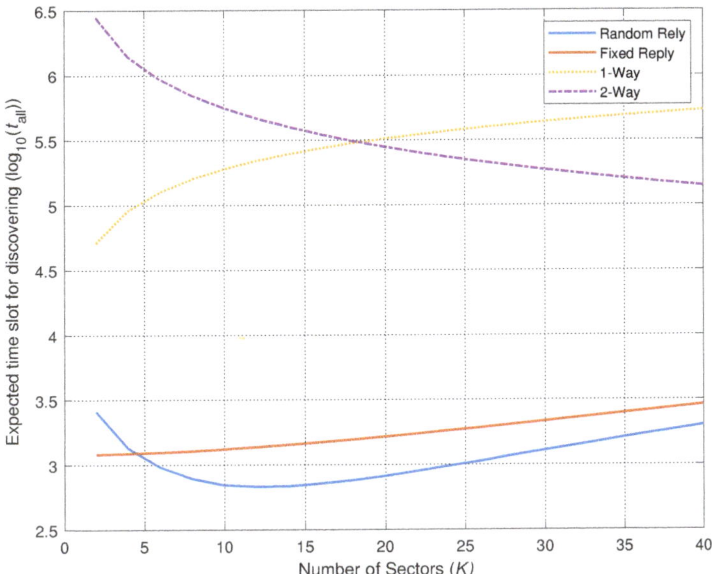

Figure 12. Time for discovery with different K. ($N = 1200$).

4.2. Ablation Experiment

In Figure 13, we set the K_t and K_r to be unequal. When the system has the multi-channel capability, the antennas' collection of "wide transmission and narrow reception" works better. Neighbor discovery time decreases along receiving antennas with the same receiving channels. However, when the number of receiving antennas increases, the number of receiving channels also increases, which will greatly improve the neighbor discovery time. When the multi-channel capability is certain, the lower number of receiving antennas occupies a certain advantage.

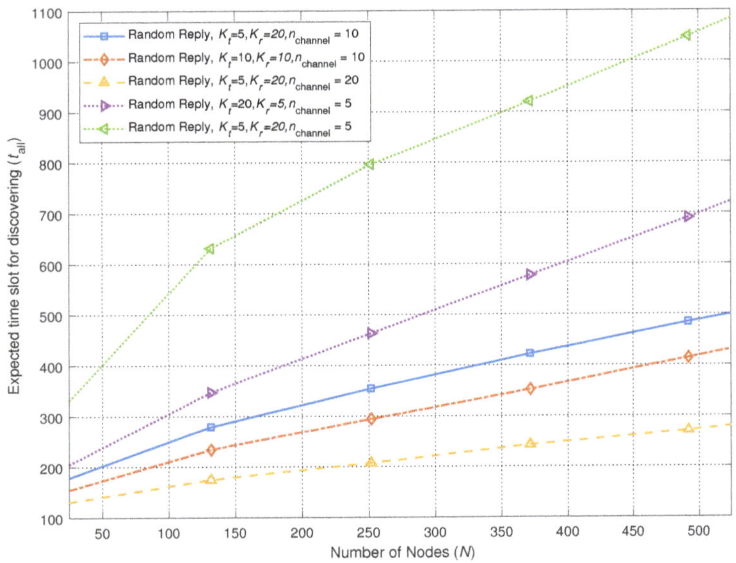

Figure 13. Symmetric K_r and K_t with different numbers of receivers.

In Figure 14, we set different α in the CSM-RR protocol. All cluster members reply to the HELLO packet with $p = 1$, so that the time of neighbor discovery can be minimum. In order to receive as many REPLY packets as possible in the next round, it needs to improve $p_{\text{reply}}(0)$, which is $p_{\text{reply}}(r) = (1 + \alpha) \cdot p_{\text{reply}}(r - 1)$.

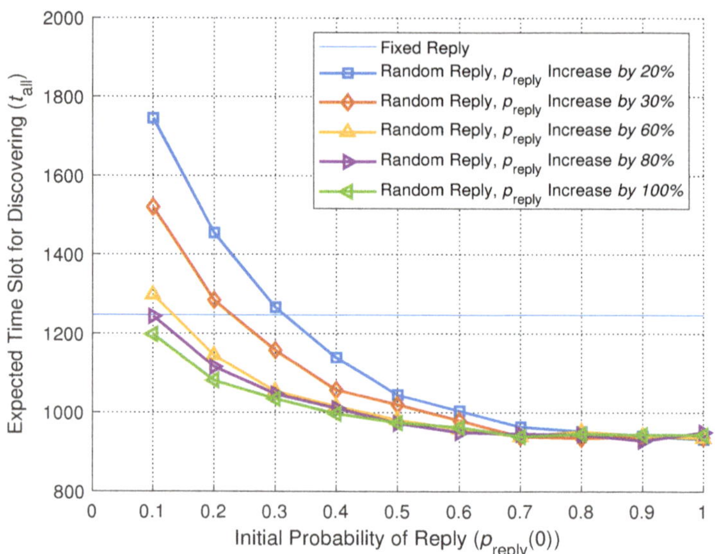

Figure 14. Discovery time vs. p_{reply}.

In Figures 15 and 16, we demonstrate how different β affects the performance of the CSM-RR. A suitable β ($\beta = 0.32$) can guarantee both the short neighbor discovery time and the neighbor discovery ratio. Since the cluster head will find new cluster members in each round of neighbor discovery, the number of undiscovered cluster members will be reduced in each round. In order to avoid the waste of reserved time slots in the reply phase, n_{reserved} should be reduced accordingly in each round, which is $n_{\text{reserved}}(r) = (1 - \beta) \cdot n_{\text{reserved}}(r - 1)$. However, the decrease in the reserved time slots will cause a large number of collisions of the REPLY packet. Thus, there are two conditions for the end of the neighbor discovery. The first one is that the cluster head discovers all cluster members, and the other is that the number of rounds of the neighbor discovery process is greater than 200, which is $r > 200$. If either of these conditions are satisfied, the neighbor discovery ends. However, when the neighbor discovery ends with $r > 200$, some of the cluster members may not be found. Once $\beta < 0.32$, the cluster head can find all cluster members. On the other hand, some neighbor nodes cannot be found when $\beta > 0.32$. This is because the large β leads to n_{reserved} which is far less than that of undiscovered cluster members, resulting in a large number of collisions in the reply packet of cluster members.

In addition to the above parameters, the position distribution of cluster members relative to the cluster head will also have a great impact on the performance of the protocol. Figure 17 shows the impact of different location distributions of cluster members on neighbor discovery time. The PMF of location distribution is shown in Figure 18. In distribution 1, distribution 2, and distribution 3, the cluster members are concentrated in different sectors. It can be seen that the different distribution of the location of cluster members will deteriorate the neighbor discovery time of the CSM-RR protocol. When the location of cluster members follows the uniform distribution, the neighbor discovery performance of the CSM-RR protocol is the best.

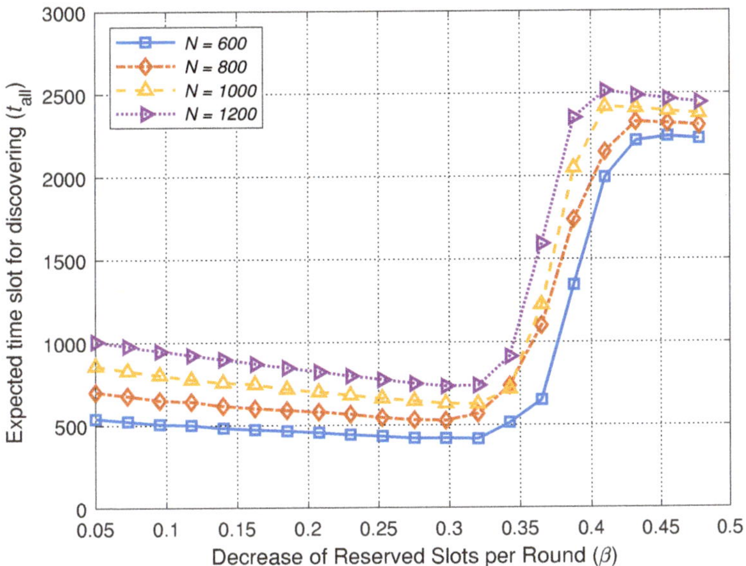

Figure 15. Time for discovery vs. β.

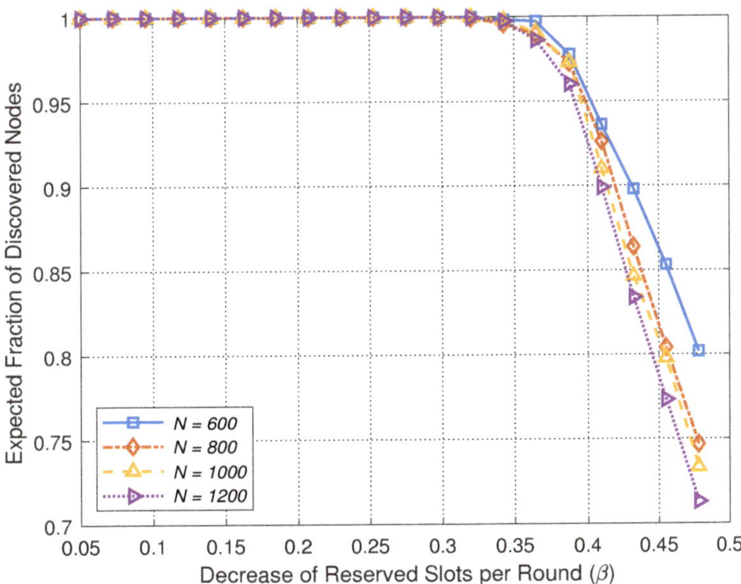

Figure 16. Fraction of discovery vs. β.

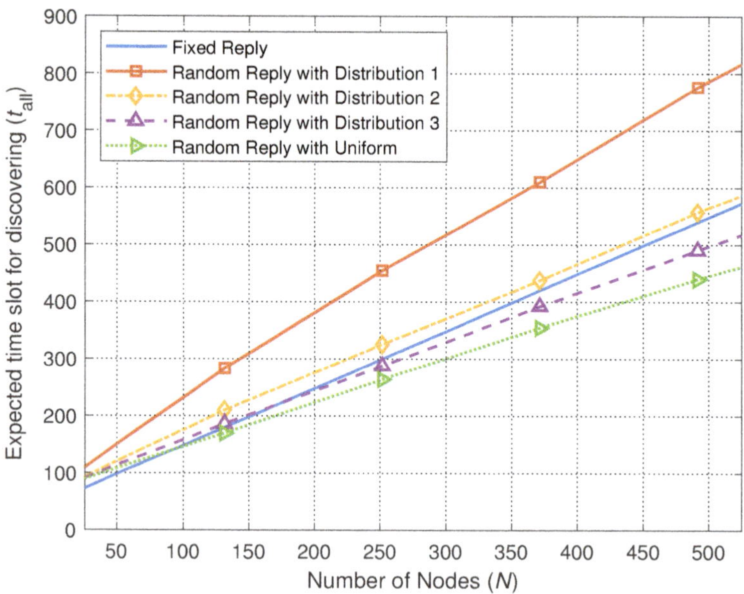

Figure 17. Time for discovery vs. different location.

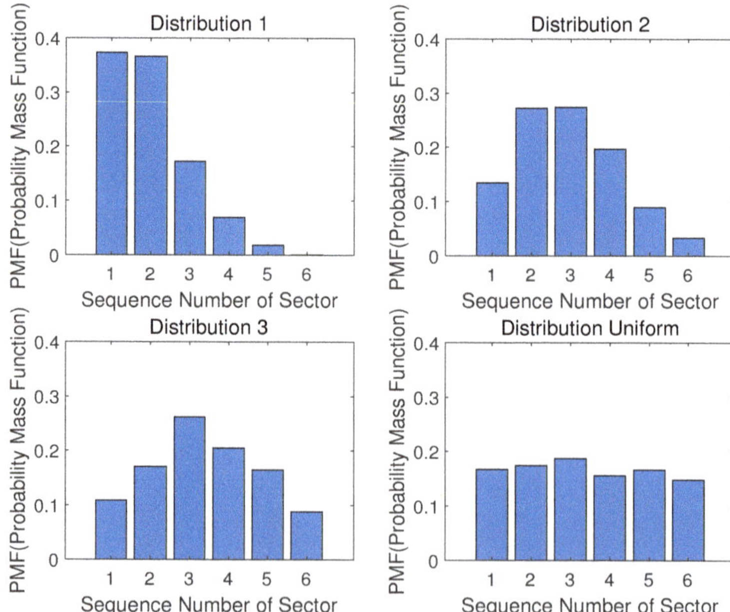

Figure 18. PMF of location distribution.

Through the simulation results from Figures 13–18, we have a good understanding of the parameter optimization of CSM-RR:

- The antennas' collection of "wide transmission and narrow reception" works better with multi-channel.
- It is necessary to increase the reply packets sent by each round of cluster members as much as possible under a certain collision probability, which is $p_{\mathrm{reply}} = 1$.

- On the premise of ensuring the neighbor discovery rate, it is necessary to reduce n_{reserved} in each round as much as possible. According to the above simulation results, the time required to discover all cluster members is the shortest with $\beta = 0.32$.
- The distribution of nodes also affects the protocol proposed in this paper. When the distribution of cluster members obeys uniform distribution, the efficiency of the random reply mechanism is the highest.

5. Conclusions

In this paper, we have proposed a neighbor discovery algorithm based on directional antennas called CSM-RR. Based on the design of our algorithm and the numberical results, we have the following key conclusions.

- The CSM-RR does not need to use the 2.4-GHz band with the omnidirectional in neighbor discovery. Therefore, the UAVs do not need additional transceivers and antennas, which will not cause link asymmetry due to different antenna gains.
- It also does not require a perfect time synchronization system to achieve initial synchronization of time slots during beacon dispatch. The UAVs do not need additional hardwares or GPS support thus reducing hardware complexity.
- The "Hidden terminal" and "deafness" problems do not arise because the CSM-RR is TDMA based.
- In addition to using the fixed reply mechanism in the case of few nodes, the CSM-RR adopts the random reply mechanism. When there are few nodes, the reply time slot of each node is very easy to plan. Therefore, the collisions do not occur.
- Since the CSM-RR uses a digital receiver, multiple channels can be used for simultaneous reception without adding additional hardware overhead. Without adding additional hardware, the CSM-RR greatly improves the efficiency of neighbor discovery.
- By optimizing N_{channel}, α, p_{reply}, β, K, the neighbor discovery time can be further reduced when the number of nodes is large. Simulations have shown that as the network scale is 100 to 500 nodes, the convergence time is 10 times higher than that of the single channel algorithm.

Consequently, the CSM-RR can be used to improve the neighbor discovery performance of FANETs. In this paper, we only consider the expected time slot for neighbor discovery as the performance. In our future work, other metrics such as the exchange of information and the number of control packets will be investigated.

Author Contributions: Y.S.: conceptualization, methodology, writing—original draft preparation; L.Z.: conceptualization, methodology, writing—review and editing; Z.L.: validation, visualization; Z.S.: conceptualization, project administration, writing—original draft preparation; J.Z.: writing—original draft preparation, writing—review and editing; J.A.: conceptualization, project administration. All authors have read and agreed to the published version of the manuscript.

Funding: This research was funded in part by the National Scientific Foundation of China under grants 62101050, 62001022, U1836201.

References

1. Xiao, Z.; Zhu, L.; Liu, Y.; Yi, P.; Zhang, R.; Xia, X.G.; Schober, R. A Survey on Millimeter-Wave Beamforming Enabled UAV Communications and Networking. *IEEE Commun. Surv. Tutor.* **2021**, *24*, 557–610. [CrossRef]
2. Shakhatreh, H.; Sawalmeh, A.H.; Al-Fuqaha, A.; Dou, Z.; Almaita, E.; Khalil, I.; Othman, N.S.; Khreishah, A.; Guizani, M. Unmanned Aerial Vehicles (UAVs): A Survey on Civil Applications and Key Research Challenges. *IEEE Access* **2019**, *7*, 48572–48634. [CrossRef]
3. Li, B.; Fei, Z.; Zhang, Y. UAV Communications for 5G and Beyond: Recent Advances and Future Trends. *IEEE Internet Things J.* **2019**, *6*, 2241–2263. [CrossRef]
4. Urama, J.; Wiren, R.; Galinina, O.; Kauppi, J.; Hiltunen, K.; Erkkila, J.; Chernogorov, F.; Etelaaho, P.; Heikkila, M.; Torsner, J.; et al. UAV-Aided Interference Assessment for Private 5G NR Deployments: Challenges and Solutions. *IEEE Commun. Mag.* **2020**, *58*, 89–95. [CrossRef]

5. Lin, J.; Cai, W.; Zhang, S.; Fan, X.; Guo, S.; Dai, J. A Survey of Flying Ad-Hoc Networks: Characteristics and Challenges. In Proceedings of the 2018 Eighth International Conference on Instrumentation Measurement, Computer, Communication and Control (IMCCC), Harbin, China, 19–21 July 2018; pp. 766–771. [CrossRef]
6. Liu, Y.; Luo, Z.; Liu, Z.; Shi, J.; Cheng, G. Cooperative Routing Problem for Ground Vehicle and Unmanned Aerial Vehicle: The Application on Intelligence, Surveillance, and Reconnaissance Missions. *IEEE Access* **2019**, *7*, 63504–63518. [CrossRef]
7. Xiao, M.; Mumtaz, S.; Huang, Y.; Dai, L.; Li, Y.; Matthaiou, M.; Karagiannidis, G.K.; Björnson, E.; Yang, K.; I, C.L.; et al. Millimeter Wave Communications for Future Mobile Networks. *IEEE J. Sel. Areas Commun.* **2017**, *35*, 1909–1935. [CrossRef]
8. Wang, X.; Kong, L.; Kong, F.; Qiu, F.; Xia, M.; Arnon, S.; Chen, G. Millimeter Wave Communication: A Comprehensive Survey. *IEEE Commun. Surv. Tutor.* **2018**, *20*, 1616–1653. [CrossRef]
9. Uwaechia, A.N.; Mahyuddin, N.M. A Comprehensive Survey on Millimeter Wave Communications for Fifth-Generation Wireless Networks: Feasibility and Challenges. *IEEE Access* **2020**, *8*, 62367–62414. [CrossRef]
10. Zhang, L.; Zhao, H.; Hou, S.; Zhao, Z.; Xu, H.; Wu, X.; Wu, Q.; Zhang, R. A Survey on 5G Millimeter Wave Communications for UAV-Assisted Wireless Networks. *IEEE Access* **2019**, *7*, 117460–117504. [CrossRef]
11. Niu, Y.; Li, Y.; Jin, D.; Su, L.; Vasilakos, A.V. A survey of millimeter wave communications (mmWave) for 5G: Opportunities and challenges. *Wirel. Netw.* **2015**, *21*, 2657–2676. [CrossRef]
12. Yang, A.; Li, B.; Yan, Z.; Yang, M. A Bi-Directional Carrier Sense Collision Avoidance Neighbor Discovery Algorithm in Directional Wireless Ad Hoc Sensor Networks. *Sensors* **2019**, *19*, 2120. [CrossRef] [PubMed]
13. Park, H.; Kim, Y.; Song, T.; Pack, S. Multiband Directional Neighbor Discovery in Self-Organized mmWave Ad Hoc Networks. *IEEE Trans. Veh. Technol.* **2015**, *64*, 1143–1155. [CrossRef]
14. Nur, F.N.; Sharmin, S.; Habib, M.A.; Razzaque, M.A.; Islam, M.S.; Almogren, A.; Hassan, M.M.; Alamri, A. Collaborative neighbor discovery in directional wireless sensor networks: Algorithm and analysis. *Eurasip J. Wirel. Commun. Netw.* **2017**, *2017*, 119. [CrossRef]
15. Zhang, Z.; Li, B. Neighbor discovery in mobile ad hoc self-configuring networks with directional antennas: Algorithms and comparisons. *IEEE Trans. Wirel. Commun.* **2008**, *7*, 1540–1549. [CrossRef]
16. Bai, W.; Xu, Y.; Wang, J.; Xu, R.; Anpalagan, A.; Chen, C.; Xu, Y.; Wang, X. Cognitive Neighbor Discovery With Directional Antennas in Self-Organizing IoT Networks. *IEEE Internet Things J.* **2021**, *8*, 6865–6877. [CrossRef]
17. Liu, B.; Rong, B.; Hu, R.Q.; Qian, Y. Neighbor discovery algorithms in directional antenna based synchronous and asynchronous wireless ad hoc networks. *IEEE Wirel. Commun.* **2013**, *20*, 106–112. [CrossRef]
18. Bilgin, Y.A.; Yılmaz, A.O. Neighbor discovery in network with directional antennas. In Proceedings of the 2013 21st Signal Processing and Communications Applications Conference (SIU), Haspolat, Turkey, 24–26 April 2013; pp. 1–4. [CrossRef]
19. Vasudevan, S.; Kurose, J.; Towsley, D. On neighbor discovery in wireless networks with directional antennas. In Proceedings of the IEEE 24th Annual Joint Conference of the IEEE Computer and Communications Societies, Miami, FL, USA, 13–17 March 2005; Volume 4, pp. 2502–2512. [CrossRef]
20. Cai, H.; Wolf, T. On 2-way neighbor discovery in wireless networks with directional antennas. In Proceedings of the 2015 IEEE Conference on Computer Communications (INFOCOM), Hong Kong, China, 26 April–1 May 2015; pp. 702–710. [CrossRef]
21. Chen, L.; Li, Y.; Vasilakos, A.V. On Oblivious Neighbor Discovery in Distributed Wireless Networks with Directional Antennas: Theoretical Foundation and Algorithm Design. *IEEE/ACM Trans. Netw.* **2017**, *25*, 1982–1993. [CrossRef]
22. Wang, Y.; Liu, B.; Gui, L. Adaptive Scan-based Asynchronous Neighbor Discovery in wireless networks using directional antennas. In Proceedings of the 2013 International Conference on Wireless Communications and Signal Processing, Hangzhou, China, 24–26 October 2013; pp. 1–6. [CrossRef]
23. Sivanantham, E.; Ramakrishnan, M. Energy-efficient sustainable cluster based neighbor discovery technique for wireless networks with directional antennas. *Clust. Comput.-J. Netw. Softw. Tools Appl.* **2017**, *20*, 1527–1534. [CrossRef]
24. Wang, Y.; Mao, S.; Rappaport, T.S. On Directional Neighbor Discovery in mmWave Networks. In Proceedings of the 2017 IEEE 37th International Conference on Distributed Computing Systems (ICDCS), Atlanta, GA, USA, 5–8 June 2017; pp. 1704–1713. [CrossRef]
25. Cai, H.; Liu, B.; Gui, L.; Wu, M.Y. Neighbor discovery algorithms in wireless networks using directional antennas. In Proceedings of the 2012 IEEE International Conference on Communications (ICC), Ottawa, ON, Canada, 10–15 June 2012; pp. 767–772. [CrossRef]
26. Tarnai, T. The observed form of coated vesicles and a mathematical covering problem. *J. Mol. Biol.* **1991**, *218*, 485–488. [CrossRef]
27. Leopardi, P. A partition of the unit sphere into regions of equal area and small diameter. *Electron. Trans. Numer. Anal.* **2006**, *25*, 2006.

 *electronics*

Article

Verification in Relevant Environment of a Physics-Based Synthetic Sensor for Flow Angle Estimation [†]

Angelo Lerro [1,*], Piero Gili [1] and Marco Pisani [2]

1   Department of Mechanical and Aerospace Engineering, Politecnico di Torino, Corso Duca degli Abruzzi, 24, 10129 Turin, Italy; piero.gili@polito.it
2   Istituto Nazionale di Ricerca Metrologica, Strada delle Cacce, 91, 10135 Turin, Italy; m.pisani@inrim.it
*   Correspondence: angelo.lerro@polito.it
†   This paper is an extended version of our paper published in Lerro, A.; Brandl, A.; Gili, P.; Pisani, M. The SAIFE Project: Demonstration of a Model-Free Synthetic Sensor for Flow Angle Estimation. In Proceedings of the 2021 IEEE 8th International Workshop on Metrology for AeroSpace (MetroAeroSpace), Virtual Conference, 22–25 June 2021.

Abstract: In the area of synthetic sensors for flow angle estimation, the present work aims to describe the verification in a relevant environment of a physics-based approach using a dedicated technological demonstrator. The flow angle synthetic solution is based on a model-free, or physics-based, scheme and, therefore, it is applicable to any flying body. The demonstrator also encompasses physical sensors that provide all the necessary inputs to the synthetic sensors to estimate the angle-of-attack and the angle-of-sideslip. The uncertainty budgets of the physical sensors are evaluated to corrupt the flight simulator data with the aim of reproducing a realistic scenario to verify the synthetic sensors. The proposed approach for the flow angle estimation is suitable for modern and future aircraft, such as drones and urban mobility air vehicles. The results presented in this work show that the proposed approach can be effective in relevant scenarios even though some limitations can arise.

Keywords: air data system; flow angle; angle-of-attack; angle-of-sideslip; flight dynamics; flight testing; synthetic sensor; analytical redundancy; model-free; physics-based

Citation: Lerro, A.; Gili, P.; Pisani, M. Verification in Relevant Environment of a Physics-Based Synthetic Sensor for Flow Angle Estimation. *Electronics* **2022**, *11*, 165. https://doi.org/10.3390/electronics11010165

Academic Editor: Bin Xu

Received: 22 November 2021
Accepted: 3 January 2022
Published: 5 January 2022

Publisher's Note: MDPI stays neutral with regard to jurisdictional claims in published maps and institutional affiliations.

1. Introduction

Following the recent aircraft crashes that occurred with the Boeing 737-MAX, the European Union Aviation Safety Agency (EASA) has become open to the use of synthetic sensors to estimate the flow angles [1]. The objective is to improve the reliability of redundant flow angle measurements, even using soft techniques starting with modern aircraft. To accommodate future applications of alternative solutions for flow angle estimations, a working group is defining the new standard AS7984 "Minimum Performance Standards, for Angle of Attack (AoA) and Angle of Sideslip (AoS)" to cover the various sensor technologies used to measure flow angles [2].

The flow angle synthetic sensor provides the same measurements as an equivalent physical sensor but without the use of dedicated equipment for this purpose. Generally speaking, a synthetic sensor, in fact, merges flight data already available on board in order to provide an additional measurement of one or more flight parameters. Therefore, a synthetic sensor can be used in replacement or combined with a conventional (physical) sensor with clear benefits from the point of view of: (i) reduction of weights and volumes; (ii) energy efficiency; (iii) dissimilarity with respect to conventional solutions.

In the field of synthetic sensors, different categories and examples are available in the literature [3–6], whereas the MIDAS project's output [7] aims to be certified. The majority of the synthetic sensors developed so far, including those of the MIDAS project, suffer from common issues: (i) they can only be used on the aircraft where calibrated; (ii) they should be tuned on the aircraft's current configuration and flight conditions; (iii) they require a tuning phase based on flight data.

State-of-the-art air data sensors are typically probes and vanes protruding externally from the aircraft fuselage, able to provide a direct measurement of air data, mainly pressures, flow angles and air temperature. In the era of digital avionics, synthetic sensors can be added to physical (or mechanical) sensors in order to analytically increase the system redundancy [8,9]. Another potential application is to use synthetic sensors to monitor physical sensors and to accommodate possible failures [10,11]. The concurrent use of dissimilar sources of the same air data (physical and synthetic ones) can be beneficial for solving some issues related to common failure modes or incorrect failure diagnosis of modern air data systems [12–14]. Moreover, synthetic sensors can be used to overcome some issues towards certification related to next generation air vehicles, for example, unmanned aerial vehicles (UAVs) [15] and urban air mobility (UAM) aircraft [16] without any limitations to the application domain [17–19].

Some examples of the synthetic estimation of flow angles can be found in [20–26]. Model-based (e.g., Kalman filter) and data-driven (e.g., neural networks) are the approaches commonly used to estimate flow angles that are designed ad hoc for a particular aircraft and, therefore, they are affected by changes of configuration and flight regime. A model-free, or physics-based, nonlinear scheme, named ASSE—Angle-of-Attack and Sideslip Estimator—is proposed in [27] and aims to have a general validity and to therefore be independent from the aircraft application or flight regime.

In fact, ASSE deals with an analytical approach that is able to provide a generic synthetic sensor for flow angle estimation applicable to any flying body independently from the flight configuration and without the need to be calibrated. ASSE is based on an analytical formulation (or scheme) that, compared to the state-of-the-art, is better suited to be certified for civil aviation. The present work is part of the project SAIFE [28]—Synthetic Air Data and Inertial Reference System—where a demonstrator of the ASSE technology is designed and manufactured to verify the Technology Readiness Level (TRL) 5. The technological demonstrator is based on "all-in-one" air data and inertial system (commonly known as ADAHRS) able to provide multiple information to pilots or to automatic control systems, partially based on synthetic sensors that are used for flow angle estimation. The proposed approach for flow angle estimation does not require dedicated physical sensors but at the same time guarantees, under recognisable circumstances, the same reliability of flow angle vanes (or probes) in order to optimise the efficiency of on board avionics for both modern and future aircraft.

The main aim of the current work, as part of the project SAIFE, is to verify the TRL 5 of the ASSE technology. For this goal, a technological demonstrator is conceived and fully characterised in order to evaluate the uncertainty budgets related to all physical sensors feeding the synthetic sensors. Therefore, results of the present work describe the flow angle estimation performance of the ASSE scheme in a relevant simulated scenario. It is worth underlining that results presented in this work have a general validity as the flight simulator is only used to generate coherent and, hence, the proposed ASSE scheme can be applied to any flying body to estimate the flow angles. Moreover, the technological demonstrator is equipped with all necessary components to be ready for flight tests for future validation in real environments.

The SAIFE project's demonstrator concept is introduced in Section 2 with a focus on the architecture and its physical sensors. The ASSE scheme is briefly presented in Section 3 in order to highlight the necessary inputs and to discuss some practical limitations emerged from the present work. The characterisation tests of the physical sensors and the consequent sensor's noisy models are defined in Section 4, whereas the approach for TRL 5 verification and results are presented in Section 5 before concluding the work.

2. The ASSE Technological Demonstrator Concept

The ASSE technological demonstrator is depicted in Figure 1a along with: (a) external power supply to be connected both to the aircraft power bus or to domestic plug; (b) global navigation satellite system (GNSS) antenna; (c) magnetometer antenna; (d) pitot boom with AoA and AoS vanes. The ASSE demonstrator, encompassing both air data, inertial and heading reference systems (ADAHRS) is able to provide the following direct measurements:

- from the Air data System (ADS):

    1. Dynamic pressure q_c (or, as alternative, total pressure);
    2. Absolute pressure p_∞;
    3. Ambient temperature T;
    4. Angle-of-attack AoA (as a reference value);
    5. Angle-of-sideslip AoS (as a reference value);

- from the Attitude and Heading Reference System (AHRS):

    6. 3-axis angular rates;
    7. 3-axis linear accelerations;
    8. 3-axis magnetometer;
    9. 2-axis inclinometer;
    10. GNSS position and velocity;

The output rate is 100 Hz. In order to provide ADS outputs, the demonstrator shall be interfaced with at least an external probe (e.g., a Pitot probe) able to measure the dynamic and absolute pressure and a temperature probe (e.g., outside ambient temperature, OAT). Whereas, for AHRS outputs, common equipment (described in Section 2.3) is used that can be installed inside the fuselage. From the aforementioned direct measurements, several parameters can be calculated, such as the attitude angles, the heading and the true airspeed, whereas, the flow angles are estimated using the ASSE technology (described in Section 3.1). Therefore, the only probes protruding externally from the fuselage are related to pressure and temperature measurements.

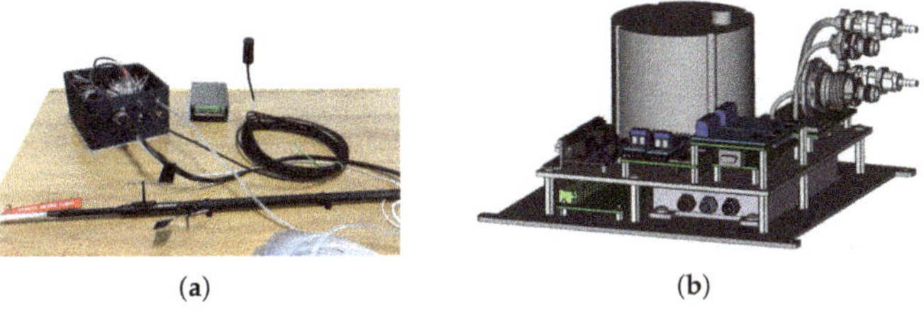

Figure 1. ASSE demonstrator design. (**a**) A view of the ASSE technological demonstrator developed under the SAIFE project, (**b**) Arrangement design. Courtesy of SELT Aerospace & Defence.

2.1. State-of-the-Art of Air Data System Sensors

As far as low speed (Mach number below 0.3) air vehicles are concerned, the air temperature is commonly measured with OAT, which is able to directly measure the ambient static temperature with a sensing element exposed to the external airflow.

As far as the pressures are concerned, two probe technologies are available: (1) the single-function probes; and (2) the multi-function probes. The conventional pressure probes (or single-function probes, SFP) considered here are total tubes (for the total pressure measurement), Pitot-static (or simply Pitot) probes (for static and total pressure measurements) or static ports (for static pressure measurement). In the majority of the examples, the conventional pressure probes are not equipped with pressure transducers, so that they have to be connected pneumatically to dedicated air data modules (ADMs), an air data computer (ADC) or a vehicle management computer (VMC). On the other side, a multi-function probe (MFP) is a probe with enhanced capabilities able to provide at least two pressure measurements and one flow angle measurement. The pressure measurements are usually related to static and total pressures, whereas the flow angle is referred to a local flow angle and only a combination of at least two MFPs can provide both AoA and AoS calculation. In the MFP category, the optical air data sensors could be considered even though there is no certified product at the moment because some issues still remain to be solved [29], for example, turbulence, vortex and wind gusts can affect the accuracy.

Considering the state-of-the-art of the current technologies, three realistic architectures for air data and inertial reference system are compared in Figure 2. An ADS based on SFP leads to a high number of LRUs to be installed protruding outside from the A/C fuselage as schematically represented in Figure 2a with a consequent increase of weight and power. However, the SFPs are mature and available worldwide. Whereas, MFPs are available only from three manufacturers [30] but it can assure a reduced number of LRUs and the absence of pneumatic tubes as represented in Figure 2b. In Figure 2c, a possible ADS based on flow angle synthetic sensors is presented. Along with the chance to use SFP probes, the other main advantages of an ADS based on flow angle synthetic sensors are: (i) a reduced number of LRUs with a consequent reduction of weights and volumes; (ii) no power required for anti-icing systems improving the overall ADS energy efficiency; (iii) improved safety due to dissimilarity with respect to other conventional flow angle vanes/probes.

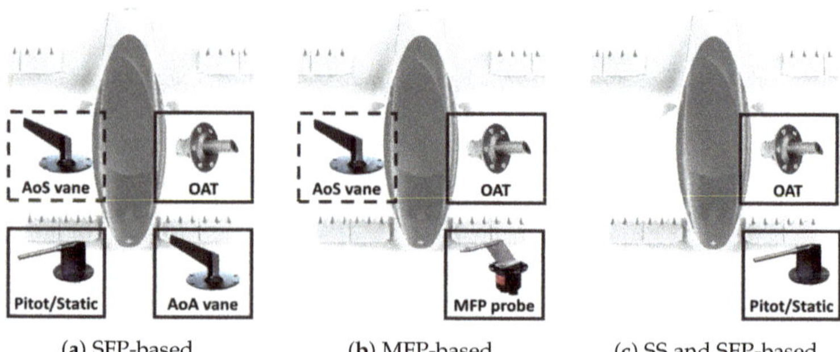

(**a**) SFP-based   (**b**) MFP-based   (**c**) SS and SFP-based

Figure 2. Generic three realistic simplex air data sensing architectures able to provide a complete set of air data. The dashed lines of AoS sensors indicate that they could not be mandatory.

It is worth underlining that the realistic sensing architectures presented in Figure 2 can provide the same air dataset in a simplex configuration. As some flight parameters can be safety critical; according to safety studies, a suitable redundancy should be assured, for example, for airspeed and AoA. Therefore, some external probes should be duplicated, or even triplicated, to satisfy the safety requirements. Although the system redundancy on board large airplanes does not represent an issue, the same can lead to some obstacles to UAVs or UAM vehicles due to the reduced fuselage surface suitable for air data probe installation. Therefore, the solutions based on fewer external LRUs can be more attractive for those categories.

2.2. Demonstrator's Architecture

The ASSE demonstrator is based on the TEENSY 4.1 board that is able to manage digital and analog interfaces as described in Figure 3. It is designed to be interfaced with AHRS sensors (described in Section 2.3), air data transducers (described in Section 2.4) and a ground computer. It is worth highlighting that the ADS is also equipped with AoA and AoS common vanes because during the demonstration they are used as reference values during actual flight tests.

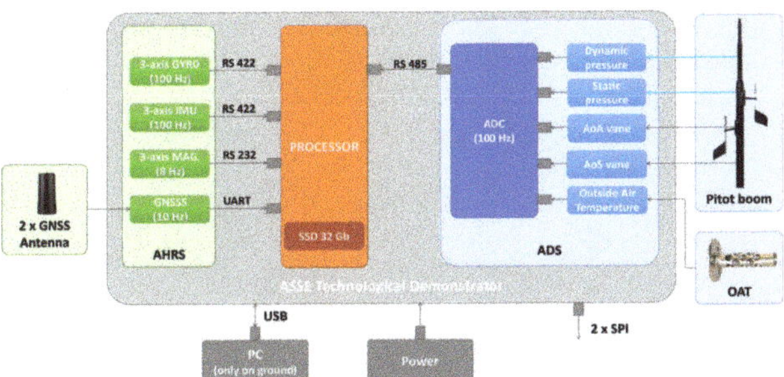

Figure 3. The architecture of the ASSE technological demonstrator. The black arrows represent data bus connections, the blue arrows represent pneumatic connections.

The AHRS and ADC are installed in a single unit (or LRU) as described in Figure 1b. The reason behind a two-layer board is to keep the center of gravity as close as possible to the gyroscope and the accelerometers.

The aircraft heading and attitudes during dynamic manoeuvres are evaluated using ad hoc Kalman filters exploiting available data from the gyroscope, accelerometers, magnetometers and GNSS. For the scope of the present work, the attitude angles are unnecessary and they are replaced with inclinometer information for integration tests.

2.3. Attitude, Heading and Reference Sub-System

The ASSE demonstrator is based on fibre optic gyroscope (FOG), a solid state accelerometer unit, a GNSS receiver and a 3-axis magnetometer as represented in Figure 3.

2.4. Air Data Sub-System

The ASSE demonstrator also includes an air data sub-system comprising an ADC with pressure transducers, calibrated interfaces for OAT and flow angle vanes. The ADC is calibrated to work in the airspeed range $[0\,\mathrm{kn}, 174\,\mathrm{kn}]$ in the altitude range $[-1800\,\mathrm{ft}, 35{,}000\,\mathrm{ft}]$.

3. Nonlinear ASSE Scheme

In this section, the ASSE scheme is presented in order to give some crucial information about the proposed technology.

Two reference frames are considered: the inertial reference frame $\mathcal{F}_I = \{X_I, Y_I, Z_I\}$ and the body reference frame $\mathcal{F}_B = \{X_B, Y_B, Z_B\}$ as described in Figure 4. The vector transformation from the inertial reference frame $\mathcal{F}_I$ to the body frame $\mathcal{F}_B$ is obtained considering the ordered sequence 3-2-1 of Euler angles: heading, elevation and bank angles. As the vector subscript denotes the reference frame where the vector is represented, the relationship between the inertial velocity $\mathbf{v}_I$, the relative velocity $\mathbf{v}_B$ and the wind velocity $\mathbf{w}_I$ can be written as:

$$\mathbf{v}_I = \mathbf{C}_{B2I}\mathbf{v}_B + \mathbf{w}_I, \tag{1}$$

where $\mathbf{C}_{B2I}$ is the direction-cosine matrix to calculate vector components in the inertial reference frame from the body reference frame [31]. It is worth recalling that the inverse transformation is $\mathbf{C}_{I2B} = \mathbf{C}_{B2I}^T$, that is, the direction-cosine matrix to calculate vector components in the body reference frame from the inertial reference frame [31]. The relative velocity $\mathbf{v}_B$ can be expressed as function of its module and flow angles, α and β, as

$$\mathbf{v}_B = V_\infty \hat{\imath}_{WB} = V_\infty \left(\cos\beta \cos\alpha \hat{\imath}_B + \sin\beta \hat{\jmath}_B + \cos\beta \sin\alpha \hat{k}_B \right), \tag{2}$$

where V_∞ is the magnitude of the relative velocity vector, or true airspeed (TAS), $\hat{\imath}_B$, $\hat{\jmath}_B$ and $\hat{k}_B$ are the three unit vectors defining the body reference axes and $\hat{\imath}_{WB}$ is the unit vector of $\mathbf{v}_B$ depending only on α and β.

Recalling that $\mathbf{\Omega}_B$ is defined as the body angular rate matrix [32], the coordinate acceleration $\mathbf{a}_B$ can be written as:

$$\mathbf{a}_B = \mathbf{C}_{I2B}\mathbf{a}_I = \dot{\mathbf{v}}_B + \mathbf{\Omega}_B\mathbf{v}_B + \mathbf{C}_{I2B}\dot{\mathbf{w}}_I = a_{X_B}\hat{\mathbf{i}}_B + a_{Y_B}\hat{\mathbf{j}}_B + a_{Z_B}\hat{\mathbf{k}}_B. \tag{3}$$

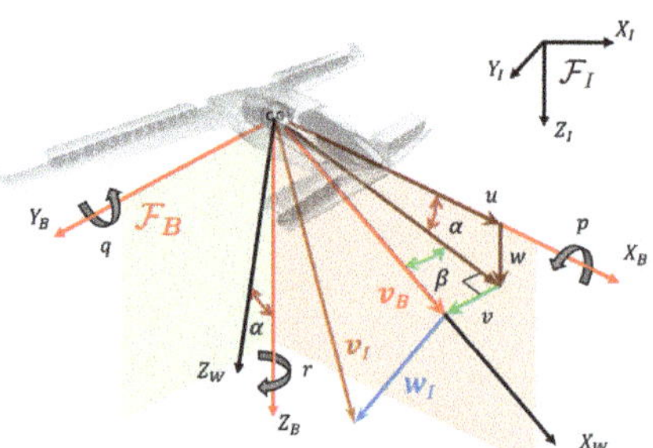

Figure 4. Representation of inertial and body reference frames with positive flow angles (α, β), linear relative velocities (u, v, w), angular rates (p, q, r) and the velocity triangle between inertial velocity $\mathbf{v}_I$, the relative velocity $\mathbf{v}_B$ and the wind velocity $\mathbf{w}_I$.

3.1. The ASSE Synopsis

As known, the time-derivative of the relative velocity's magnitude can be expressed as:

$$\dot{V}_\infty = \frac{\mathbf{v}_B^T\dot{\mathbf{v}}_B}{V_\infty} \Rightarrow \dot{V}_\infty V_\infty = \mathbf{v}_B^T(\mathbf{a}_B - \mathbf{C}_{I2B}\dot{\mathbf{w}}_I), \tag{4}$$

where all terms are measured at the same time instant. Moreover, the relative velocity vector $\mathbf{v}_B$ at time t can be expressed starting from $\mathbf{v}_B$ at a generic time τ, with $t \geq \tau$, as

$$\mathbf{v}_B(t) = \mathbf{v}_B(\tau) + \int_\tau^t \dot{\mathbf{v}}_B(\mathcal{T})\,d\mathcal{T}. \tag{5}$$

Henceforth, in order to ease the notation, the independent variable of the integrand function is omitted and the time of the measurement is reported as subscript. For example, the relative velocity evaluated at time τ is denoted as $\mathbf{v}_{B,\tau}$.

Recalling Equation (3), Equation (5) can be rewritten as:

$$\mathbf{v}_{B,t} = \mathbf{v}_{B,\tau} + \int_\tau^t (\mathbf{a}_B - \mathbf{\Omega}_B\mathbf{v}_B - \mathbf{C}_{I2B}\dot{\mathbf{w}}_I)\,d\mathcal{T} \tag{6}$$

and

$$\mathbf{v}_{B,\tau} = \mathbf{v}_{B,t} - \int_\tau^t \mathbf{a}_B\,d\mathcal{T} + \int_\tau^t \mathbf{\Omega}_B\mathbf{v}_B\,d\mathcal{T} + \int_\tau^t \mathbf{C}_{I2B}\dot{\mathbf{w}}_I\,d\mathcal{T}. \tag{7}$$

Replacing $\mathbf{v}_{B,\tau}$ with Equation (7), Equation (4) can be written at time τ as:

$$V_{\infty,\tau}\dot{V}_{\infty,\tau} = \left[\mathbf{v}_{B,t} + \int_\tau^t \mathbf{\Omega}_B\mathbf{v}_B\,d\mathcal{T}\right]^T (\mathbf{a}_B - \mathbf{C}_{I2B}\dot{\mathbf{w}}_I)_\tau, \tag{8}$$

where all terms depending on $\mathbf{v}_B$, and hence on the flow angles, are collected on the right hand side.

The ASSE scheme based on the zero-order approximation [27] assumes that the integral term $\int_\tau^t \mathbf{\Omega}_B\mathbf{v}_B\,d\mathcal{T}$ of Equation (8) is constant in the generic time interval $[\tau, t]$, therefore

$$\int_\tau^t \mathbf{\Omega}_B\mathbf{v}_B\,d\mathcal{T} = (\mathbf{\Omega}_B\mathbf{v}_B)_t\Delta t, \tag{9}$$

where $\Delta t = t - \tau$. Considering the latter expression, Equation (8) can be rewritten as:

$$V_{\infty,\tau}\dot{V}_{\infty,\tau} + \left[\int_{\tau}^{t} \mathbf{a}_B \, d\mathcal{T} - \int_{\tau}^{t} \mathbf{C}_{I2B}\dot{\mathbf{w}}_I \, d\mathcal{T}\right]^{T} (\mathbf{a}_B - \dot{\mathbf{w}}_B \mathbf{C}_{I2B}\dot{\mathbf{w}}_I)_{\tau} =$$
$$= V_{\infty,t}\hat{\mathbf{i}}_{WB,t}^{T}(\mathbf{I} - \mathbf{\Omega}_{B,t}\Delta t)(\mathbf{a}_B - \mathbf{C}_{I2B}\dot{\mathbf{w}}_I)_{\tau}. \tag{10}$$

Equation (10) is denoted in [27] as the basic expression of the zero-order ASSE scheme referred to the generic time τ where the flow angles, $\alpha(t)$ and $\beta(t)$, are the only unknowns and all other terms are supposed to be measured. For the latter reason, the unknown variables are always referred to as current time t, henceforth, the flow angles are represented without subscripts related to time.

Considering the previous notations, Equation (10) can be presented as:

$$n_{\tau} = \hat{\mathbf{i}}_{WB,t}^{T}\mathbf{m}_{\tau}, \tag{11}$$

where

$$n_{\tau} = V_{\infty,\tau}\dot{V}_{\infty,\tau} + \left[\int_{\tau}^{t} \mathbf{a}_B \, d\mathcal{T} - \int_{\tau}^{t} \mathbf{C}_{I2B}\dot{\mathbf{w}} \, d\mathcal{T}\right]^{T} (\mathbf{a}_B - \mathbf{C}_{I2B}\dot{\mathbf{w}})_{\tau} \tag{12}$$

and

$$\mathbf{m}_{\tau} = V_{\infty,t}(\mathbf{I} - \mathbf{\Omega}_{B,t}\Delta t)(\mathbf{a}_B - \mathbf{C}_{I2B}\dot{\mathbf{w}}_I)_{\tau} = h_{\tau}\hat{\mathbf{i}}_B + l_{\tau}\hat{\mathbf{j}}_B + m_{\tau}\hat{\mathbf{k}}_B. \tag{13}$$

It is worth underlining that, in Equation (13), all parameters are referred to as time τ, whereas the true airspeed $V_{\infty,t}$ is always referred to as time t and, hence, AoA and AoS as well. Rewriting Equation (11) back in time starting from t to n-th generic τ_i with $i \in [0, 1, \ldots, n]$ where $\tau_0 \equiv t$ and AoA and AoS are always referred to at the same time t. Therefore, the following system of $n + 1$ nonlinear equations, or ASSE scheme, is obtained:

$$\begin{cases} n_t = \hat{\mathbf{i}}_{WB,t}^{T}\mathbf{m}_t \\ n_{\tau_1} = \hat{\mathbf{i}}_{WB,t}^{T}\mathbf{m}_{\tau_1} \\ \vdots \\ n_{\tau_n} = \hat{\mathbf{i}}_{WB,t}^{T}\mathbf{m}_{\tau_n}, \end{cases} \tag{14}$$

where the AoA and AoS at time t are assumed to be the only unknowns, the wind acceleration $\dot{\mathbf{w}}$ is considered to be negligible in the time interval and all other parameters can be measured. Therefore, the flow angle estimation based on the ASSE scheme requires direct measurements of: (1) TAS; (2) body angular rates p, q and r and (3) coordinate acceleration vector $\mathbf{a}_B$. It is worth highlighting that the AHRS does not measure the coordinate acceleration vector but it can be calculated from the inertial acceleration directly measured by the AHRS in addition to the aircraft attitudes (or Euler angles).

Equation (14) is the generic form of the proposed zero-order ASSE scheme based on $n + 1$ equations. In this work, an expansion in the past is considered ($\tau_{i+1} < \tau_i$). The most suitable solver can be adopted to solve the system of Equation (14) for AoA and AoS estimation.

3.2. Practical Implementation

As discussed in [33], solving the ASSE nonlinear scheme can be challenging dealing with real signals that are affected by uncertainties mainly related to random noise, bias, drift of the sensors. In fact, more robust solvers can be considered to solve the ASSE scheme that are more tolerant to the input noise. In this work, the nonlinear ASSE scheme [34] with 200 time steps (or equations) is adopted to validate the ASSE demonstrator. The number of equations indicates that the ASSE scheme is applied considering a 2 s time observation. The latter choice is based on the experience gained with the present project to improve the ASSE performance characteristics in the presence of noisy input with respect to results presented in [33]. In this work, the system of nonlinear equations is solved using an iterative method based on the Levenberg–Marquardt algorithm [35]. The first guess in iteration methods is crucial but it is hardly realistic considering a reliable first guess in practical applications; therefore, the proposed scheme is applied, adopting null values as the initial condition.

As known from [27], the ASSE scheme can be applied only during dynamic flight conditions. Moreover, from [34], the linear formulation can lead to understanding where ASSE can be applied or not. In order to introduce the reliability criteria, the linearised ASSE scheme is discussed. Considering Taylor series expansions of trigonometric functions, the versor of Equation (2) can be approximated at first order (i.e., $\cos(x) \approx 1$ and $\sin(x) \approx x$) as:

$$\hat{\mathbf{i}}_{WB,t} \approx \tilde{\mathbf{i}}_{WB,t} = [1, \beta, \alpha]^T. \tag{15}$$

Therefore, writing Equation (14) for time t and a generic previous time τ (with $\tau < t$), the following system of two linear equations is obtained:

$$\begin{cases} n_t = \hat{\mathbf{i}}_{WB,t}^T \mathbf{m}_t \approx h_t + l_t \beta_{lin} + m_t \alpha_{lin} \\ n_\tau = \hat{\mathbf{i}}_{WB,t}^T \mathbf{m}_\tau \approx \tilde{\mathbf{i}}_{WB,t}^T \mathbf{m}_\tau = h_\tau + l_\tau \beta_{lin} + m_\tau \alpha_{lin}. \end{cases} \tag{16}$$

Moreover, considering that the wind acceleration vector $\dot{\mathbf{w}}$ is negligible, for example, if the two observed time steps are sufficiently close to consider the wind vector constant in the time interval $[\tau, t]$, Equation (16) can be rewritten as:

$$\begin{cases} \alpha_{lin,\dot{\mathbf{w}}\approx 0} = \frac{a_{Y,t}\left(\dot{V}_{\infty,\tau}-a_{X,\tau}\right)-a_{Y,\tau}\left(\dot{V}_{\infty,t}-a_{X,t}\right)}{D_{\dot{\mathbf{w}}\approx 0}} \\ \beta_{lin,\dot{\mathbf{w}}\approx 0} = \frac{a_{Z,\tau}\left(\dot{V}_{\infty,t}-a_{X,t}\right)-a_{Z,t}\left(\dot{V}_{\infty,\tau}-a_{X,\tau}\right)}{D_{\dot{\mathbf{w}}\approx 0}}, \end{cases} \tag{17}$$

where

$$D_{\dot{\mathbf{w}}\approx 0} = l_{t,\dot{\mathbf{w}}\approx 0} m_{\tau,\dot{\mathbf{w}}\approx 0} - m_{t,\dot{\mathbf{w}}\approx 0} l_{\tau,\dot{\mathbf{w}}\approx 0} \tag{18}$$

is the determinant of the system evaluated with negligible wind accelerations and it shall be nonzero to guarantee the existence of the closed-form solution.

In order to evaluate possible flight conditions leading to a null determinant, it is worth noting that the determinant $D_{\dot{\mathbf{w}}\approx 0}$ does not depend on the time derivative of the airspeed but only on the true airspeed, body accelerations and angular rates; the latter weighted by the chosen time interval Δt. In fact, when the time interval tends to zero, the $\mathbf{m}$ vector of Equation (13) can be approximated as $\mathbf{m}_\tau = [h_\tau, l_\tau, m_\tau]^T \approx V_{\infty,t}[a_{X,\tau}, a_{Y,\tau}, a_{Z,\tau}]^T$ yielding to the following determinant approximation:

$$D_{\dot{\mathbf{w}}\approx 0, \Delta t \to 0} = V_{\infty,t}^2 (a_{Y,t} a_{Z,\tau} - a_{Z,t} a_{Y,\tau}). \tag{19}$$

Along with the existence condition discussion presented in [34], from Equation (19) it emerges that a null simplified determinant could lead to very large errors. Therefore, some thresholds are defined according to a trial and error approach where the nonlinear ASSE scheme would lead to very large errors (larger than 5°). The proposed values can be possible values for real applications even though they should be verified in an extended flight envelope. In this work, the following thresholds are considered:

1. vertical/lateral inertial acceleration:

   - $a_Z > a_{thr} = 0.5\,\mathrm{m\,s^{-2}}$ for AoA estimation
   - $a_Y > a_{thr} = 0.5\,\mathrm{m\,s^{-2}}$ for AoS estimation

2. determinant of Equation (18): $\tilde{D} > D_{thr} = 0.2\,\mathrm{m^4/s^6}$,

where the determinant $D_{\dot{\mathbf{w}}\approx 0}$ of Equation (18) is denoted henceforth as $\tilde{D}$ for simplicity of notation. The latter conditions are used to drive the integration tests of the demonstrator's algorithms. In fact, even in the case of larger time intervals (i.e., $\Delta t > 0$), if lateral or vertical inertial accelerations are very small (i.e., $a_Y \to 0$ or $a_Z \to 0$, respectively) the determinant tends to zero and the condition number will be very large leading to a very noise-sensitive scheme. It is worth underlining that, in this work, the thresholds do not influence the results because they are only applied after the ASSE solution is computed.

Moreover, considering that there are no metrological standards to evaluate the synthetic sensor performance [36], some criteria are defined to evaluate the flow angle estimation performed by the nonlinear ASSE scheme. In this work, the flow angle estimations are analysed using the mean, maximum, 1σ and 2σ errors. The latter choice, inspired by industrial experience, is useful for providing an overall overview of the flow angle estimation performances. In the present work, 1σ and 2σ intended the value such that the probability $\Pr(-\sigma \leq X \leq \sigma) = 68.3\%$ and $\Pr(-2\sigma \leq X \leq 2\sigma) = 95.4\%$ also in the case the error is not normally distributed. If the mean error can be easily removed from the synthetic sensor estimations, the maximum and the 2σ errors should be specified according to their operative scopes. In fact, if AoA and AoS are used for monitoring or displaying purposes, maximum absolute errors up to 5° could be accepted, whereas, for control and navigation applications the performance requirements could be more demanding. However, considering the stage of the present project, the following thresholds are considered in the current work:

- maximum absolute error $< 5°$
- 2σ error $< 2°$.

4. Characterisation of the ASSE Demonstrator's Sensors

The section introduces the methods and results of the characterisation activities for the ASSE demonstrator's sensors. Calculated uncertainties are used to corrupt simulated flight data with realistic errors that are added to reproduce a relevant testing scenario as defined in this section.

The three-axes gyroscope and the three-axes accelerometer are tested separately to verify their performances independently from manufacturer's data sheets using a rotating platform to verify the angular rates, whereas tilting and sliding platforms are used to verify the linear acceleration performances. Once the AHRS sensors are installed in the ASSE demonstrator, the same characterisation tests are performed to verify possible misalignment issues. For the sake of clarity, the characterisation presented in this section is related to the technological demonstrator.

In more detail, three dedicated facilities are used to calibrate the inertial sensor at the best accuracy level. The fibre optic gyroscope (FOG) is calibrated against a rotating platform. The platform is based on a precision air-bearing table driven by a microstep motor. The platform is able to generate smooth and accurate rotation ranging from fractions of a degree per second to the full range of the device, which is greater than $400°\,\mathrm{s}^{-1}$.

The FOG is mounted in three orthogonal positions in order to calibrate the three axes and to check for the orthogonality of the measurement axis. The accuracy of the orthogonality is checked by a precision autocollimator and the accuracy of the rotating table is checked by comparison with the National Angle Standard (REAC) [37].

The accelerometers, or the inertial measurement unit (IMU), is calibrated for very low accelerations with a tilt table platform where the projection of the acceleration gravity vector parallel to the platform is used as a reference acceleration. The technique has been already used for the calibration of the accelerometers on board the ESA spacecrafts BepiColombo and JUICE with destinations of Mercury and Jupiter [38]. For higher accelerations, the device is calibrated using a facility built to the purpose making use of dynamic laser interferometry having an accuracy of at least $1 \times 10^{-4}\,\mathrm{m\,s}^{-2}$. In this case, the accelerometers are also calibrated at the three orthogonal axes.

The ADS is stimulated using the pressure test bench. Thanks to suitable probe adapters, the air data probe is connected pneumatically to the pressure test bench that is able to generate both the static and the total pressures. Moreover, the pressure test bench is able to simulate both constant and dynamic flight conditions according to predefined steps that can be set by the user. The temperature sensor is connected and left to the ambient temperature.

4.1. Gyroscope

The FOG is tested to characterise the uncertainties on the measurement for several constant angular rates and periodic angular rates with several frequencies in order to evaluate both steady-state and dynamic errors and, hence, to evaluate the FOG's expanded uncertainty. An example of the steady-state gyroscope measurements is represented in Figure 5a with respect to the reference values on the Z-axis. From the Figure 5a it is clear that at 0 Hz, the gyroscope is characterised by a linear behaviour; in fact, the regression slope is 1.00 and the $R^2 = 1.00$. Moreover, a cross-coupling between the three axes is visible even though it is less than 0.05 %. Including dynamic analysis, the expanded uncertainty on the angular rates measured by the ASSE demonstrator is evaluated as $Q(0.05, 5 \times 10^{-4}\nu)$. The notation $Q(0.05, 5 \times 10^{-4}\nu)$ is intended to be the quadratic sum of two terms: the first is the constant, the second is proportional to the measured value ν. For example, for $100\,°\,s^{-1}$, the uncertainty is $U = \sqrt{0.05^2 + \left(100 \cdot 5 \times 10^{-4}\right)^2} = \sqrt{0.0025 + 0.0025} \approx 0.071\,°\,s^{-1}$. If the error distribution is normal, the $U = 2\sigma$. In this work, the latter assumption is adopted. Therefore, the single axis gyroscope measurements, p, q, r, are corrupted with a white noise whose 1σ depends on the measurement itself. For example, the pitch rate q is corrupted using a white noise with 1σ calculated as $q_{1\sigma} = \frac{1}{2}\sqrt{0.05^2 + q^2\left(5 \times 10^{-4}\right)^2}\,°\,s^{-1}$.

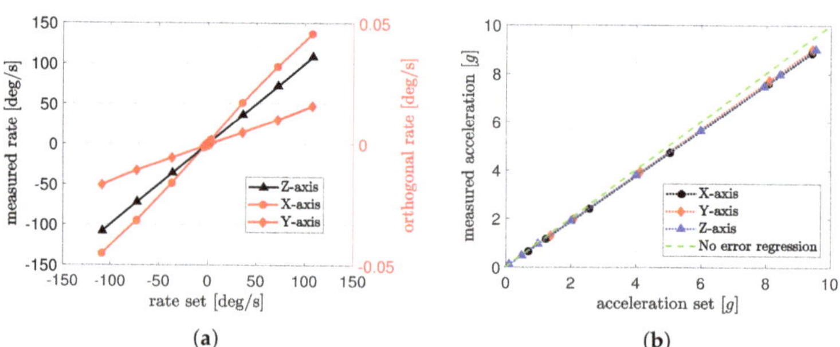

Figure 5. Example of AHRS data analysis. (**a**) Constant angular rate on the Z-axis, (**b**) Dynamic accelerations at 80 Hz.

4.2. Inertial Measurement Unit

The IMU is tested to characterise the uncertainties on the measurement for several constant accelerations and periodic accelerations with several frequencies in order to evaluate both steady-state and dynamic errors and, therefore, to evaluate the expanded uncertainty. An example of dynamic measurements with periodic reference accelerations at 80 Hz are represented in Figure 5b. From the Figure 5b it is clear that at 80 Hz, the inertial measurement unit loses accuracy. In fact, even though the linearity is maintained, because with a linear regression $R^2 = 1.00$, the slope is 0.941, whereas for very low frequencies (less than 5 Hz) the regression slope is 1.00 for each of the three axis. This particular behaviour is taken into account with a very large expanded uncertainty of $Q(0.007, 0.02\nu)$ up to 10 g and up to 80 Hz. It is worth underlying that for realistic aircraft applications, limiting the frequency to 10 Hz as discussed in Section 4.8, the expanded uncertainty can be recalculated as $Q(0.007, 1 \times 10^{-3}\nu)$. In this work, the largest values are considered for conservative reasons. The single axis acceleration measurements, a_X, a_Y, a_Z, are corrupted with a white noise whose 1σ depends on the measurements itself. For example, the longitudinal accelerations a_X are corrupted using a white noise with 1σ calculated as $a_{X,1\sigma} = \frac{1}{2}\sqrt{0.007^2 + a_X^2 0.02^2}\,m\,s^{-2}$.

4.3. Inclinometer

The two-axes inclinometer is tested to characterise the uncertainties on the measurement for several constant inclinations in order to evaluate the steady-state errors and, therefore, to evaluate its expanded uncertainty. As the inclinometer is only used for initialisation purposes, it is not considered in the uncertainty chain for the ASSE scheme. In fact, the attitude angles are calculated with a dedicated algorithm but they are not involved in the verification activities of the present work. However, the inclinometer is used to verify the ASSE algorithm during integration tests where the attitude angle is used to derive the inertial acceleration a from the proper acceleration $a + g$ measured by the IMU, where the vector g is the gravity vector. The measured expanded uncertainty is Q$(0.2, 0.01\nu)$ in the range $[-60°,\ 60°]$.

4.4. Calibrated Airspeed

Some airspeed profiles are simulated in terms of velocity and altitude using the pressure test bench. As said before, the pressure test bench has two independent pressure ports to provide both static and dynamic pressures. The air data probe is connected pneumatically to the pressure test bench using a suitable probe adapter. With the latter experimental setup, the air data test set is able to stimulate the ADC with predefined airspeed and pressure altitude according to a predefined rate that can be programmed by the user. Preliminarily, the ADC of the ASSE demonstrator is calibrated using reference values of airspeed generated using the pressure test bench.

The 2nd order calibration polynomial is derived fitting the error measured during the bench tests at ambient temperature as represented in Figure 6a. The residual maximum error after the calibration is about $0.05\,\mathrm{m\,s^{-1}}$ as can be noted in Figure 6a.

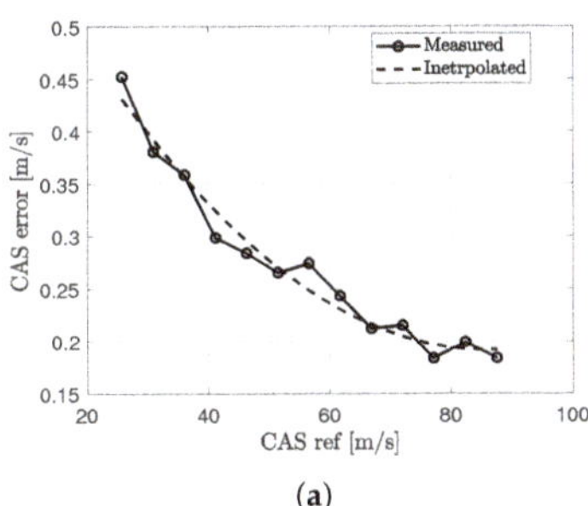
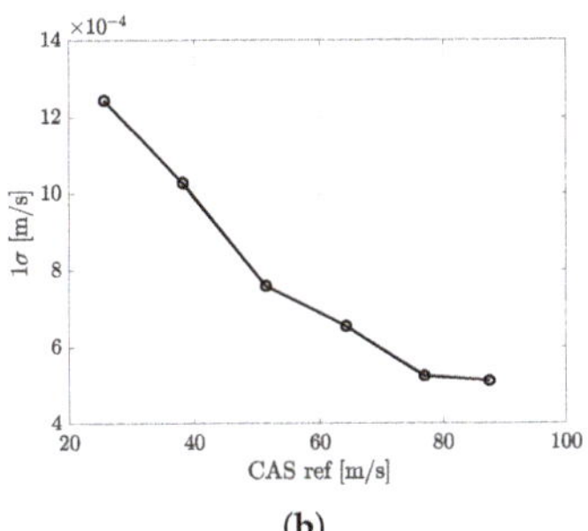

(a)   (b)

Figure 6. Calibrated airspeed analysis. (**a**) Measured and calibration of the IAS calculated by the ASSE demonstrator, (**b**) Noise on the CAS measurement.

A high-pass filter with $1\,\mathrm{Hz}$ cutoff frequency is used to evaluate the random noise on the CAS measurement. From Figure 6b, it is clear that the noise on the measurement of the CAS is less than $1.3 \times 10^{-3}\,\mathrm{m\,s^{-1}}$.

The air data test set can maintain a constant value of airspeed with internal controllers with a total uncertainty of $0.26\,\mathrm{m\,s^{-1}}$. This leads to the assumption that the CAS measurement uncertainty is dominated by unknown bias errors and less influenced by the random noise on the measurement. Therefore, the uncertainty of the CAS measurement is modelled as the sum of: (1) a bias not depending on the CAS itself calculated as the sum of the residual error after the calibration plus the contribution of the reference values generated; (2) white noise whose 1σ value is derived from the maximum of Figure 6b to be conservative. Therefore, in this work, the realistic CAS (or the CAS with uncertainties) is obtained using $\mathrm{CAS}_{bias} = 3.1 \times 10^{-2}\,\mathrm{m\,s^{-1}}$ and $\mathrm{CAS}_{1\sigma} = 1.3 \times 10^{-3}\,\mathrm{m\,s^{-1}}$.

As far as the dynamic pressure error is concerned, the main uncertainty can be evaluated considering the sole CAS_{bias} as $q_{c,bias} = \sqrt{\frac{1}{2}\rho_{SL}\left|(\text{CAS} \pm \text{CAS}_{bias})^2 - \text{CAS}^2\right|}$, where $\rho_{SL} = 1.225\,\text{kg/m}^3$.

4.5. Altitude

Some constant altitudes are simulated in terms of static pressure using the pressure test bench. As said before, using suitable probe adapters, the air data probe is connected pneumatically to the pressure test bench that is able to stimulate the ADC's absolute pressure transducer. As a first step, the ADC of the ASSE demonstrator is calibrated using reference values of static pressure (or altitudes) generated using the pressure test bench.

A linear regression is used to fit the error measured during the bench tests at ambient temperature as represented in Figure 7a. The residual maximum error after the calibration is about 3 Pa, as can be seen from Figure 7a.

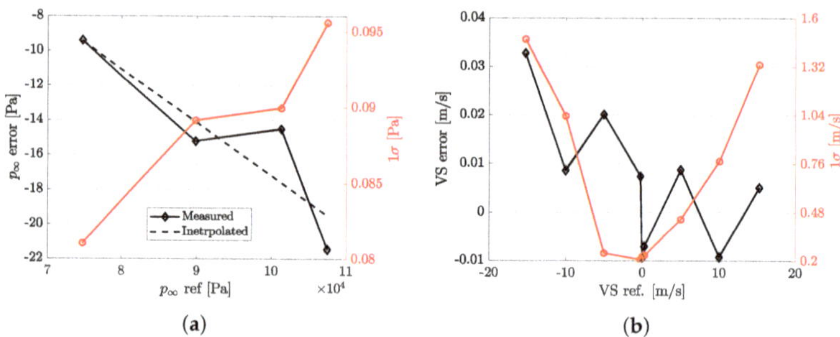

Figure 7. Altitude and vertical speed uncertainty analysis. (**a**) Static pressure, (**b**) Vertical speed.

Considering 30 s time windows, the random noise is allocated to the 1σ error of the measured static pressure. The noise on the measurement of the static pressure is less than 0.1 Pa and it is practically constant if compared with the mean errors. For conservative reasons, the latter value is used in this work.

From the latter analysis, the static pressure measurement uncertainty is dominated by unknown bias errors and less influenced by the random noise of the measurement. Therefore, adopting the same model of the CAS presented in Section 4.4, the uncertainty on the static pressure measurement is modelled as the sum of a constant bias and white noise. Therefore, the uncertainty model of the static pressure is obtained using $p_{\infty,bias} = 3.0\,\text{Pa}$ and $p_{\infty,1\sigma} = 0.1\,\text{Pa}$. Applying the proposed static pressure error model in a realistic altitude range between $-500\,\text{m}$ and $3000\,\text{m}$ for ultralight aircraft applications, the maximum uncertainty is enclosed in $\pm 3\,\text{Pa}$ (or $\pm 0.36\,\text{m}$) as can be noted from Figure 7a. It is worth underlining that the latter result is related to the sole ASSE demonstrator without considering any position errors due to the aircraft installation.

4.6. Vertical Speed

In this section, the vertical speed uncertainty is evaluated even though it is not used in the ASSE scheme. In fact, it is reported here to provide some additional information of the ADC performance. Some constant vertical speeds are simulated in terms of static pressure using the pressure test bench connected to the ADC using suitable probe adapters. Results are presented in Figure 7b and it is clear that the VS uncertainty is dominated by the random noise. Therefore, the calculated expanded uncertainty for the vertical speed is $Q(-0.00551, 0.00518v)$.

4.7. True Airspeed

The same uncertainty model of the CAS can be applied to the TAS measurement with higher values due to uncertainty related to the temperature and pressure uncertainties. In fact, the error in terms of the TAS calculation can be estimated as:

$$\text{TAS} = \sqrt{\frac{2(q_c \pm q_{c,bias})R(T \pm \Delta T)}{p_\infty \pm p_{\infty,bias}}} = V_\infty + \text{TAS}_{bias}, \tag{20}$$

where $R = 287.06\,\text{J}\,\text{kg}^{-1}\,\text{K}^{-1}$ is the air specific gas constant. The quantity $\Delta T = \pm 2.5\,^\circ\text{C}$, whereas $q_{c,bias}$ and $p_{\infty,bias}$ are the largest bias error calculated according to models presented in Section 4.4 and Section 4.5 respectively. Therefore, in this work, the resulting $\text{TAS}_{bias} = \pm 0.47\,\text{m}\,\text{s}^{-1}$, whereas $\text{TAS}_{1\sigma} = \text{CAS}_{1\sigma}$.

4.8. Time Derivative of the Airspeed

The airspeed time derivative is a crucial measurement for ASSE applications. As the direct measurement cannot be taken, several schemes [39] are considered in this work: (i) backward two point 1st order; (ii) backward three point 2nd order; (iii) backward four point 3rd order; (iv) backward five point 4th order; (v) backward six point 5th order; (vi) backward seven point 6th order; (vii) centred three point 2nd order; (iix) centred five point 4th order. The numerical derivative is denoted as $\tilde{\dot{y}}$. In order to evaluate the numerical estimation error at several frequency of several derivative schemes, a linear increasing frequency cosine function and its exact derivative are defined as

$$y = \cos 2\pi f t$$
$$\dot{y} = -4\pi f_{max}\frac{t}{t_{end}} \sin 2\pi f t, \tag{21}$$

where $t \in [0, t_{end}]$ is the time in seconds and $f = f_{max}\frac{t}{t_{end}}$ is the increasing frequency and $f_{max} = 10\,\text{Hz}$ is the maximum frequency. The maximum frequency is chosen considering the target application of the ASSE demonstrator during flight trials. In fact, as an ultralight motorised aircraft is chosen for flight tests, the highest dynamic mode frequencies are not greater than $5\,\text{Hz}$. The same can be applicable for UAV, UAM, general aviation and civil aircraft. Considering a constant sampling time, the absolute error $|\tilde{\dot{y}} - \dot{y}|$ is reported in Figure 8a as a function of the frequency for several schemes. It is clear that: (1) the lower the frequency, the lower the error; (2) the higher the scheme order, the lower the error. The same conclusions can be obtained from another analysis that are summarised in Figure 8b. In the latter analysis, only five frequencies are selected (1 Hz, 2.5 Hz, 5 Hz, 7.5 Hz and 10 Hz) and the 1σ error (i.e., $\tilde{\dot{y}} - \dot{y}$) is calculated over an entire period. As noted before, the 7-point backward scheme shows the best performance among the backward schemes with comparable performances to those achieved with the 5-point centred scheme.

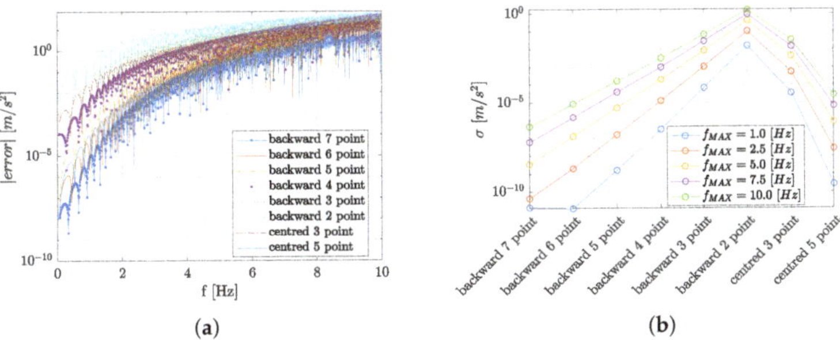

Figure 8. Error estimation of the numerical derivative estimation of Equation (21) at different frequencies. (a) Numerical derivative estimation at different frequencies, (b) 1σ error for different frequencies.

The 1σ error analysis reported in Figure 8b is very important to build the error model for the airspeed time derivative as shown at the end of the current section.

Although the best solution could be a centred derivative scheme, in this work, the backward schemes are preferred as the refresh rate of the estimated flow angles is considered aligned to all other parameters. Of course, other implementations with one or two step delays (i.e., 3-point and 5-point centred schemes respectively) can be accepted according to the specific final application (e.g., if it does not affect the control logics). However, the choice of the differential scheme is not made at the moment as the objective of the present analysis is to characterise the numerical derivative uncertainty.

The ASSE demonstrator's sampling rate is not constant due to the hardware implementation of several devices with different output rate and computational load. In fact, the ASSE demonstrator sampling rate is $0.01\,\text{s} \pm 0.002\,\text{s}$ as experienced in a log file record of more than 2 min.

Using the pressure test set, constant airspeed rates are generated at a constant altitude in order to avoid interference of the total and static pressure controllers. However, from the frequency analysis of Figure 9a, it emerged that the positive airspeed rates are affected by the pressure test set's controller (between 0.5 Hz and 20 Hz), whereas the negative rates are only dominated by high frequency dynamics as can be seen from Figure 9b. Therefore, only the negative airspeed rates are considered in order to isolate the effect of the derivative scheme.

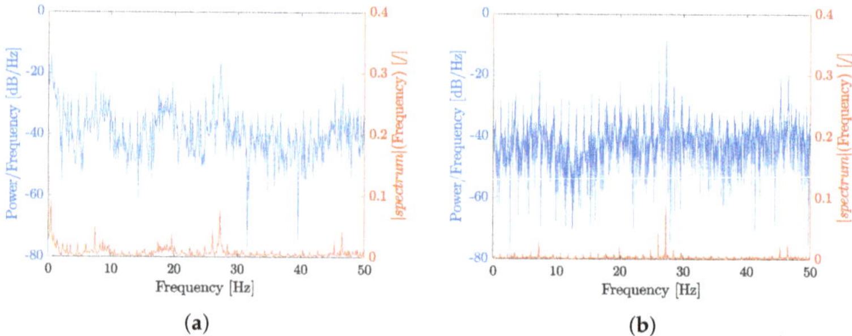

Figure 9. Frequency analysis of the numerical airspeed time derivative using the backward three-point scheme. (**a**) Positive TAS rate, $\dot{V}_\infty = 0.50\,\text{m s}^{-2}$, (**b**) Negative TAS rate, $\dot{V}_\infty = -0.51\,\text{m s}^{-2}$.

Thirty second time windows with constant airspeed rates are considered in order to evaluate the numerical errors of the airspeed time derivative with respect to the scheme adopted. As can be noted in Figure 10, the more points that are considered in the numerical scheme, the higher the errors. The latter phenomenon is due to the sampling time of the ASSE demonstrator that is not perfectly constant as mentioned before.

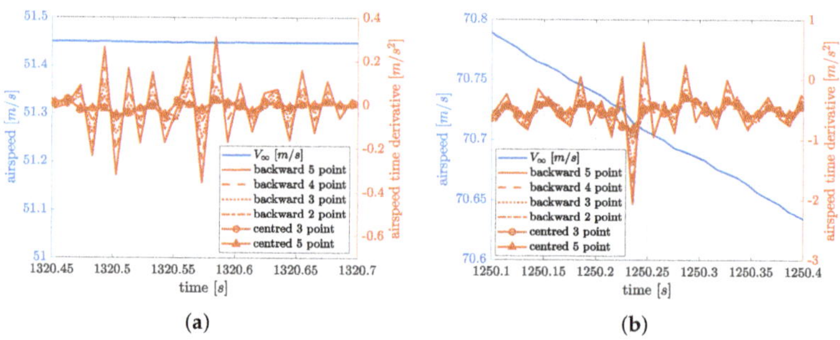

Figure 10. Airspeed time derivative using different numerical schemes. (**a**) Null TAS rate, $\dot{V}_\infty = 0.01\,\text{m s}^{-2}$, (**b**) Negative TAS rate, $\dot{V}_\infty = -0.51\,\text{m s}^{-2}$.

In Table 1 the error distributions for all the schemes considered in this work are reported.

Table 1. Numerical error analysis using different numerical schemes for airspeed time derivative calculation. With 1σ, 2σ and 3σ is intended the value such that the probability $\Pr(-\sigma \leq X \leq \sigma) = 68.3\%$, $\Pr(-2\sigma \leq X \leq 2\sigma) = 95.4\%$ and $\Pr(-3\sigma \leq X \leq 3\sigma) = 99.7\%$ also in case the error is not normally distributed.

Mean $\dot{V}_\infty$	Error (m s^{-2})	5 Point Backward	4 Point Backward	3 Point Backward	2 Point Backward	5 Point Centred	3 Point Centred
0.01 m s^{-2}		0.12	0.08	0.06	0.04	0.03	0.03
−0.51 m s^{-2}	σ	0.35	0.28	0.21	0.13	0.11	0.09
−1.00 m s^{-2}		0.69	0.56	0.40	0.25	0.23	0.18
0.01 m s^{-2}		0.23	0.17	0.12	0.08	0.07	0.06
−0.51 m s^{-2}	2σ	0.67	0.51	0.37	0.23	0.20	0.17
−1.00 m s^{-2}		1.15	0.91	0.72	0.46	0.41	0.35
0.01 m s^{-2}		0.35	0.25	0.18	0.12	0.11	0.09
−0.51 m s^{-2}	3σ	1.52	1.21	0.95	0.58	0.38	0.31
−1.00 m s^{-2}		3.11	2.55	2.00	1.24	0.76	0.62

Overall, in order to establish the optimal scheme and, hence, the related uncertainty model, a trade-off between static performance (reported in Table 1) and dynamic performance (described in Figure 8) should be carried out. As far as the backward schemes are concerned, the lowest numerical errors are obtained with the 5 point stencil considering the 1σ error due to the frequency effects of Figure 8b but it shows the largest errors when applied to the ASSE technological demonstrator (Figure 10) for the reasons mentioned before. On the other hand, the 2 point stencil shows the opposite performance: the best performance during steady derivative conditions (Figure 10) and the worst with dynamic derivative conditions (Figure 8). As the present characterisation aims to define an uncertainty budget for the airspeed time derivative, the numerical scheme selection is out of the scope. However, considering the backward schemes, the 3 point backward scheme can be a good compromise between steady state error ($\approx 0.4|\dot{V}_\infty|$) from Table 1 and the dynamic error (≈ 0.073 m s^{-2} at 10 Hz) from Figure 8b. Therefore, the total 1σ uncertainty of the time derivative of the airspeed can be approximated as $1\sigma_{\dot{V}_\infty} = 0.073 + 0.4|\dot{V}_\infty|$.

The frequency limit of 10 Hz is applicable to the majority of modern aircraft, whereas higher frequencies are only limited to very high performance aircraft (e.g., military ones) that should be investigated separately. Moreover, the use of a low-pass filter would also be beneficial to reducing the steady state errors but, at the same time, it would introduce higher delays during dynamic manoeuvres. As said before, the best trade-off should be studied but the latter analysis is out of the present work's scope.

5. TRL 5 ASSE Verification

As both accelerometers and gyroscope sensors cannot be excited at the same time, the TRL 5 verification of the proposed flow angle synthetic estimator is made up of two complementary activities: (1) integration verification stimulating the accelerometers or the gyroscope along with the air data reproduction, aiming to verify the correct implementation of all necessary algorithms; (2) flight simulations using corrupted data to feed the flow angle synthetic estimator according to noise characterisation introduced in Section 4.

5.1. Integration Verification

The integration verification, or single point verification, aims to verify the correct implementation of all necessary routines and mainly to verify the correct communication interface between the ASSE software module and all other physical sensors. To this purpose, the linearised ASSE scheme of Equation (17) is used because more adequate to the scope. The nonlinear scheme would have required longer time histories of flight manoeuvres that are not feasible in the laboratory environment. In fact, the nonlinear ASSE scheme is tested using flight simulations as reported in Section 5.2.

The test points are designed according to criteria presented in Section 3.2. A first test dataset is conceived, exploiting both the inertial acceleration and the angular rate at the same time as reported in Table 2. The value reported in Table 2 shall be considered as reference values and, therefore, the test results are reported according to the mean values recorded for 1 s in order to limit the effect of the measurement noise. Moreover, when the flow angle is denoted as N/A it means that the linear scheme of Equation (17) leads to unrealistic values.

Comparing the reference values and the measured values of the flow angles, it is clear that the estimations are implemented correctly even though a residual error can be noted. The latter errors mainly arise because of the deviation from the reference values and those actually realised during the integration tests. For example, considering the test case 1, the value of the reference airspeed time derivative is $1\,\mathrm{m\,s^{-2}}$ whereas the mean value obtained during the test with the pressure test set is $0.96\,\mathrm{m\,s^{-2}}$ because of the limitation of the experimental setup itself. The deviation of 4% on the $\dot{V}_\infty$ leads to the $0.2°$ error on the AoS measurement. On the other hand, the acceleration and the angular rates are affected by very low error as described in Section 4.

Table 2. Numerical ASSE estimation exciting both the accelerometer and the gyroscope with steady state values.

Test #	TAS $(\mathrm{m\,s^{-1}})$	$\dot{V}_\infty$ $(\mathrm{m\,s^{-2}})$	a $[a_X, a_Y, a_Z]^T$ (g)	ω $[p, q, r]^T$ $(° \mathrm{s}^{-1})$	Flow Angle Ref. (AoA, AoS) (°)	Flow Angle Meas. AoA, AoS (°)
1	10	1	$[0,1,0]^T$	$[0.707, -0.707, 0]^T$	N/A, 5.80	N/A, 5.6
2	10	2	$[0,1,0]^T$	$[0.707, -0.707, 0]^T$	N/A, 11.6	N/A, 11.3
3	10	2.5	$[0,1,0]^T$	$[0.707, -0.707, 0]^T$	N/A, 14.6	N/A, 14.0
4	10	1.5	$[0,1,0]^T$	$[0.707, -0.707, 0]^T$	N/A, 8.8	N/A, 8.4
5	10	-0.5	$[0,0,1]^T$	$[-0.707, -0.707, 0]^T$	-2.92, N/A	-3.1, N/A
6	10	0.25	$[0,0,1]^T$	$[-0.707, -0.707, 0]^T$	1.46, N/A	1.3, N/A
7	10	1	$[0,0,1]^T$	$[-0.707, -0.707, 0]^T$	5.84, N/A	5.84, N/A

A second test dataset is prepared, exploiting both the inertial acceleration and the attitudes at the same time as reported in Table 3. The attitudes are generated using a single axis tilting table. As mentioned before, the main errors arise from the deviation between the air data reference values (TAS and its time derivative) and those actually realised in the laboratory.

Table 3. Numerical ASSE estimation exciting the accelerometer with steady state values.

Test #	TAS $(\mathrm{m\,s^{-1}})$	$\dot{V}_\infty$ $(\mathrm{m\,s^{-2}})$	Attitudes (Pitch, Roll) (°)	Flow Angle Ref. (AoA, AoS) (°)	Flow Angle Meas. AoA, AoS (°)
8	10	0.25	96, 0	N/A, 14.3	N/A, 13.7
9	10	-0.5	102, 0	N/A, -14.3	N/A, -15.7
10	10	0.5	114, 0	N/A, 7.16	N/A, 6.9
11	10	0.25	0, 84	14.4, N/A	13.0, N/A
12	10	-0.5	0, 78	-14.2, N/A	-15.1, N/A
13	10	0.5	0, 66	7.39, N/A	7.7, N/A

Considering the results obtained and reported in Tables 2 and 3, the communication is correctly implemented between the ASSE software module and all other physical sensors.

5.2. Verification by Simulation

Flight simulated data are obtained using a coupled 6 degree of freedom nonlinear (ultralight) aircraft model equipped with nonlinear aerodynamic and thrust models designed accordingly to flight test results and the engine data sheet. The simulation is run using the explicit Euler scheme with a fixed time step of 10 ms and, therefore, $\tau_1 = t - 10$ ms aiming to simulate the ASSE demonstrator output rate. As the simulator does not implement any sensor noise, all simulated signals are noise-free and synchronised. In order to evaluate preliminarily the ASSE sensitivity to noise, the input signals are corrupted using error models described in Section 4.

A stall manoeuvre, described in Figure 11a, is performed to excite the AoA up to maximum values. After a short dive, the stall manoeuvre is performed producing initially an increase of airspeed and then a smooth deceleration leading to high angle-of-attack, as can be seen in Figure 11a, with limited changes in angle-of-sideslip and lateral acceleration a_Y. The angle-of-sideslip sweep manoeuvre is performed, exciting the angle-of-sideslip in a large range whereas the angle-of-attack is almost constant as can be seen in Figure 11b as well as the vertical acceleration a_Z.

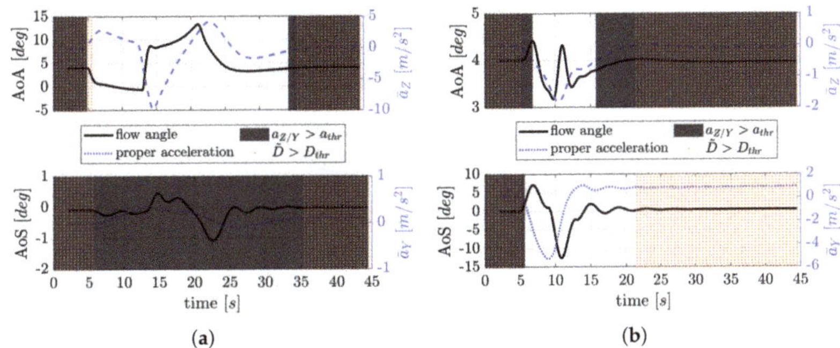

Figure 11. Verification manoeuvres. The shaded areas indicate the regions where the corresponding inertial acceleration modules are below the predefined threshold a_{thr} and dotted areas indicate where the determinant is below a predefined threshold D_{thr}. (**a**) Stall manoeuvre, (**b**) AoS sweep manoeuvre.

In Figure 11 those flight conditions where the ASSE is not reliable or applicable are indicated. As presented in Section 3.2, some thresholds exist where the analytical ASSE solution cannot be reliable. The criteria of Section 3.2 are sequentially applied. If the absolute value of the inertial acceleration is below the threshold a_{thr}, the ASSE scheme leads to unrealistic solutions characterised by very large errors (higher than 5°). The latter condition is represented using dark shaded areas in the Figure 11. Whereas, if $a_Y > a_{thr}$ and/or $a_Z > a_{thr}$, the determinant $\tilde{D}$ is evaluated: if the determinant absolute value is below the threshold D_{thr}, the ASSE scheme leads to unreliable solutions where errors up to 5° can be produced. This latter condition is represented using dotted areas in Figure 11. In order to guarantee the flow angle estimation during dynamic manoeuvres, each criteria is considered satisfied only if it is verified for at least 100 consecutive samples, that is, equivalent to 1 s in this work.

From Figure 11a, it is clear that the AoS cannot be estimated during the stall manoeuvre using the ASSE scheme because the lateral acceleration is below the defined threshold a_{thr} and the AoS is around the null value. Whereas, the AoA can be estimated using the ASSE scheme in the time window $[5\,\text{s}, 34\,\text{s}]$. As far as the AoS sweep manoeuvre is concerned, the AoA can be estimated in the time window $[6\,\text{s}, 17\,\text{s}]$. On the other hand the AoS could be estimated in the time window $[5\,\text{s}, 45\,\text{s}]$ but accepting less reliability from 21 s onwards.

Results of the AoA and AoS estimations using the nonlinear ASSE scheme are presented in Figure 12. First of all, it can be noted that the sign of the CAS bias does not play a significant role as both the dashed red ($\text{TAS}_{bias} = 0.47\,\text{m}\,\text{s}^{-1}$) and blue ($\text{TAS}_{bias} = -0.47\,\text{m}\,\text{s}^{-1}$) lines are not distinguishable.

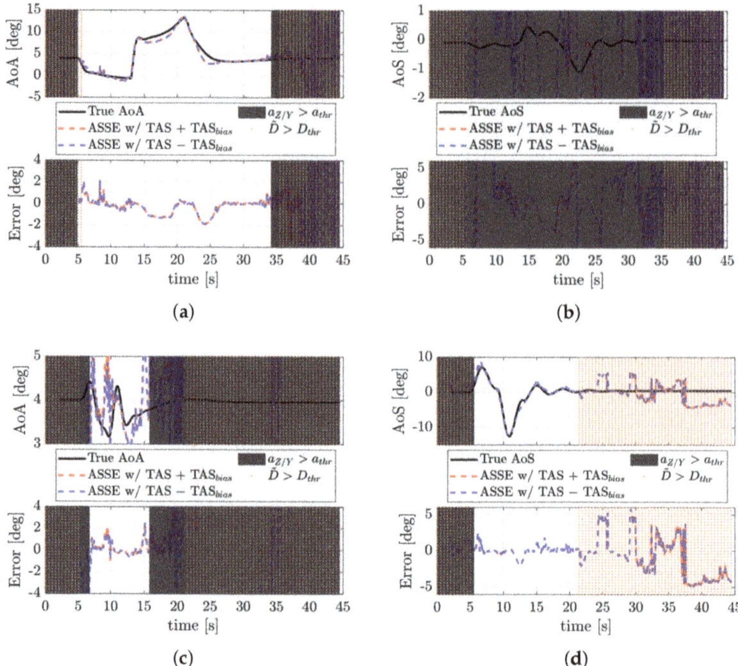

Figure 12. AoA and AoS estimation using the nonlinear ASSE scheme during the verification manoeuvres. The shaded areas indicate the regions where the corresponding inertial acceleration modules are below the predefined threshold a_{thr} and the dotted areas indicate where the determinant is below a predefined threshold D_{thr}. (**a**) AoA estimation during the stall manoeuvre, (**b**) AoS estimation during the stall manoeuvre, (**c**) AoA estimation during the sweep manoeuvre, (**d**) AoS estimation during the sweep manoeuvre.

As said before, during the stall simulation, the AoS cannot be estimated using the nonlinear ASSE scheme as can be observed in Figure 12b. In fact, except for a very limited time range between 12 s and 20 s where the error estimation is below $\pm 5°$, the error is not acceptable. On the other hand, the AoA is estimated with adequate accuracy in the time window $[5\,\text{s}, 34\,\text{s}]$ as the error is always below $\pm 2.5°$. However, the AoA estimation is less accurate at the beginning of the manoeuvre (time ≈ 5 s) where the determinant criteria are not satisfied even if the acceleration criteria are satisfied. These latter considerations, as also observed in [27], limit the application of the ASSE scheme at the beginning of the manoeuvre because the scheme relies on previous time steps (related to steady-state conditions) and, hence, not significant equations are considered in the scheme to be solved. Therefore, a preliminary conclusion leads to consider the determinant condition crucial, that is, as important as the acceleration criteria, when the manoeuvre begins.

As far as the AoS sweep manoeuvre is concerned, the vertical acceleration a_Z is small, even though it is beyond the threshold in the time window $[6\,\text{s}, 17\,\text{s}]$ where the ASSE scheme is used to estimate the AoA as in Figure 12c. Even though the estimation is not very accurate, the error is bounded in $\pm 2.5°$, which is acceptable for the scope of the present work. On the other hand, the AoS estimation can be performed with the nonlinear ASSE scheme for the entire manoeuvre except for the initial steady state conditions where the acceleration criteria are not met. However, from ≈ 22 s the determinant criteria are not met and, in fact, large errors (up to $\pm 5°$) can be noted. It is worth noting that, when the $\tilde{D} > D_{thr}$ the AoS estimation is biased with respect to the true value.

The error analysis proposed in Table 4 does not have a general validity on the performance of the ASSE scheme for AoA and AoS estimation because it would require the simulation of a whole flight envelope that is practically impossible. Moreover, a better strategy should be defined to take into account the conclusions of the present work. In fact, the proposed error analysis is intended to provide an overview of the ASSE possible performance in a realistic simulated environment (as required for TRL 5 validation) along with the integration verification tests.

At first sight, the OR condition between the determinant and the acceleration criteria would lead to more accurate results only because the ASSE scheme is applied to less flight data where the conditions are more suitable. However, it should be noted that the very large AoA maximum error is mainly due to the transition at the beginning of the stall manoeuvre (Figure 12a). On the other hand, the large AoS maximum error is experienced when the AoS values are quite constant and null as can be noted in Figure 12d, leading to a large 2σ as well.

Therefore, from the proposed results and error analysis, the OR condition should be preferred to achieve very low errors ($2\sigma < 2°$). Moreover, more stringent performance requirements on the flow estimation would require other solvers as mentioned before, rather than an iterative approach to solve the nonlinear ASSE scheme.

Table 4. ASSE estimation error analysis of the mean, absolute maximum, 1σ and 2σ errors. With 1σ and 2σ the value is intended such that the probability $\Pr(-\sigma \leq X \leq \sigma) = 68.3\%$ and $\Pr(-2\sigma \leq X \leq 2\sigma) = 95.4\%$ also in case the error is not normally distributed.

Flow Angle	ASSE Exclusion Criteria	Mean Error (°)	Max Abs. Error (°)	1σ Error (°)	2σ Error (°)
AoA	$\tilde{D} < D_{thr}$ OR $a_{Y,Z} < a_{thr}$	-0.19	3.02	0.60	1.66
	$a_{Y,Z} < a_{thr}$	0.18	3.02	0.61	1.67
AoS	$\tilde{D} > D_{thr}$ OR $a_{Y,Z} > a_{thr}$	0.04	2.52	0.41	1.74
	$a_{Y,Z} > a_{thr}$	-0.52	5.80	2.11	4.73

6. Conclusions

Within the scenario of flow angle synthetic estimators, the project SAIFE's scope is to design and manufacture a suitable technological demonstrator in order to verify at TRL 5 a model-free approach for flow angle estimation. The proposed approach is based on the rearrangement of classical flight mechanic equations in order to obtain a set of nonlinear equations, or the ASSE scheme. As the proposed scheme is only applicable when the aircraft is manoeuvring, practical thresholds are used to identify the flight conditions where the flow angle estimation is more reliable. The technological demonstrator is able to provide all necessary inputs to the ASSE scheme: true airspeed, angular rates, inertial accelerations and aircraft attitudes. In the present work, the inputs provided by physical sensors are characterised in order to evaluate the uncertainty budget on the performed measurements. This latter aspect is crucial for testing the nonlinear ASSE scheme in a realistic scenario for the TRL 5 verification. Firstly, the technological demonstrator is tested in the laboratory both to evaluate the uncertainty budget of the physical sensors and to verify the correct implementation of the required algorithms. Secondly, the ASSE scheme is tested using flight simulations data that are corrupted with realistic uncertainties. In order to tolerate the uncertainties of the input signals, 200 nonlinear equations are used to define the ASSE scheme, that is, data collected for 2 s. The latter choice highly depends on the physical sensors used and, therefore, on the particular aircraft application. The numerical results show low errors with $2\sigma < 2°$ both for AoA and AoS that are within the initial objectives. It is worth noting that results of the present work can be applied to any flying body to estimate the flow angles as the proposed ASSE scheme is model-free. On the other hand, the proposed setup relies on iterative methods to solve a scheme of 200 nonlinear equations and is unlikely to fit with a practical implementation. Further investigations are required to solve the ASSE scheme using alternative solvers that, for example, may contribute to reducing the number of nonlinear equations.

Author Contributions: Conceptualization, A.L. and P.G.; methodology, A.L.; software, A.L. and M.P; validation, A.L., P.G. and M.P.; formal analysis, A.L. and M.P.; investigation, A.L. and M.P.; resources, A.L.; data curation, A.L. and M.P.; writing—original draft preparation, A.L.; writing—review and editing, A.L., P.G. and M.P.; visualization, A.L.; supervision, P.G.; project administration, A.L.; funding acquisition, A.L. and P.G. All authors have read and agreed to the published version of the manuscript.

Funding: This research is supported by the PoC Instrument initiative funded by the "Compagnia di San Paolo" Foundation, promoted by LINKS with the support of LIFTT.

Conflicts of Interest: The authors declare no conflict of interest.

Abbreviations

The following abbreviations are used in this manuscript:

A/C	Aircraft
ADAHRS	Air Data System, Attitude and Heading Reference System
AHRS	Attitude and Heading Reference System
ADC	Air Data Computer
ADS	Air Data System or Sub-system
AoA	Angle-of-Attack
AoS	Angle-of-Sideslip
ASSE	Angle-of-Attack and -Sideslip Estimator
CAS	Calibrated Airspeed
FOG	Fibre Optical Gyroscope
GNSS	Global Navigation Satellite System
IMU	Inertial Measurement Unit
LRU	Line Replaceable Unit
MFP	Multi-Function Probe
OAT	Outside Air Temperature
SAIFE	Synthetic Air Data and Inertial Reference System
SFP	Single-Function Probe
SL	Sea level
SS	Synthetic Sensor
TAS	True Airspeed
TAT	Total Air Temperature
TRL	Technology Readiness Level
UAM	Urban Air Mobility
UAV	Unmanned Aerial Vehicles
VMC	vehicle management computer

References

1. Norris, G. Enhanced angle-of-attack system set for 737-10 flight tests. *Aviat. Week Space Technol.* **2021**. Available online: https://aviationweek.com/shownews/dubai-airshow/enhanced-angle-attack-system-set-737-10-flight-tests (accessed on 30 November 2021).
2. SAE International. *Minimum Performance Standards, for Angle of Attack (aoa) and Angle of Sideslip (AOS)*; SAE International: Warrendale, PA, USA, 2019.
3. Vitale, A.; Corraro, F.; Genito, N.; Garbarino, L.; Verde, L. An Innovative Angle of Attack Virtual Sensor for Physical-Analytical Redundant Measurement System Applicable to Commercial Aircraft. *Adv. Sci. Technol. Eng. Syst. J.* **2021**, *6*, 698–709. [CrossRef]
4. Ariante, G.; Ponte, S.; Papa, U.; Del Core, G. Estimation of Airspeed, Angle of Attack, and Sideslip for Small Unmanned Aerial Vehicles (UAVs) Using a Micro-Pitot Tube. *Electronics* **2021**, *10*, 2325. [CrossRef]
5. Prem, S.; Sankaralingam, L.; Ramprasadh, C. Pseudomeasurement-aided estimation of angle of attack in mini unmanned aerial vehicle. *J. Aerosp. Inf. Syst.* **2020**, *17*, 603–614. [CrossRef]
6. Valasek, J.; Harris, J.; Pruchnicki, S.; McCrink, M.; Gregory, J.; Sizoo, D.G. Derived angle of attack and sideslip angle characterization for general aviation. *J. Guid. Control. Dyn.* **2020**, *43*, 1039–1055. [CrossRef]

7. Lerro, A.; Battipede, M.; Gili, P.; Ferlauto, M.; Brandl, A.; Merlone, A.; Musacchio, C.; Sangaletti, G.; Russo, G. The clean sky 2 MIDAS project—An innovative modular, digital and integrated air data system for fly-by-wire applications. In Proceedings of the 2019 IEEE 6th International Workshop on Metrology for AeroSpace (MetroAeroSpace), Turin, Italy, 19–21 June 2019; Volume 1, pp. 714–719. [CrossRef]

8. Gertler, J. Analytical redundancy methods in fault detection and isolation - survey and synthesis. *IFAC Proc. Vol.* **1991**, *24*, 9–21. [CrossRef]

9. Perhinschi, M.; Campa, G.; Napolitano, M.; Lando, M.; Massotti, L.; Fravolini, M. Modelling and simulation of a fault-tolerant flight control system. *Int. J. Model. Simul.* **2006**, *26*, 1–10. [CrossRef]

10. Pouliezos, A.D.; Stavrakakis, G.S. *Analytical Redundancy Methods*; Springer: Amsterdam, The Netherlands, 1994; Volume 12, pp. 93–178. [CrossRef]

11. Rhudy, M.B.; Fravolini, M.L.; Porcacchia, M.; Napolitano, M.R. Comparison of wind speed models within a Pitot-free airspeed estimation algorithm using light aviation data. *Aerosp. Sci. Technol.* **2019**, *86*, 21–29. [CrossRef]

12. Eubank, R.; Atkins, E.; Ogura, S. Fault detection and fail-safe operation with a multiple-redundancy air-data system. In *AIAA Guidance, Navigation, and Control Conference*; American Institute of Aeronautics and Astronautics: Reston, VA, USA, 2010; pp. 1–14. [CrossRef]

13. Lu, P.; Van Eykeren, L.; Van Kampen, E.J.; Chu, Q. Air data sensor fault detection and diagnosis with application to real flight data. In *AIAA Guidance, Navigation, and Control Conference*; AIAA SciTech Forum: Reston, VA, USA, 2015; pp. 1–18. [CrossRef]

14. Komite Nasional Keselamatan Transportasi Republic of Indonesia. *Aircraft Accident Investigation Report - PT. Lion Mentari Airlines Boeing 737-8 (MAX)*; KNKT.18.10.35.04; 2018. Available online: http://knkt.dephub.go.id/knkt/ntsc_aviation/baru/2018-035-PK-LQPFinalReport.pdf (accessed on 30 November 2021).

15. European Aviation Safety Agency. *Easy Access Rules for Unmanned Aircraft Systems*; EASA: Cologne, Germany, 2015.

16. European Aviation Safety Agency. *Proposed Means of Compliance with the Special Condition VTOL*; MOC SC-VTOL Issue 1; EASA: Cologne, Germany, 2020.

17. Estrada, M.A.R.; Ndoma, A. The uses of unmanned aerial vehicles—UAV's- (or drones) in social logistic: Natural disasters response and humanitarian relief aid. *Procedia Comput. Sci.* **2019**, *149*, 375–383. [CrossRef]

18. Radoglou-Grammatikis, P.; Sarigiannidis, P.; Lagkas, T.; Moscholios, I. A compilation of UAV applications for precision agriculture. *Comput. Networks* **2020**, *172*, 107148. [CrossRef]

19. Baniasadi, P.; Foumani, M.; Smith-Miles, K.; Ejov, V. A transformation technique for the clustered generalized traveling salesman problem with applications to logistics. *Eur. J. Oper. Res.* **2020**, *285*, 444–457. [CrossRef]

20. Dendy, J.; Transier, K. *Angle-of-Attack Computation Study*; Technical Report AFFDL-TR-69-93; Air Force Flight Dynamics Laboratory: Wright-Patterson AFB, OH, USA, 1969.

21. Freeman, D.B. *Angle of Attack Computation System*; Technical Report AFFDL-TR-73-89; Air Force Flight Dynamics Laboratory: Wright-Patterson AFB, OH, USA, 1973.

22. Rohloff, T.J.; Whitmore, S.A.; Catton, I. Air data sensing from surface pressure measurements using a neural network method. *AIAA J.* **1998**, *36*, 2094–2101. [CrossRef]

23. Samara, P.A.; Fouskitakis, G.N.; Sakellariou, J.S.; Fassois, S.D. Aircraft angle-of-attack virtual sensor design via a functional pooling narx methodology. *IEEE Eur. Control. Conf.* **2003**, *1*, 1816–1821. [CrossRef]

24. Wise, K.A. Computational Air Data System for Angle-of-Attack and Angle-of-Sideslip. U.S. Patent 6,928,341 B2, 2005.

25. Langelaan, J.W.; Alley, N.; Neidhoefer, J. Wind field estimation for small unmanned aerial vehicles. *J. Guid. Control. Dyn.* **2011**, *34*, 1016–1030. [CrossRef]

26. Lu, P.; Van Eykeren, L.; van Kampen, E.; de Visser, C.C.; Chu, Q.P. Adaptive three-step kalman filter for air data sensor fault detection and diagnosis. *J. Guid. Control. Dyn.* **2016**, *39*, 590–604. [CrossRef]

27. Lerro, A.; Brandl, A.; Gili, P. Model-free scheme for angle-of-attack and angle-of-sideslip estimation. *J. Guid. Control. Dyn.* **2021**, *44*, 595–600. [CrossRef]

28. Lerro, A.; Brandl, A.; Gili, P.; Pisani, M. The SAIFE Project: Demonstration of a Model-Free Synthetic Sensor for Flow Angle Estimation. In Proceedings of the 2021 IEEE 8th International Workshop on Metrology for AeroSpace (MetroAeroSpace), Naples, Italy, 23–25 June 2021; pp. 98–103. [CrossRef]

29. Lacondemine, X.; Barbier, D. In-Flight Demonstration of a LiDAR Based Air Data System DANIELA Project. In Proceedings of the Sixth European Aeronautics Days (Aerodays), Madrid, Spain, 30 March–1 April 2011.

30. Lerro, A. Survey of certifiable air data systems for urban air mobility. In Proceedings of the 2020 AIAA/IEEE 39th Digital Avionics Systems Conference (DASC), San Antonio, TX, USA, 11–15 October 2020; pp. 1–10. [CrossRef]

31. Schmidt, D. *Modern Flight Dynamics*; McGraw-Hill: New York, NY, USA, 2011.

32. Etkin, B.; Reid, L. *Dynamics of Flight: Stability and Control*, 3rd ed.; Wiley: Hoboken, NJ, USA, 1995.

33. Lerro, A.; Brandl, A.; Gili, P. Neural network techniques to solve a model-free scheme for flow angle estimation. In Proceedings of the 2021 International Conference on Unmanned Aircraft Systems (ICUAS), Athens, Greece, 15–18 June 2021; pp. 1187–1193. [CrossRef]

34. Lerro, A. Physics-based modelling for a closed form solution for flow angle Estimation. *Adv. Aircr. Spacecr. Sci.* **2021**, *8*, 273–287. [CrossRef]

35. Marquardt, D.W. An algorithm for least-squares estimation of nonlinear parameters. *J. Soc. Ind. Appl. Math.* **1963**, *11*, 431–441. [CrossRef]
36. Lerro, A.; Musacchio, C. Preliminary definition of metrological guidelines for synthetic sensor verification. In Proceedings of the 2020 IEEE 7th International Workshop on Metrology for AeroSpace (MetroAeroSpace), Pisa, Italy, 22–24 June 2020; pp. 187–192. [CrossRef]
37. Pisani, M.; Astrua, M. The new INRIM rotating encoder angle comparator (REAC). *Meas. Sci. Technol.* **2017**, *28*, 045008. [CrossRef]
38. Pisani, M.; Astrua, M.; Iafolla, V.; Santoli, F.; Lucchesi, D.; Lefevre, C.; Lucente, M. On-ground actuator calibration for ISA—BepiColombo. In Proceedings of the 2015 IEEE Metrology for Aerospace (MetroAeroSpace), Benevento, Italy, 4–5 June 2015; pp. 312–317. [CrossRef]
39. Chung, T.J. Solution methods of finite difference equations. In *Computational Fluid Dynamics*; Cambridge University Press: Cambridge, UK, 2002; pp. 63–105. [CrossRef]

Article

Iterative Learning Sliding Mode Control for UAV Trajectory Tracking

Lanh Van Nguyen [1,*], **Manh Duong Phung** [1,2] **and Quang Phuc Ha** [1]

[1] School of Electrical and Data Engineering, University of Technology Sydney (UTS), 15 Broadway, Ultimo, NSW 2007, Australia; manhduong.phung@uts.edu.au (M.D.P.); quang.ha@uts.edu.au (Q.P.H.)

[2] VNU University of Engineering and Technology (VNU-UET), Vietnam National University, Hanoi (VNU), 144 Xuan Thuy, Cau Giay, Hanoi 100000, Vietnam

* Correspondence: vanlanh.nguyen@uts.edu.au

Abstract: This paper presents a novel iterative learning sliding mode controller (ILSMC) that can be applied to the trajectory tracking of quadrotor unmanned aerial vehicles (UAVs) subject to model uncertainties and external disturbances. Here, the proposed ILSMC is integrated in the outer loop of a controlled system. The control development, conducted in the discrete-time domain, does not require a priori information of the disturbance bound as with conventional SMC techniques. It only involves an equivalent control term for the desired dynamics in the closed loop and an iterative learning term to drive the system state toward the sliding surface to maintain robust performance. By learning from previous iterations, the ILSMC can yield very accurate tracking performance when a sliding mode is induced without control chattering. The design is then applied to the attitude control of a 3DR Solo UAV with a built-in PID controller. The simulation results and experimental validation with real-time data demonstrate the advantages of the proposed control scheme over existing techniques.

Keywords: iterative learning; sliding mode control; unmanned arial vehicles; trajectory tracking

Citation: Nguyen, L.V.; Phung, M.D.; Ha, Q.P. Iterative Learning Sliding Mode Control for UAV Trajectory Tracking. *Electronics* **2021**, *10*, 2474. https://doi.org/10.3390/electronics 10202474

Academic Editor: Umberto Papa

Received: 19 September 2021
Accepted: 5 October 2021
Published: 12 October 2021

Publisher's Note: MDPI stays neutral with regard to jurisdictional claims in published maps and institutional affiliations.

1. Introduction

In recent years, the interest in developing and utilizing unmanned aerial vehicles (UAVs) has been growing, with numerous applications in practice, such as mapping [1,2], inspection, search and rescue [3,4], construction automation [5], and agricultural surveillance [6]. When a quadrotor drone performs a desired trajectory, accurate tracking is highly necessary. In trajectory tracking control, feedback linearization (FL) has been widely used [7,8]. This control method works well under the assumption of known system dynamics. When faced with large uncertainties and disturbances, FL-based approaches may lead to poor tracking performance, and other advanced control laws are required. An adaptive feedback linearization controller was applied in [9], allowing for adjustments of the control parameters to enhance control performance. Robust control methods were also developed to improve control performance [10]. A backstepping controller was introduced in [11] to improve the tracking accuracy and robustness of UAVs' attitude control, wherein the external disturbances are estimated using a nonlinear disturbance observer.

Sliding mode control (SMC), a well-known control method for improving system robustness, has been successfully applied to various control systems [12,13] in general and particularly to UAVs [14,15]. However, information on disturbance bounds is required in these techniques. Adaptive SMC was developed to overcome this requirement [16]. This approach still reveals the main disadvantage of SMC, i.e., control signals usually present a chattering behavior, especially when dealing with large uncertainties and disturbances, which often require excessively high control gains. Various techniques have been suggested as a remedy, mostly using an approximation of the sign function to avoid or reduce chattering with some trade-offs with system robustness for control signal smoothing. Deep-learning-based techniques with convolutional neural networks have recently shown promise in learning systems that are subject to highly complex and time-varying uncertainties, for example, in vision-based applications such as in robust face tracking [17]. For control in robotic manipulators, a deep convolutional neural network was developed to integrate learning schemes into a fractional-order terminal sliding-mode controller for enhancing tracking accuracy [18]. Although the deep learning methods can achieve high-quality performance with continuous control signals, a major disadvantage is the high computational cost incurred for implementing a deep neural network.

Iterative learning control (ILC) is an effective technique for dealing with repetitive tasks. It allows for learning from system data to update the control input repeatedly to improve system performance [19]. Through trial-based learning, ILC is able to achieve highly accurate tracking performance even with large model uncertainties and disturbances. Unlike nonlearning control techniques, the system in ILC is reset to zero after the system reaches the final time, and then repeatedly follows the same reference again. Thereby, the control input can be adjusted through the repetitions to result in perfect tracking. Since ILC can learn from the system response to provide feedforward control in the iteration domain, it is more robust and can effectively compensate for model uncertainties and unknown disturbances, particularly iteration-invariant disturbances [20].

The application of iterative cybernetics, first proposed in [21], has emerged for iterative learning control in robotic systems [22], later developed for industrial control [23]. In the past decades, ILC has become an effective tool in various control systems, such as robot arm manipulators [24], chemical batch processes [25], wafer scanner systems [26], and video-rate atomic force microscopy [27]. Unlike other learning techniques such as artificial neural networks, which obtain the inverse dynamics from a training set [28] or adaptive controller, which tunes the control parameters [29,30], requiring a time-consuming process, ILC updates the control input from the information from previous executions and can hence converge quickly after a limited number of repetitions [19]. Moreover, as ILC does not require a system model, it is quite beneficial in practical applications that deal with unknown characteristics.

In this paper, integrating the learning capacity of ILC with the strong robustness of SMC, we propose a novel iterative learning sliding mode controller (ILSMC) to achieve high-accuracy trajectory tracking for UAVs while retaining strong robustness as well as alleviating control chattering. In terms of iterative control, several existing techniques have been introduced for UAVs. In [31], a plug-in controller was designed and implemented in aerial robots. Although the average tracking error is reduced for periodic reference trajectories, the technique does not concern the effect of disturbances. In [32], fuzzy PID-typed ILC was introduced; where fuzzy logic is used to tune the control parameters, high-accuracy tracking performance is hard to attain in comparison to other rigorous control strategies. In [33], optimization-based ILC was developed to improve the UAV trajectory tracking performance. In this approach, learning and filtering schemes are formulated into convex optimal problems. Although the two-step convex optimization problem can be solved using software, it involves high computational complexity.

The proposed ILSMC offers a simpler design, and thus is more robust and effective. The main contributions of this work include (i) the comprehensive development of an iterative learning term with fast convergence after several iterations to compensate for system uncertainties and unknown disturbances, and (ii) the integration of ILC and SMC schemes to a built-in PID controller in cascade to yield high performance for quadcopter trajectory tracking.

This paper is structured as follows: The control development of ILSMC is presented in Section 2. The convergence and stability analysis of the proposed learning algorithm is also provided in this section. Next, Section 3 presents system modeling, including kinematic and quadcopter dynamics. Then, the integration of ILSMC with PID control for the UAV is described in Section 4. Section 5 provides numerical simulation results, and Section 6 presents experimental validation with real-time data. Finally, a conclusion is drawn in Section 7.

2. Iterative Learning Sliding Mode Control

Iterative learning control (ILC) is a tracking control strategy for systems performing repetitive tasks, which are commonly required in industry. This technique aims to generate a feedforward control signal so that the system can learn from the previous responses to improve tracking performance and repeatedly eliminate disturbance after each iteration. The basic structure of an ILC is depicted in Figure 1 for an iteration number j. At this iteration, the input $u_{(j)}(k)$ and the deviation $e_{(j)}(k)$ between the reference $y_d(k)$ and the output $y_{(j)}(k)$ are stored to compute the control signal for the next iteration, with k starting from an initial time instant ($k = 0$). In this section, an iterative learning sliding mode control scheme is designed to deal with large uncertainties and disturbances.

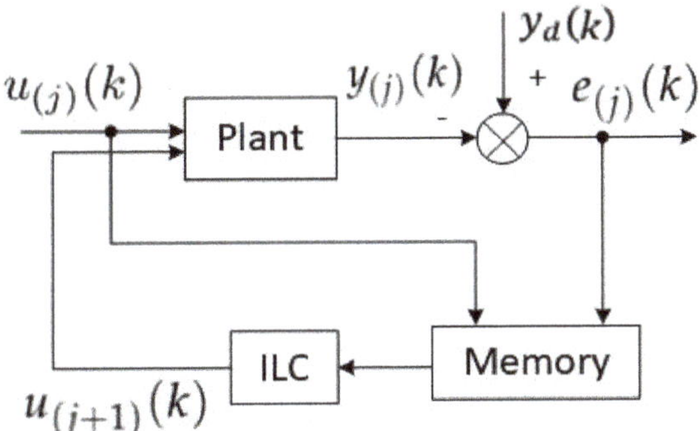

Figure 1. Basic structure of ILC.

2.1. ILSMC Design

Consider the following general discrete-time control system:

$$\begin{cases} x_{1_{(j)}}(k+1) = x_{1_{(j)}}(k) + \Delta_t x_{2_{(j)}}(k), \\ x_{2_{(j)}}(k+1) = f(x_{(j)}(k)) + \Delta_t B\left[u_{(j)}(k) + d_{(j)}(k)\right], \end{cases} \tag{1}$$

where k is the time instant and Δ_t is the sampling period. The subscript j denotes the iteration index, also called trial, run, cycle, or repetition in the ILC literature. The system state vector is $x_{(j)}(k) = \left[x_{1_{(j)}}(k) \quad x_{2_{(j)}}(k)\right]^T \in \mathbb{R}^{2m}$, where m is the dimension of state $x_{1_{(j)}}(k)$, $x_{2_{(j)}}(k)$ is its derivative, $u_{(j)}(k) \in \mathbb{R}^m$ is the control signal, $B \in \mathbb{R}^{m \times m}$ is a positive definite matrix, and $f(.)$ is a vector function. The influence of parameter variations and loading conditions, model uncertainties and external disturbances can be lumped into a vector $d_{(j)}(k)$. In each iteration, the input and state variables comprise an N-sample sequence each, where N is a finite number of samples.

Definition 1. *At iteration j, an exogenous input $\delta_{(j)}(k)$ is called iteration-invariant if it occurs repeatedly over iterations, or persistent within the iteration domain, that is, $\delta_{(1)}(k) = \delta_{(2)}(k) = \ldots = \delta_{(j)}(k)$ for all $k = \{0, 1, ..., N-1\}$.*

To proceed with the ILC methodology, the following assumption [20,27,34] is made:

Assumption 1. *In this paper, the lumped disturbance $d_{(j)}(k)$ is assumed to be iteration-invariant.*

In this paper, an ILSMC law is developed aiming to drive the tracking error asymptotically to zero from any initial condition and under external disturbances $d_{(j)}(k)$. The control algorithm consists of two steps. The first step is to induce a desired sliding surface to drive a learning sliding function to zero after some iterations regardless of external disturbance and system uncertainties. In the second step, the tracking error of the system is driven to zero in the sliding mode associated with the control sliding function. Instead of using a discontinuous gain as in the conventional SMC methodology, here, an iteration learning process is involved; hence, a priori information of the disturbance bound is not required, and chattering is fully alleviated.

The control design is initiated by considering the tracking error in an iteration as

$$e_{(j)}(k) = \begin{bmatrix} e_{1(j)}(k) & e_{2(j)}(k) \end{bmatrix}^T = x_{(j)}(k) - x_d(k), \tag{2}$$

where $x_d(k) = \begin{bmatrix} x_{1d}(k) & x_{2d}(k) \end{bmatrix}^T$ is the desired trajectory vector, which is also iteration-invariant during the execution of repetitive tasks.

Let us define the control sliding function as:

$$\sigma_{(j)}(k) = e_{2(j)}(k) + ce_{1(j)}(k), \tag{3}$$

where $c = diag(c_i) \in \mathbb{R}^{m \times m}$, $c_i > 0$.

By denoting

$$\Delta_{\sigma_{(j)}}(k) = \begin{bmatrix} \sigma_{(j)}(k+1) - \sigma_{(j)}(k) \end{bmatrix} \Delta_t^{-1}, \tag{4}$$

we have from (3):

$$\begin{aligned} \sigma_{(j)}(k+1) &= e_{2(j)}(k+1) + ce_{1(j)}(k+1) \\ &= x_{2(j)}(k+1) - x_{2d}(k+1) + ce_{1(j)}(k+1). \end{aligned} \tag{5}$$

Substituting (1), (3), and (5) into (4) yields:

$$\begin{aligned} \Delta_{\sigma_{(j)}}(k) = &\Big[f(x_{(j)}(k)) + \Delta_t B \Big[u_{(j)}(k) + d_{(j)}(k) \Big] - x_{2d}(k+1) \\ &+ c \Big[e_{1(j)}(k+1) - e_{1(j)}(k) \Big] - e_{2(j)}(k) \Big] \Delta_t^{-1}. \end{aligned} \tag{6}$$

By using the forward Euler method for discretization, we also have

$$e_{1(j)}(k+1) = e_{1(j)}(k) + \Delta_t e_{2(j)}(k) \Rightarrow e_{1(j)}(k+1) - e_{1(j)}(k) = \Delta_t e_{2(j)}(k). \tag{7}$$

Applying (7) into (6) produces:

$$\begin{aligned} \Delta_{\sigma_{(j)}}(k) = &[f(x_{(j)}(k)) + \Delta_t B(u_{(j)}(k) + d_{(j)}(k)) \\ &- x_{2d}(k+1) + (c\Delta_t - 1)e_{2(j)}(k)] \Delta_t^{-1}. \end{aligned} \tag{8}$$

2.1.1. Equivalent Control

Now let us consider the following dynamics to be induced by the learning process:

$$S_{(j)}(k) = \Delta_{\sigma_{(j)}}(k) + \mu\sigma_{(j)}(k) = 0, \tag{9}$$

where $\mu > 0$ is a control parameter. Substituting (8) into (9) yields:

$$\begin{aligned} [f(x_{(j)}(k)) &+ \Delta_t B(u_{(j)}(k) + d_{(j)}(k)) - x_{2d}(k+1) \\ &+ (c\Delta_t - 1)e_{2(j)}(k)] \Delta_t^{-1} + \mu\sigma_{(j)}(k) = 0. \end{aligned} \tag{10}$$

In nominal conditions of the system under no model error or disturbance, the equivalent control of the system is described by:

$$u_{eq_{(j)}}(k) = (\Delta_t B)^{-1} \Big[-f(x_{(j)}(k)) + x_{2d}(k+1) - (c\Delta_t - 1)e_{2(j)}(k) - \Delta_t \mu\sigma_{(j)}(k) \Big]. \tag{11}$$

2.1.2. Learning Control

In the learning step, to drive the system trajectories toward the sliding surface (9) regardless of disturbances, an iterative learning scheme is introduced using the stored data from previous iterations as:

$$u_{ilc_{(j)}}(k) = (\Delta_t B)^{-1} \sum_{i=0}^{j-1} \lambda S_{(i)}(k)$$

$$= u_{ilc_{(j-1)}}(k) - (\Delta_t B)^{-1} \lambda S_{(j-1)}(k), \tag{12}$$

where the initial iterations $u_{ilc_{(0)}}(k) = 0$ and $\lambda > 0$ are a design parameter for the learning rate.

From (11) and (12), the ILSMC law is described by:

$$u_{(j)}(k) = u_{eq_{(j)}}(k) + u_{ilc_{(j)}}(k). \tag{13}$$

We summarize the ILSMC design in the following theorem:

Theorem 1. *For a discrete-time system* (1) *with sampling period Δ_t subject to iteration-invariant disturbance $d_{(j)}(k)$, under the iterative learning sliding mode control* (13) *comprising the equivalent control* (11) *and learning control* (12)*, if the control parameter μ and learning rate λ are respectively chosen such that $0 < \mu < 2/\Delta_t$, $0 < \lambda < 2$ and $\lambda \neq 1$, then the tracking error* (2) *is driven to zero at a sufficiently large number of iterations and the control system is asymptotically stable.*

Proof. By substituting (8) and (11)–(13) into (9), we obtain:

$$S_{(j)}(k) = -\sum_{i=0}^{j-1} \lambda S_{(i)}(k) + \Delta_t B d_{(j)}(k). \tag{14}$$

Similarly,

$$S_{(j-1)}(k) = -\sum_{i=0}^{j-2} \lambda S_{(i)}(k) + \Delta_t B d_{(j-1)}(k). \tag{15}$$

According to the Assumption, as $d_{(j)}(k)$ is iteration-invariant, from (14) and (15), we have:

$$S_{(j)}(k) - S_{(j-1)}(k) = -\lambda S_{(j-1)}(k)$$

$$\Leftrightarrow S_{(j)}(k) = (1-\lambda)S_{(j-1)}(k) = (1-\lambda)^2 S_{(j-2)}(k)$$

$$= \ldots = (1-\lambda)^j S_{(0)}(k). \tag{16}$$

From (16), the iterative learning algorithm converges to 0 at large values of the iteration number under the condition $|1-\lambda| < 1$. Therefore, if the learning rate λ is selected to satisfy $0 < \lambda < 2$ and $\lambda \neq 1$, we have

$$\lim_{j \to \infty} S_{(j)}(k) = \lim_{j \to \infty} (1-\lambda)^j S_{(0)}(k) = 0. \tag{17}$$

By substituting $\Delta_{\sigma_{(j)}}(k)$ into $S_{(j)}(k)$, Equation (9) can be rewritten as

$$S_{(j)}(k) = \Delta_t^{-1} \sigma_{(j)}(k+1) - (1-\mu\Delta_t)\Delta_t^{-1}\sigma_{(j)}(k), \tag{18}$$

whereby $S_{(j)}(k) \to 0$, with a proper selection of the learning rate λ and at an adequate number of iterations j, the sliding function (3) becomes:

$$\sigma_{(j)}(k) \rightarrow (1 - \mu\Delta_t)\sigma_{(j)}(k-1)$$
$$= (1 - \mu\Delta_t)^2\sigma_{(j)}(k-2) = \ldots = (1 - \mu\Delta_t)^k\sigma_{(j)}(0), \tag{19}$$

where $\sigma_{(j)}(0)$ is the initial value of $\sigma_{(j)}(k)$ at the jth iteration. Therefore, given a positive constant μ with $0 < \mu < 2/\Delta_t$, the sliding function $\sigma_{(j)}(k)$ in (19) approaches zero at a sufficiently large value of k. Thus, since the sliding function $\sigma_{(j)}(k)$ as defined in (3) is driven to zero after some iterations j, a sliding mode is induced from the selection of parameter $c > 0$. It follows that

$$\lim_{j,k \rightarrow \infty} e_{(j)}(k) = 0. \tag{20}$$

Notably, the asymptotic convergence of the tracking error $e_{(j)}(k)$ here does not come from the switching of the control signal with a high discontinuous gain as in conventional SMC, but is a result of the proposed learning process (12). Hence, the sliding mode (3) induced for the tracking error can retain system robustness in the face of uncertainties and disturbances while avoiding the high-frequency switching of the control signal. This is the reason why the proposed ILSMC can achieve highly accurate tracking without control chattering. The tracking performance then depends on the convergence of the learning process, governed by the learning rate λ.

To verify the system stability, let us consider the control sliding function $\sigma_{(j)}(k)$ at iteration j. According to [35], the discrete-time control system will be asymptotically stable if, for all its entries $[\sigma_{(j)}(k)]$:

$$\begin{cases} [\sigma_{(j)}(k+1) - \sigma_{(j)}(k)]Sign([\sigma_{(j)}(k)]) < 0, \\ [\sigma_{(j)}(k+1) + \sigma_{(j)}(k)]Sign([\sigma_{(j)}(k)]) \geq 0, \end{cases} \tag{21}$$

where $Sign(\cdot)$ is the signum function.

To verify the above conditions, from (18), we have,

$$\sigma_{(j)}(k+1) = (1 - \mu\Delta_t)\sigma_{(j)}(k). \tag{22}$$

We obtain, accordingly

$$\sigma_{(j)}(k+1) - \sigma_{(j)}(k) = -\mu\Delta_t\sigma_{(j)}(k). \tag{23}$$

Therefore, the first condition of (21) is satisfactory as

$$-\mu\Delta_t[\sigma_{(j)}(k)]Sign([\sigma_{(j)}(k)]) < 0. \tag{24}$$

From (22), we also have

$$\sigma_{(j)}(k+1) + \sigma_{(j)}(k) = (2 - \mu\Delta_t)\sigma_{(j)}(k), \tag{25}$$

and with the choice $0 < \mu < 2/\Delta_t$, the second condition of (21) is also satisfactory since

$$(2 - \mu\Delta_t)[\sigma_{(j)}(k)]Sign([\sigma_{(j)}(k)]) \geq 0. \tag{26}$$

Therefore, the control system is asymptotically stable. The proof is completed. $\square$

Remark 1. *From (16), to quickly induce a sliding surface, a high rate of convergence is required, subject to the condition $|1 - \lambda| < 1$. This condition is similar to the ILC convergence condition presented in the frequency domain [36]. On one hand, the closer λ is to 1, the faster the convergence in the learning step. On the other hand, under the effect of noise and nonrepeating disturbances, a rapid learning rate could affect robustness. In practice, one can choose λ as close to 1 for fast convergence and gradually lower this value if required to reduce the system sensitivity.*

Remark 2. *The learning process can be terminated upon satisfaction of a required tracking performance index (TPI), e.g., when the integral time absolute error (ITAE) of the control error satisfies the requirement for tracking performance for a specific task of the system.*

3. System Description and Modeling

The UAV employed in this study was a quadcopter with a symmetric rigid structure and driven by four motors, as shown in Figure 2. For the quadcopter, the pitch angle, varied in accordance with the quadcopter's longitudinal motion, is controlled by adjusting the front and rear propellers' velocities, which generate the force F_1 and F_3. Meanwhile, its lateral displacement is governed by the roll angle, which is controlled through the right and left rotors' speeds, resulting in the forces F_2 and F_4. Finally, the yaw angle, associated with the UAV yaw motion, is regulated by the difference between torques generated by these pairs of rotors. In this work, we focused on the attitude tracking control, and thus, only the quadcopter orientation is considered here. The torques acting on the quadcopter include the thrust forces τ, the gyroscopic torques caused by the rotation of the quadcopter's rigid body τ_b and of four propellers τ_p, as well as the torque due to aerodynamic friction τ_a. Here, the propellers' gyroscopic effects and the drag from air resistance are considered external disturbances.

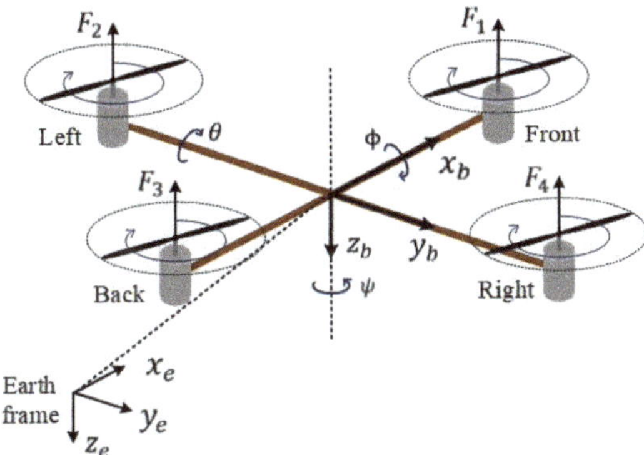

Figure 2. Configuration of a quadcopter.

3.1. Kinematics

As shown in the configuration in Figure 2, an earth frame, $\{x_e, y_e, z_e\}$, is fixed at the ground and a body frame, $\{x_b, y_b, z_b\}$, is attached to the CoG of the quadcopter, both with the z axis pointing downward. The position of the UAV's mass center in the earth frame is defined by a vector $P = (x, y, z)^T$. The UAV orientation is represented by angles $(\phi, \theta, \psi)^T$, corresponding to roll, pitch, and yaw motion, respectively. For attitude control, these angles are limited as $\phi \in [-\pi/2, \pi/2]$, $\theta \in [-\pi/2, \pi/2]$ and $\psi \in [-\pi, \pi]$. With respect to the earth frame, the orientation of the quadcopter is obtained a rotation transformation resulting from successively rotating around the x_b, y_b and z_b axes, and characterized by an orthonormal rotation matrix R [37]:

$$R = \begin{bmatrix} c_\psi c_\theta & c_\psi s_\theta s_\phi - s_\psi c_\theta & c_\psi s_\theta c_\phi + s_\psi s_\theta \\ s_\psi c_\theta & s_\psi s_\theta s_\phi + c_\psi c_\theta & s_\psi s_\theta c_\phi - c_\psi s_\theta \\ -s_\theta & c_\theta s_\phi & c_\theta c_\phi \end{bmatrix}, \tag{27}$$

where s_x and c_x denote $\sin x$ and $\cos x$, respectively.

Denoting the angular velocity vector of the quadcopter in the body frame as $(\omega_\phi \ \omega_\theta \ \omega_\psi)^T$, the rotational kinematics can be obtained as follows [38]:

$$\begin{bmatrix} \dot{\phi} & \dot{\theta} & \dot{\psi} \end{bmatrix}^T = W^{-1} \begin{bmatrix} \omega_\phi & \omega_\theta & \omega_\psi \end{bmatrix}^T, \tag{28}$$

where

$$W^{-1} = \begin{bmatrix} 1 & \sin\phi\tan\theta & \cos\phi\tan\theta \\ 0 & \cos\phi & -\sin\phi \\ 0 & \sin\phi\sec\theta & \cos\theta\sec\theta \end{bmatrix}.$$

3.2. Quadcopter Dynamics

From the quadcopter description, the components of torque vector $\tau = [\tau_\phi\ \tau_\theta\ \tau_\psi]^T$, corresponding to rotation in the roll, pitch, and yaw directions, respectively, are calculated as

$$\tau_\phi = l(F_2 - F_4), \tag{29}$$

$$\tau_\theta = l(-F_1 + F_3), \tag{30}$$

$$\tau_\theta = l(-F_1 + F_3), \tag{31}$$

where l is the distance from each rotor to the CoG, and β is the apparent radius for converting the force into the yaw torque.

From (29)–(31), the control inputs are described as:

$$\begin{bmatrix} u_\phi \\ u_\theta \\ u_\psi \\ u_z \end{bmatrix} = \begin{bmatrix} \tau_\phi \\ \tau_\theta \\ \tau_\psi \\ F \end{bmatrix} = \begin{bmatrix} 0 & l & 0 & -l \\ -l & 0 & l & 0 \\ -\beta & \beta & -\beta & \beta \\ l & l & l & l \end{bmatrix} \begin{bmatrix} F_1 \\ F_2 \\ F_3 \\ F_4 \end{bmatrix}, \tag{32}$$

where $u_\phi, u_\theta,$ and u_ψ are the roll, pitch, and yaw torques, respectively; $F = \sum_{n=1}^{4} F_n$ is the lift force, representing the total thrust utilizing from the four motors. As the only attitude of the quadcopter is controlled, u_z is assumed to balance with gravity.

The gyroscopic torque due to the rotation of the symmetric body of the quadcopter is described by [29]:

$$\tau_b = -SI \begin{bmatrix} \omega_\phi & \omega_\theta & \omega_\psi \end{bmatrix}^T, \tag{33}$$

where

$$S = \begin{bmatrix} 0 & -\omega_\psi & \omega_\theta \\ \omega_\psi & 0 & -\omega_\phi \\ -\omega_\theta & \omega_\phi & 0 \end{bmatrix}$$

is a skew-symmetric matrix. As shown in the configuration in Figure 2 with the body frame assigned to the quadcopter, given a mass point m_i with its coordinates (x_i, y_i, z_i) in the body, the quadcopter's inertia can be obtained as a diagonal matrix:

$$I = \begin{bmatrix} \sum_i (y_i^2 + z_i^2)m_i & 0 & 0 \\ 0 & \sum_i (x_i^2 + z_i^2)m_i & 0 \\ 0 & 0 & \sum_i (x_i^2 + y_i^2)m_i \end{bmatrix} = \begin{bmatrix} I_{xx} & 0 & 0 \\ 0 & I_{yy} & 0 \\ 0 & 0 & I_{zz} \end{bmatrix}. \tag{34}$$

Accordingly, Equation (33) can be rewritten as

$$\tau_b = \begin{bmatrix} (I_{yy} - I_{zz})\omega_\theta\omega_\psi & (I_{zz} - I_{xx})\omega_\phi\omega_\psi & (I_{xx} - I_{yy})\omega_\phi\omega_\theta \end{bmatrix}^T. \tag{35}$$

The gyroscopic torque due to the rotation of four propellers is determined as [29]:

$$\tau_p = \begin{bmatrix} I_r\omega_r\omega_\theta & -I_r\omega_r\omega_\phi & 0 \end{bmatrix}^T, \tag{36}$$

where I_r is the moment of inertia of the rotor of each motor, and $\omega_r = -\omega_{r1} + \omega_{r2} - \omega_{r3} + \omega_{r4}$ is the residual angular velocity, in which $\omega_{r1}, \dots, \omega_{r4}$ are the corresponding angular velocities of the propellers.

The air drag torque is calculated as [29]:

$$\tau_a = \begin{bmatrix} k_{ax}\omega_\theta^2 & k_{ay}\omega_\phi^2 & k_{az}\omega_\psi^2 \end{bmatrix}^T, \tag{37}$$

where k_{ax}, k_{ay}, and k_{az} are aerodynamic friction factors.

The dynamics of the quadcopter in attitude control can then be represented as:

$$\begin{bmatrix} \dot{\omega}_\phi & \dot{\omega}_\theta & \dot{\omega}_\psi \end{bmatrix}^T = I^{-1}(\tau_b + \tau + \tau_p - \tau_a). \tag{38}$$

Now, if the propeller gyroscopic and aerodynamic torques are considered as external disturbances, i.e.,

$$d = \begin{bmatrix} d_\phi & d_\theta & d_\psi \end{bmatrix}^T = \tau_p - \tau_a, \tag{39}$$

where d_ϕ, d_θ, and d_ψ are disturbance components, then substituting (32), (35), and (39) into (38) yields:

$$\dot{\omega}_\phi = I_{xx}^{-1}\left[(I_{yy} - I_{zz})\omega_\theta\omega_\psi + u_\phi + d_\phi\right], \tag{40}$$

$$\dot{\omega}_\theta = I_{yy}^{-1}\left[(I_{zz} - I_{xx})\omega_\phi\omega_\psi + u_\theta + d_\theta\right] \tag{41}$$

$$\dot{\omega}_\theta = I_{yy}^{-1}\left[(I_{zz} - I_{xx})\omega_\phi\omega_\psi + u_\theta + d_\theta\right]. \tag{42}$$

To express the quadcopter dynamics via the orientation angles, the model can be simplified by considering $[\omega_\phi, \omega_\theta, \omega_\psi] \approx [\dot{\phi}, \dot{\theta}, \dot{\psi}]$. This approximation is acceptable since a minor model error can be adequately addressed by a good controller. Accordingly, the quadcopter model is obtained as:

$$\ddot{\phi} = I_{xx}^{-1}\left[(I_{yy} - I_{zz})\dot{\theta}\dot{\psi} + u_\phi + d_\phi\right], \tag{43}$$

$$\ddot{\theta} = I_{yy}^{-1}[(I_{zz} - I_{xx})\dot{\phi}\dot{\psi} + u_\theta + d_\theta], \tag{44}$$

$$\ddot{\psi} = I_{zz}^{-1}\left[(I_{xx} - I_{yy})\dot{\phi}\dot{\theta} + u_\psi + d_\psi\right]. \tag{45}$$

3.3. Discrete-Time Model

In the discrete-time domain, by considering the difference approximation for first and second derivatives using the forward Euler method, the transformed discrete-time model can be obtained as below:

$$\begin{aligned}
\phi(k+2) &= 2\phi(k+1) - \phi(k) + I_{xx}^{-1}(I_{yy} - I_{zz})[\theta(k+1) - \theta(k)][\psi(k+1) - \psi(k)] \\
&\quad + \Delta_t^2 I_{xx}^{-1}\left[u_\phi(k) + d_\phi(k)\right],
\end{aligned} \tag{46}$$

$$\begin{aligned}
\theta(k+2) &= 2\theta(k+1) - \theta(k) + I_{yy}^{-1}(I_{zz} - I_{xx})[\phi(k+1) - \phi(k)][\psi(k+1) - \psi(k)] \\
&\quad + \Delta_t^2 I_{yy}^{-1}[u_\theta(k) + d_\theta(k)],
\end{aligned} \tag{47}$$

$$\begin{aligned}
\psi(k+2) &= 2\psi(k+1) - \psi(k) + I_{zz}^{-1}(I_{xx} - I_{yy})[\phi(k+1) - \phi(k)][\theta(k+1) - \theta(k)] \\
&\quad + \Delta_t^2 I_{zz}^{-1}\left[u_\psi(k) + d_\psi(k)\right].
\end{aligned} \tag{48}$$

Now, consider the system state vector $x(k) = [x_1(k) \ \ x_2(k)]^T$ defined by:

$$x_1(k) = \begin{bmatrix} \phi_1(k) & \theta_1(k) & \psi_1(k) \end{bmatrix}^T = \begin{bmatrix} \phi(k) & \theta(k) & \psi(k) \end{bmatrix}^T, \tag{49}$$

$$x_2(k) = \begin{bmatrix} \phi_2(k) & \theta_2(k) & \psi_2(k) \end{bmatrix}^T = \begin{bmatrix} \frac{\phi(k+1)-\phi(k)}{\Delta_t} & \frac{\theta(k+1)-\theta(k)}{\Delta_t} & \frac{\psi(k+1)-\psi(k)}{\Delta_t} \end{bmatrix}^T. \tag{50}$$

From (46)–(50), we obtain the UAV state equation in discrete time of the form (1) as:

$$x_1(k+1) = x_1(k) + \Delta_t x_2(k), \tag{51}$$

$$x_2(k+1) = f(x(k)) + \Delta_t B[u(k) + d(k)], \tag{52}$$

where $f(x(k)) = x_2(k) + \Delta_t I^{-1}\tau_b(k)$ and $B = I^{-1}$.

4. Integrated ILSMC for UAV Attitude Control

The proposed ILSMC was then applied to the outer loop of a quadcopter with a built-in PID controller in the inner loop for flight control. Here, the ILSMC is integrated in cascade control to improve the performance of the UAV trajectory tracking in dealing with noise, nonrepeating uncertainties, and disturbances. Figure 3 shows the block diagram of the proposed controller wherein the reference signal of a feedback controller is generated by the ILSMC signal $\hat{u}_{(j)}(k)$ at a time instant k.

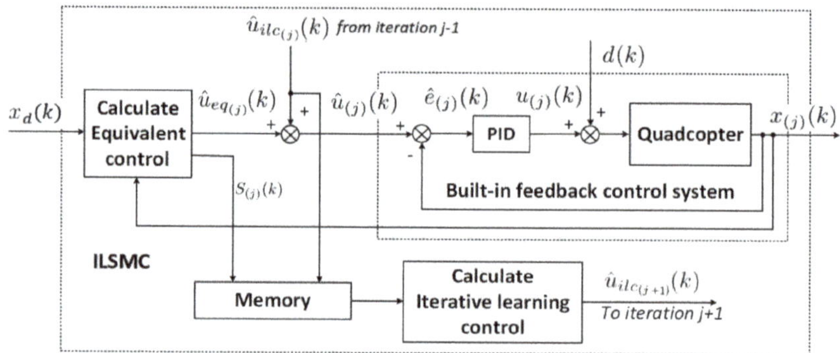

Figure 3. ILSMC in a cascade PID-controlled quadcopter.

4.1. Inner-Loop PID Controller

As shown in Figure 3, the output of the quadcopter PID controller is computed as

$$u_{(j)}(k) = K_p \circ \hat{e}_{(j)}(k) + K_i \Delta_t \circ \sum_{\kappa=1}^{k} \hat{e}_{(j)}(\kappa) + K_d \Delta_t^{-1} \circ [\hat{e}_{(j)}(k) - \hat{e}_{(j)}(k-1)], \tag{53}$$

where $\circ$ denotes the elementwise Hadamard product; $K_p = \begin{bmatrix} K_{p_\phi} & K_{p_\theta} & K_{p_\psi} \end{bmatrix}^T$, $K_i = \begin{bmatrix} K_{i_\phi} & K_{i_\theta} & K_{i_\psi} \end{bmatrix}^T$, and $K_d = \begin{bmatrix} K_{d_\phi} & K_{d_\theta} & K_{d_\psi} \end{bmatrix}^T$ are PID control parameters. The error of the PID feedback loop $\hat{e}_{(j)}(k)$ is defined as

$$\hat{e}_{(j)}(k) = \hat{u}_{(j)}(k) - x_{(j)}(k), \tag{54}$$

where $\hat{u}_{(j)}(k)$ is the ILSMC control signal.

Substituting (54) into (53) yields:

$$u_{(j)}(k) = D \circ \hat{u}_{(j)}(k) + H(k), \tag{55}$$

where

$$D = K_p + K_i \Delta_t + K_d / \Delta_t, \tag{56}$$

$$H(k) = -D \circ x_{(j)}(k) + K_i \Delta_t \circ \sum_{\kappa=1}^{k-1} \hat{e}_{(j)}(\kappa) - K_d \circ \hat{e}_{(j)}(k-1) \Delta_t^{-1}. \tag{57}$$

4.2. Outer-Loop ILSMC

As mentioned above, the proposed ILSMC was then added to the predesigned PID controller for improving the UAV tracking performance. Substituting (55) into (10) yields:

$$\Delta_t^{-1}[f(x_{(j)}(k)) + \Delta_t B(D \circ \hat{u}_{(j)}(k) + H(k) + d(k)) - x_{2d}(k+1) \\ + (c\Delta_t - 1)e_{2(j)}(k)] + \mu\sigma_{(j)}(k) = 0. \tag{58}$$

The equivalent control of the outer loop is then given by:

$$\hat{u}_{eq_{(j)}}(k) = D^{-1} \circ (\Delta_t B)^{-1} \{ -f(x_{(j)}(k)) - \Delta_t BH(k) + x_{2d}(k+1)$$
$$- (c\Delta_t - 1)e_{2(j)}(k) - \Delta_t \mu \sigma_{(j)}(k) \}. \tag{59}$$

In the learning step, the iterative learning term is computed as

$$\hat{u}_{ilc_{(j)}}(k) = \hat{u}_{ilc_{(j-1)}}(k) - D^{-1} \circ (\Delta_t B)^{-1} \lambda S_{(j-1)}(k), \tag{60}$$

where $S_{(j-1)}(k)$ is obtained from the learning process at a previous iteration $(j-1)$ as per (15).

This finally leads to the integrated iterative learning sliding mode control law (13) for the quadcopter:

$$\hat{u}_{(j)}(k) = \hat{u}_{eq_{(j)}}(k) + \hat{u}_{ilc_{(j)}}(k). \tag{61}$$

4.3. Implementation Procedure

In summary, a step-by-step procedure to implement the proposed control scheme is summarized as:

- Step 1: Declare $I_{xx}, I_{yy}, I_{zz}, K_p, K_i, K_d, c, \mu, \lambda$.
- Step 2: Set $x_d(k)$, $j = 0$, and $\hat{u}_{ilc_{(j)}}(k) = 0$.
- Step 3: Compute the ILSMC $\hat{u}_{(j)}(k)$ from (61) as a reference to the inner loop.
- Step 4: Compute, from the measured states $x_{(j)}(k)$, $e_{(j)}(k)$, $\sigma_{(j)}(k)$, $S_{(j)}(k)$, and the selected TPI.
- Step 5: Check if the tracking performance requirement is met to terminate the learning process. Otherwise, proceed to Step 6.
- Step 6: Set $j = j + 1$, update $\hat{u}_{ilc_{(j)}}(k)$ from (60), then return to Step 3.

5. Simulation Results

This section provides the simulation results of the proposed ILSMC design. The parameters used for simulation were obtained from the 3DR Solo drone [29], as listed in Table 1. The selected control parameters are described in Table 2. Here, in the learning process, a suitable value for λ was chosen to obtain a fast convergence rate, so $S_{(j)}(k)$ was driven to zero quickly. In the control phase, coefficients c_ϕ, c_θ, c_ψ ,and μ were chosen by the desired error dynamics described in (3). Initially, the iterative learning control signal was set to zero, $u_{ilc_{(0)}}(k) = 0$, and then updated after each iteration. To evaluate performance of the proposed controller, we compared it with other available techniques including the PD feedback controller, PD-type ILC [19], adaptive twisting sliding mode controller (ATSMC) [29], and adaptive finite-time control scheme (AFTC) [30].

Table 1. Parameters of the 3DR Solo drone.

Parameter	Value	Unit
m	1.5	kg
l	0.205	m
g	9.81	m/s^2
I_{xx}	9.1×10^{-3}	kg m^2
I_{yy}	16.4×10^{-3}	kg m^2
I_{zz}	24.1×10^{-3}	kg m^2

Table 2. Control parameters.

Parameter	Value	Parameter	Value
c_ϕ	50	μ	10
c_θ	50	λ	0.9
c_ψ	20	-	-

5.1. Step Response in Nominal Conditions

In this section, the performance of the proposed controller is evaluated via the step responses in nominal conditions where disturbances and uncertainties were set to zero. The desired reference attitude angles were set to $\phi_d = -20°$, $\theta_d = 20°$, and $\psi_d = 60°$ at 1 s. The simulation results of the step responses and control signals are shown in Figure 4, in which the black step signal is the desired angle, and responses of ATSMC, AFTC, PD, PD-ILC, and the proposed ILSMC controllers are depicted in cyan, green, magenta, blue, and red, respectively. It can be seen from Figure 4 for all three orientation angles that while ATSMC and AFTC provide some oscillations in the control, and PD presents a slow response, both iterative-learning-based techniques, ILSMC and PD-ILC, exhibit fast responses with zero steady-state error. PDILC, however, incurs a large overshoot, whereas ILSMC is able to maintain the desired dynamics without overshoot owing to the merits of sliding mode control. Notably, the fast response of ILSMC in comparison to ATSMC, AFTC, PD, ILC, and PD is attributed to the choice of $c = \mathrm{diag}(c_\phi, c_\theta, c_\psi)$ and μ. A faster transient response, however, requires more control efforts that may exceed the physical limits imposed by the motors and power supply for the drone. Moreover, the proposed controller is chattering-free in the steady state.

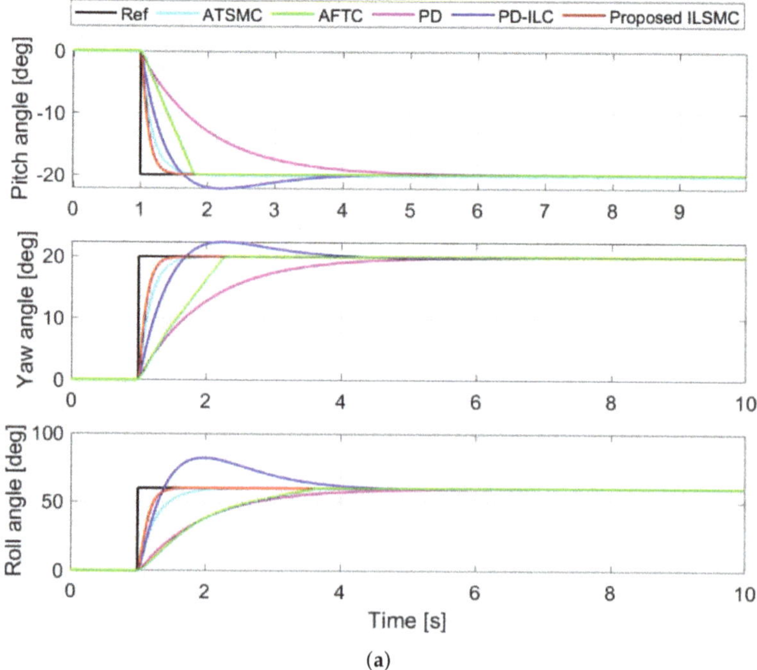

Figure 4. *Cont.*

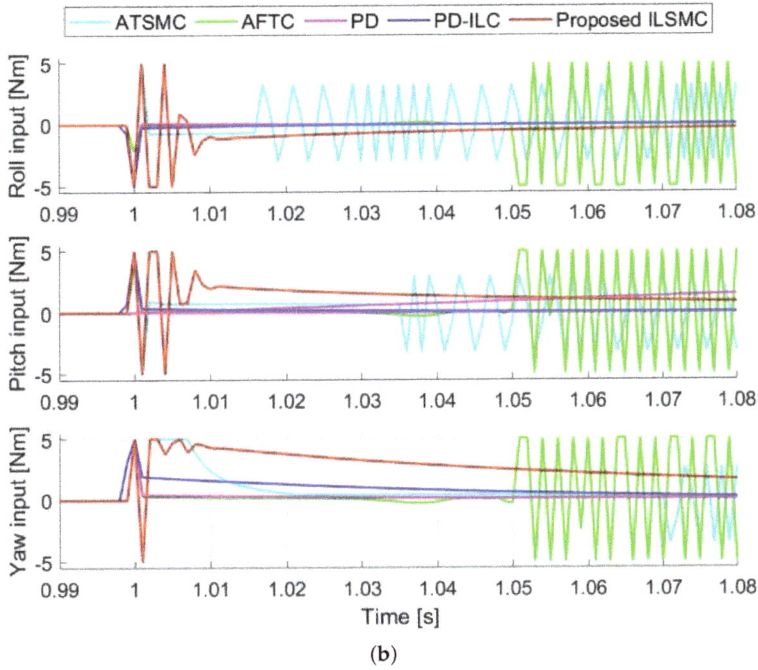

Figure 4. Step response in nominal conditions: (**a**) step response and (**b**) control effort.

5.2. Trajectory Tracking Performance under Disturbances and Uncertainties

To evaluate the tracking performance of ILSMC with the presence of uncertainties and disturbances caused by load variations, the reference attitude angles in this simulation were set, in degrees, as below:

$$
\begin{aligned}
\phi_d(k) &= 20 - 20\sin(2k), \\
\theta_d(k) &= 20 + 20\sin(2k), \\
\psi_d(k) &= 20 + 60\sin(2k).
\end{aligned}
\tag{62}
$$

For performance evaluation, the system was injected with a disturbance at $t = 10$ s whose components were:

$$
d_\phi = d_\theta = d_\psi = -0.2.
\tag{63}
$$

Considering 20% loading conditions, the model uncertainties were introduced by setting:

$$
\hat{I}_{xx} = 1.2 I_{xx}, \quad \hat{I}_{yy} = 1.2 I_{yy}, \quad \hat{I}_{zz} = 1.2 I_{zz},
\tag{64}
$$

where $\hat{I}_{xx}$, $\hat{I}_{yy}$, and $\hat{I}_{zz}$ are the estimation of I_{xx}, I_{yy}, and I_{zz}, respectively.

Figure 5 shows the tracking performance of the attitude angles, while Table 3 presents the TPI, for which the integral time absolute error (ITAE) is adopted here for all angles. It can be seen that the PD controller cannot cope with disturbances with large tracking errors at $t = 10$ s and high values of ITAE. As with PDILC, it can suppress disturbances but suffers from control overshoot and a noticeable error. Both the ATSMC and AFTC techniques present relatively accurate tracking performance with a small ITAE, between 1.40 and 5.03. The proposed ILSMC presents a relatively large tracking error at the first iteration (since $u_{ilc_{(0)}} = 0$), but owing to the learning ability, the tracking error decreases with increasing iterations by updating the iterative learning control term after each iteration. As the tracking performance is improved significantly, at the last iteration, ILSMC has the smallest ITAE among the considered techniques. Additionally, the absolute error is also the smallest, almost zero in steady state, as shown in the magnified figure, demonstrating the advantage of the proposed ILSMC.

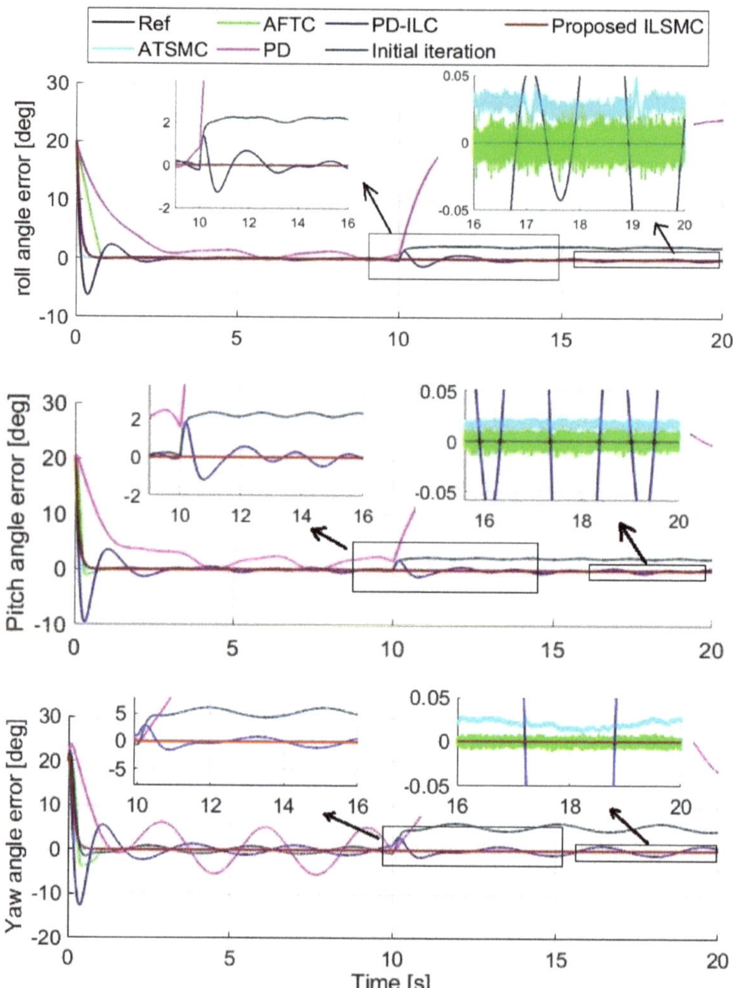

Figure 5. Performance in the presence of disturbances and uncertainties.

Table 3. ITAE of UAV attitude control angles.

UAV Angle (degrees)	ATSMC	AFTC	PD	PD-ILC	ILSMC
Roll	5.03	2.03	3344.2	39.91	0.255
Pitch	3.87	1.40	3468.4	55.60	0.261
Yaw	3.47	2.42	3323.0	149.89	0.444

Figure 6 shows the control efforts in the presence of disturbances and uncertainties. It can be seen that its magnitude increases after 10 s, which implies that more energy is required to compensate for the external disturbances. More importantly, the control efforts of ILSMC display oscillation only in the transient state, but no chattering in the steady state, which is beneficial for practical implementation.

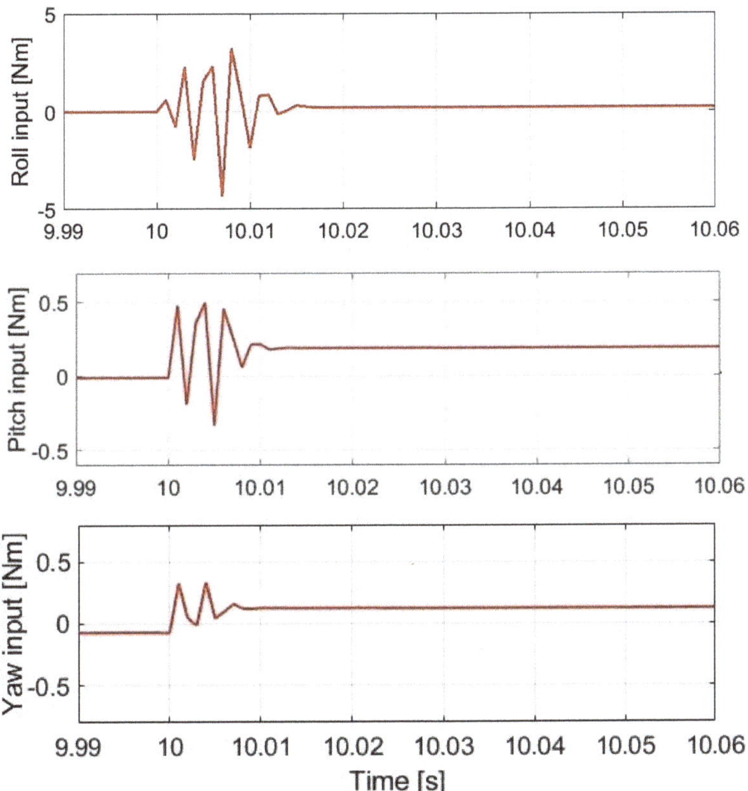

Figure 6. Control efforts in the presence of disturbances and uncertainties.

To evaluate the effect of the proposed learning mechanism, the ITAE values were computed for ILSMC after each iteration with different values of λ. The results up to 15 iterations are presented in Figure 7. They indicate that the ITAE of the all three attitude angle errors quickly decreases and converges to zero after several iterations. To induce a fast system sliding mode, a higher rate of convergence must be selected. It can be seen in Figure 7 that this can be obtained when λ is close to one. In this work, $\lambda = 0.9$ was chosen to achieve the desired control performance.

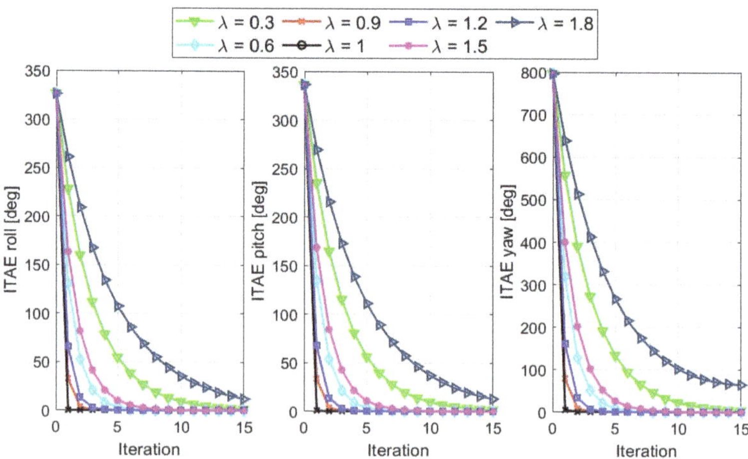

Figure 7. ITAEs of the tracking errors after each iteration.

6. Experimental Validation

In this section, we evaluate the performance of the combined ILSMC and PID control algorithm in the trajectory tracking problem for our UAV testbed, in which a built-in PID was already employed.

6.1. Experimental Setup

The setup for the experiments is shown in Figure 8, using a 3DR Solo drone with its parameters described in Table 1 [39]. It consists of two Cortex M4 168 MHz processors used for low-level control and one ARM Cortex A9 processor used for running the Arducopter flight operating system. The drone is equipped with a camera, a laser scanner, and environmental sensors for data acquisition. During the experiments, communication data, including control reference signals and drone sensor outputs, were transmitted to the ground control station via the local network established by the drone system. Mission Planner software was connected to the network to upload the flight plan to the drone and log flight data for analysis. In the experiments, the PID gains were set to their default values implemented in the 3DR Solo. From the desired and actual roll, pitch, and yaw angles, the tracking error was computed.

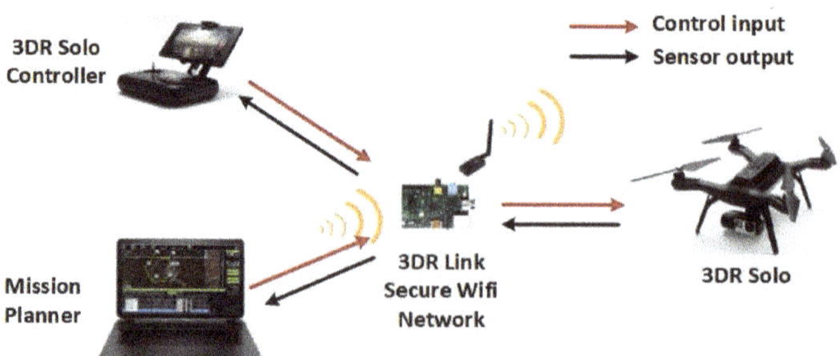

Figure 8. System architecture.

6.2. Real-Time Data Validation Results

The steps for conducting the experiments to validate the trajectory tracking performance of the proposed ILSMC were as follows: First, a trajectory was predefined with a starting point being set at the home position of the drone in an absolute frame of reference, as depicted in Figure 9. After that, the longitude, latitude, and altitude of the waypoints forming the trajectory were imported into Mission Planner, as depicted in Figure 10. Next, those waypoints were uploaded to the 3DR Solo to fly automatically, as shown in Figure 11. Then, the reference and actual attitude angles were logged by Mission Planner, as shown in Figure 12, for comparison. The errors between those angles were used to update the iterative learning term. Finally, the trajectories of the 3DR Solo drone obtained by using the built-in PID controller were compared with the results obtained using ATSMC, AFTC, PD-ILC, and the proposed ILSMC.

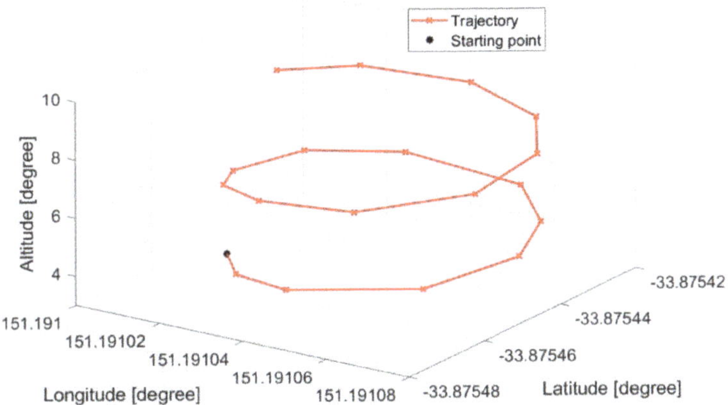

Figure 9. Predefined trajectory.

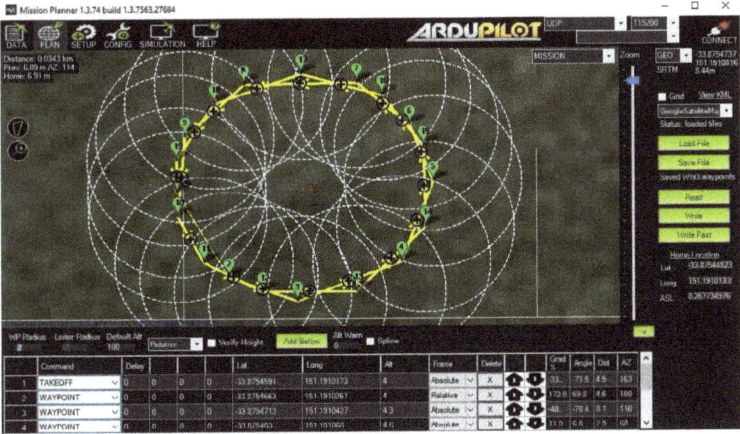

Figure 10. Imported trajectory.

The comparison was performed by setting the references obtained from the 3DR Solo drone under similar control settings as in the simulation. Figure 13 shows typical comparison results for the UAV roll, pitch, and yaw responses. It can be seen that the deviation between the reference and the actual roll angle controlled by the built-in PID is relatively high due to disturbances. Advanced techniques can improve the tracking performance in which ILSMC remains the best, as indicated by its smallest tracking error, as can be seen clearly in the insets in the figure. The results obtained confirm the validity and efficiency of the proposed approach.

Figure 11. Flying 3DR Solo drone.

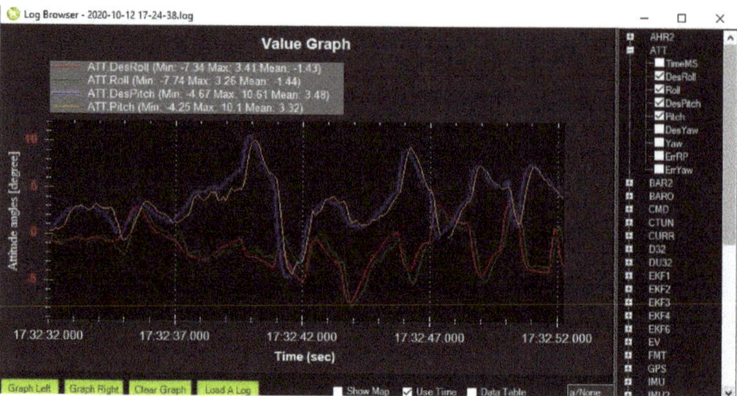

Figure 12. Logged flight data.

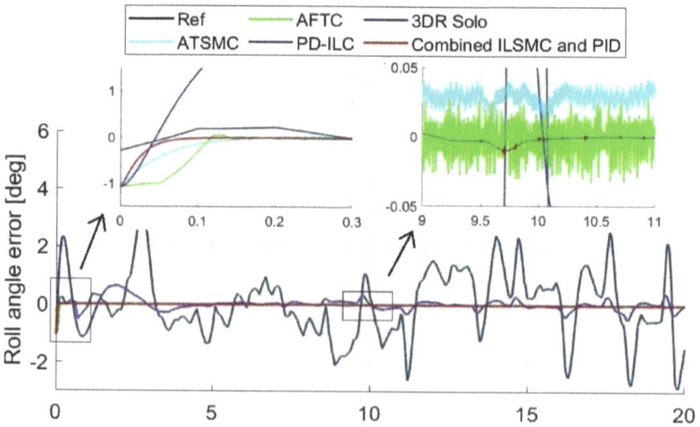

Figure 13. *Cont.*

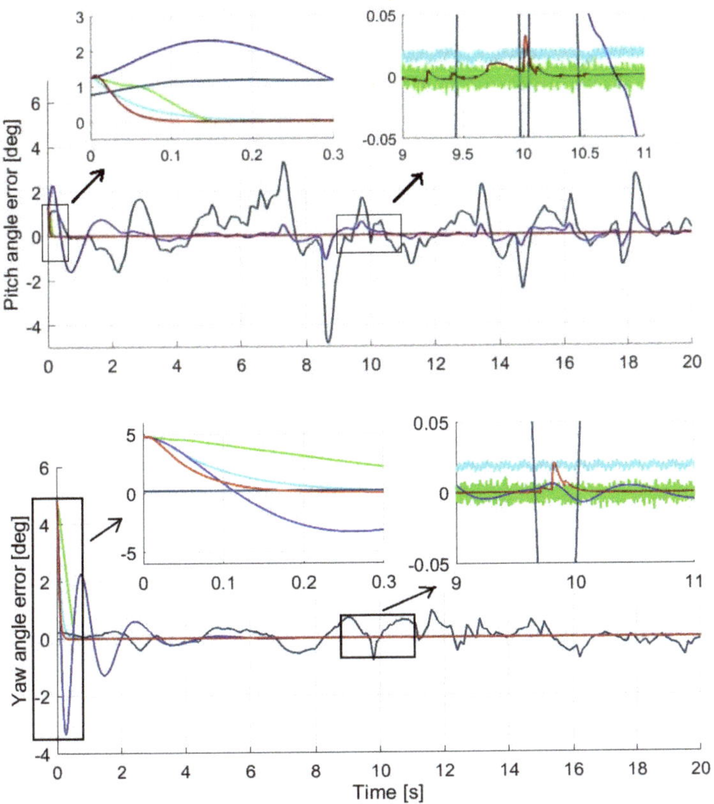

Figure 13. Tracking performance with real-time data.

From the real-time experiments, the recorded control efforts are shown in Figure 14, where the steady-state yaw torque is a constant as the quadcopter was controlled to lift up with a linearly increasing height while performing a circular trajectory during the test.

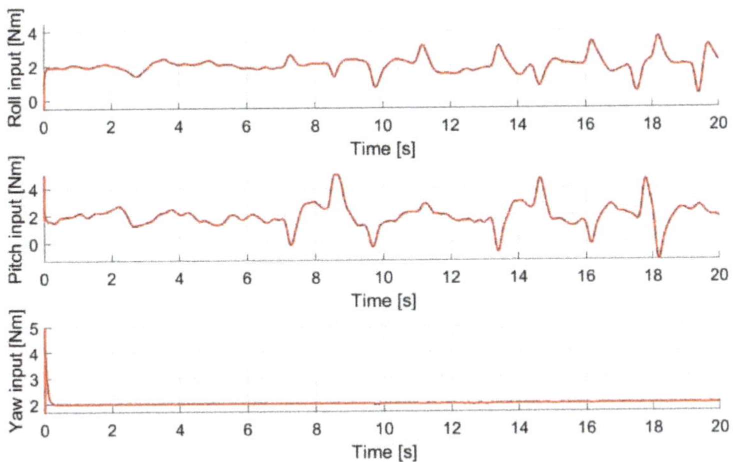

Figure 14. Control efforts with real-time data.

7. Conclusions

We proposed an effective control technique called ILSMC to address the tracking control problem experienced by quadcopters when subject to disturbances and uncertainties. The control signal consists of an equivalent term to control the system states within the desired sliding surface, and an iterative learning term to drive the system states toward the sliding surface and then remain in the sliding surface despite the presence of uncertainties and disturbances. The iterative learning signal is updated following some iterations to improve the tracking performance by using the data acquired from previous iterations. The simulation results showed, in the case of disturbances and uncertainties, that the iterative learning sliding mode controller presented the smallest tracking errors compared to some other existing control techniques used for quadcopter control. For UAVs with built-in PID controllers, the proposed control scheme can be integrated in a cascade structure to improve the trajectory tracking accuracy and robustness. Field tests were performed, and validation with real-time experimental data was conducted to confirm the advantages of the proposed approach. Our future work will focus on extending the learning mechanism to enable the control of multiple UAVs for real-time formation.

Author Contributions: L.V.N. and Q.P.H. proposed the main idea and the methodology of the research; L.V.N. performed the simulation, conducted the experiments, and wrote the original manuscript; M.D.P. and Q.P.H. reviewed the manuscript; the manuscript was directed and edited by Q.P.H. All authors have read and agreed to the published version of the manuscript.

Funding: This research received no external funding.

Acknowledgments: This work was supported by the Vingroup Science and Technology Scholarship (VSTS) Program for Overseas Study for Master's and Doctoral Degrees. The VSTS Program is managed by VinUniversity and sponsored by Vingroup.

Conflicts of Interest: The authors declare no conflict of interest.

Abbreviations

Abbreviations

The following abbreviations are used in this manuscript:

SMC	Sliding mode control
ILC	Iterative learning control
ILSMC	Iterative learning sliding mode control
UAV	Unmanned aerial vehicle
PID	Proportional–integral–derivative
FL	Feedback linearization
CoG	Center of gravity
TPI	Tracking performance index
ITAE	Integral time absolute error
ATSMC	Adaptive twisting sliding mode control
AFTC	Finite-time control scheme.

References

1. Grzonka, S.; Grisetti, G.; Burgard, W. A fully autonomous indoor quadrotor. *IEEE Trans. Robot.* **2012**, *28*, 90–100. [CrossRef]
2. Grisetti, G.; Stachniss, C.; Burgard, W. Non-linear constraint network optimization for efficient map learning. *IEEE Trans. Intell. Transp. Syst.* **2009**, *10*, 428–439. [CrossRef]
3. Flint, M.; Polycarpou, M.; Fernandez-Gaucherand, E. Cooperative control for multiple autonomous UAV's searching for targets. In Proceedings of the 41st IEEE Conference on Decision and Control, Las Vegas, NV, USA, 10–13 December 2002.
4. Brown, A.; Anderson, D. Trajectory optimization for high-altitude long endurance UAV maritime radar surveillance. *IEEE Trans. Aerosp. Electron. Syst.* **2020**, *56*, 2406–2421. [CrossRef]
5. Metni, N.; Hamel, T. A UAV for bridge inspection: Visual servoing control law with orientation limits. *Autom. Constr.* **2007**, *17*, 3–10. [CrossRef]

6. Herwitz, S.R.; Johnson, L.F.; Dunagan, S.E.; Higgins, R.G.; Sullivan, D.V.; Zheng, J.; Lobitz, B.M.; Leung, J.G.; Gallmeyer, B.A.; Aoyagi, M.; et al. Imaging from an unmanned aerial vehicle: Agricultural surveillance and decision support. *Comput. Electron. Agric.* **2004**, *44*, 49–61. [CrossRef]

7. Choi, Y.C.; Ahn, H.S.; Nonlinear control of quadrotor for point tracking: Actual implementation and experimental tests. *IEEE/ASME Trans. Mechatron.* **2015**, *20*, 1179–1192. [CrossRef]

8. Park, J.; Kim, Y.; Kim, S. Landing site searching and selection algorithm development using vision system and its application to quadrotor. *IEEE Trans. Control Syst. Technol.* **2015**, *23*, 488–503. [CrossRef]

9. Prasenjit, M.; Steven, W. Direct adaptive feedback linearization for quadrotor control. In Proceedings of the AIAA Guidance, Navigation, and Control Conference, Minneapolis, MN, USA, 13–16 August 2012; American Institute of Aeronautics and Astronautics: Minneapolis, MN, USA, 2012.

10. L'afflitto, A.; Anderson, R.B.; Mohammadi, K. An introduction to nonlinear robust control for unmanned quadrotor aircraft. *IEEE Control Syst. Mag.* **2018**, *38*, 102–121. [CrossRef]

11. Qingtong, W.; Honglin, W.; Qingxian, W.; Mou, C. Backstepping-based attitude control for a quadrotor UAV using nonlinear disturbance observer. In Proceedings of the 2015 34th Chinese Control Conference (CCC), Hangzhou, China, 28–30 July 2015.

12. Chincholkar, S.H.; Jiang, W.; Chan, C.-Y. Continuous nonsingular terminal sliding mode control of DC–DC boost converters subject to time-varying disturbances. *IEEE Trans. Circuits Syst. II Exp. Briefs* **2020**, *67*, 92–96. [CrossRef]

13. Ma, H.; Xiong, Z. Sliding mode control for uncertain discrete-time systems using an adaptive reaching law. *IEEE Trans. Circuits Syst. II Exp. Briefs* **2021**, *68*, 722–726. [CrossRef]

14. Ali, S.U.; Samar, R.; Shah, M.Z.; Bhatti, A.I.; Munawar, K.; Al-Sggaf, U.M. Lateral guidance and control of UAVs using second-order sliding modes. *Aerosp. Sci. Technol.* **2016**, *49*, 88–100. [CrossRef]

15. Sankaranarayanan, V.; Mahindrakar, A.D. Control of a class of underactuated mechanical systems using sliding modes. *IEEE Trans. Robot.* **2009**, *25*, 459–467. [CrossRef]

16. Tian, B.; Yin, L.; Wang, H. Finite-time reentry attitude control based on adaptive multivariable disturbance compensation. *IEEE Trans. Ind. Electron.* **2015**, *62*, 5889–5898. [CrossRef]

17. Qi, Y.; Zhang, S.; Jiang, F.; Zhou, H.; Tao, D.; Li, X. Siamese local and global networks for robust face tracking. *IEEE Trans. Image Process.* **2020**, *29*, 9152–9164. [CrossRef] [PubMed]

18. Zhou, M.; Feng, Y.; Xue, C.; Han, F. Deep convolutional neural network based fractional-order terminal sliding-mode control for robotic manipulators. *Neurocomputing* **2020**, *416*, 143–151. [CrossRef]

19. Bristow, D.A.; Tharayil, M.; Alleyne, A.G. A survey of iterative learning control. *IEEE Control Syst. Mag.* **2006**, *26*, 96–114.

20. Yu, M.; Li, C. Robust Adaptive Iterative Learning Control for Discrete-Time Nonlinear Systems with Time-Iteration-Varying Parameters. *IEEE Trans. Syst. Man Cybern. Syst.* **2017**, *47*, 1737–1745. [CrossRef]

21. Uchiyama, M. Formation of high speed motion pattern of mechanical arm by trial. *Trans. Soc. Instrum. Control. Eng.* **1978**, *14*, 706–712. [CrossRef]

22. Arimoto, S.; Kawamura, S.; Miyazaki, F. Iterative learning control for robot systems. In Proceedings of the IECON, Tokyo, Japan, 22–26 October 1984; pp. 393–398.

23. Moore, K.L. Iterative learning control for deterministic systems. In *Advances in Industial Control*; Springer: London, UK, 1993.

24. Tayebi, A.; Abdul, S.; Zaremba, M.B.; Ye, Y. Robust iterative learning control design: Application to a robot manipulator. *IEEE/ASME Trans. Mechatron.* **2008**, *13*, 608–613. [CrossRef]

25. Mezghani, M.; Roux, G.; Cabassud, M.; Lann, M.V.L.; Dahhou, B.; Casamatta, G. Application of iterative learning control to an exothermic semibatch chemical reactor. *IEEE Trans. Control Syst. Technol.* **2002**, *10*, 822–834. [CrossRef]

26. Zhu, Q.; Song, F.; Xu, J.; Liu, Y. An internal model based iterative learning control for wafer scanner systems. *IEEE/ASME Trans. Mechatron.* **2019**, *24*, 2073–2084. [CrossRef]

27. Nikooienejad, N.; Maroufi, M.; Moheimani, R. Iterative learning control for video-rate atomic force microscopy. *IEEE/ASME Trans. Mechatron.* **2021**, *26*, 2127–2138. [CrossRef]

28. Madani, T.; Benallegue, A. Adaptive control via backstepping technique and neural networks of a quadrotor helicopter. *IFAC Proc. Vol.* **2008**, *41*, 6513–6518. [CrossRef]

29. Hoang, V.T.; Phung, M.D.; Ha, Q.P. Adaptive twisting sliding mode control for quadrotor unmanned aerial vehicles. In Proceedings of the 2017 Asian Control Conference (ASCC 2017), Gold Coast, Australia, 17–20 December 2017; pp. 671–676.

30. Tian, B.; Lu, H.; Zuo, Z.; Zong, Q. Adaptive finite-time attitude tracking of quadrotors with experiments and comparisons. *IEEE Trans. Ind. Electron.* **2019**, *66*, 9428–9438. [CrossRef]

31. He, X.; Guo, D.; Leang, K.K. Repetitive control design and implementation for periodic motion tracking in aerial robots. In Proceedings of the 2017 American Control Conference (ACC), Seattle, WA, USA, 24–26 May 2017; pp. 5101–5108.

32. Dong, J.; He, B. Novel fuzzy PID-type iterative learning control for quadrotor UAV. *Sensors* **2019**, *19*, 24. [CrossRef] [PubMed]

33. Adlakha, R.; Zheng, M. An Optimization-Based Iterative Learning Control Design Method for UAV's Trajectory Tracking. In Proceedings of the 2020 American Control Conference (ACC), Denver, CO, USA, 1–3 July 2020; IEEE: Piscataway, NJ, USA, 2020; pp. 1353–1359.

34. Chen, Y.; Moore, K.L. Harnessing the nonrepetitiveness in iterative learning control. In Proceedings of the 41st IEEE Conference on Decision and Control, Las Vegas, NV, USA, 10–13 December 2002; IEEE: Piscataway, NJ, USA, 2002; Volume 3, pp. 3350–3355.

35. Sarpturk, S.Z.; Istefanopulos, Y.; Kaynak, O. On the stability of discrete-time sliding mode control systems. *IEEE Trans. Autom. Control* **1987**, *32*, 930–932. [CrossRef]
36. Norrlöf, M.; Gunnarsson, S. Time and frequency domain convergence properties in iterative learning control. *Int. J. Control* **2002**, *75*, 1114–1126. [CrossRef]
37. Guilherme, V.R.; Manuel, G.O.; Francisco, R.R. An integral predictive/nonlinear H_∞ control structure for a quadrotor helicopter. *Automatica* **2010**, *46*, 29–39.
38. Craig, J.J. *Introduction to Robotics—Mechanics and Control*, 2nd ed.; Addison-Wesley Publishing Company, Inc.: Reading, MA, USA, 1989.
39. Hoang, V.T.; Phung, M.D.; Dinh, T.H.; Ha, Q.P. System architecture for real-time surface inspection using multiple UAVs. *IEEE Sens. J.* **2020**, *40*, 4430–4441. [CrossRef]

 *electronics*

Article

A Hybrid-Driven Optimization Framework for Fixed-Wing UAV Maneuvering Flight Planning

Renshan Zhang [1], Su Cao [2,*], Kuang Zhao [1], Huangchao Yu [2] and Yongyang Hu [1]

[1] Nanjing Telecommunication Technology Research Institute, National University of Defense Technology, Nanjing 210007, China; zhangrenshan16@nudt.edu.cn (R.Z.); zhaokuang12@nudt.edu.cn (K.Z.); huyongyang17@nudt.edu.cn (Y.H.)
[2] Institute of Unmanned Systems, National University of Defense Technology, Changsha 410073, China; yuhuangchao@nudt.edu.cn
* Correspondence: caosu18@nudt.edu.cn

Abstract: Performing autonomous maneuvering flight planning and optimization remains a challenge for unmanned aerial vehicles (UAVs), especially for fixed-wing UAVs due to its high maneuverability and model complexity. A novel hybrid-driven fixed-wing UAV maneuver optimization framework, inspired by apprenticeship learning and nonlinear programing approaches, is proposed in this paper. The work consists of two main aspects: (1) Identifying the model parameters for a certain fixed-wing UAV based on the demonstrated flight data performed by human pilot. Then, the features of the maneuvers can be described by the positional/attitude/compound key-frames. Eventually, each of the maneuvers can be decomposed into several motion primitives. (2) Formulating the maneuver planning issue into a minimum-time optimization problem, a novel nonlinear programming algorithm was developed, which was unnecessary to determine the exact time for the UAV to pass by the key-frames. The simulation results illustrate the effectiveness of the proposed framework in several scenarios, as both the preservation of geometric features and the minimization of maneuver times were ensured.

Keywords: flight maneuvers; hybrid data and model driven; key-frame; motion primitives

Citation: Zhang, R.; Cao, S.; Zhao, K.; Yu, H.; Hu, Y. A Hybrid-Driven Optimization Framework for Fixed-Wing UAV Maneuvering Flight Planning. *Electronics* **2021**, *10*, 2330. https://doi.org/10.3390/electronics 10192330

Academic Editors: Umberto Papa, Marcello Rosario Napolitano, Giuseppe Del Core and Salvatore Ponte

Received: 11 August 2021
Accepted: 16 September 2021
Published: 23 September 2021

Publisher's Note: MDPI stays neutral with regard to jurisdictional claims in published maps and institutional affiliations.

1. Introduction

Recent times have witnessed a wide range of applications for unmanned aerial vehicles (UAVs) including the commercial, military, and research fields [1–5]. Most of the autonomous UAV flight missions are limited to cruise on a predefined path with steady flight states. However, in some scenarios, such as dog fights and high-speed obstacle avoidance, UAVs are required to perform fast and agile maneuver flights. During maneuvers, drastic changes in position and attitude will hinder UAVs from maintaining trim conditions. Therefore, developing maneuvering flight techniques is of great importance, and the current research mainly focuses on the quadrotor UAV [6–9]. It must be noted that fixed-wing UAVs have longer endurance and a larger payload capacity compared to those of the quadrotors. Yet the maneuvers of fixed-wing UAVs have not been thoroughly studied due to the fact of its sophisticated movement features [10]. Meanwhile, it is harder to conduct real flight tests also.

The research on autonomous maneuverable flight is assumed to be hierarchical including several topics such as maneuver decision-making, maneuver planning, and tracking control [11]. The planner first decides the category of the maneuver, then generates a specific trajectory for the tracking control. Among them, maneuver planning is essential for the maneuverability of UAVs. In this article, we were mainly concerned with maneuver planning issues. When validating the complex maneuver representation approaches, appropriate maneuver generation algorithms are also considered.

Generally, state-of-the-art maneuver generation algorithms incorporate data-driven [12] and model-driven approaches. One category of maneuver generation algorithms that have been successfully utilized in UAV maneuver generation is learning from demonstration [13]. By collecting the flight data taught by experts with a high-level planning layer, imitating learning methods can be applied to extract the maneuver features [14]. The resulting algorithm is practically capable of reproducing and generalizing the learned motions to some extent. However, due to the limitations of viewing distance, communication delay, and external disturbances, such as wind gusts, it is hard for the pilot to perform their flight skills optimally [11]. Therefore, optimality is essential in the learning procedure and few of the data-driven algorithms in the recent literature take the optimization into consideration.

In the robotic literature, model-driven planning has been widely studied and the typical approaches include polynomial interpolation, Dubins curve [10,15,16]. For some special scenarios, it is essential to plan the position and attitude trajectories simultaneously. Therefore, many researchers have also made explorations in SE(3) space [17–19]. Most of the methods are based on the differential flatness principle which can greatly simplify the calculation complexity. However, for fixed-wing UAV flight maneuvers, it has to be stressed that those approaches did not take the complicated physical models with complex aerodynamic features into consideration. Therefore, the non-differential flatness property of fixed-wing UAV [20] hinders the application of the aforementioned methods. Solving the optimal control problems is also an effective maneuver planning method and has attracted growing attention. In particular, both the direct optimization method and indirect optimization methods have been applied in UAV flights in recent years [21–23]. Through taking the physical model into consideration, dynamical feasibility is ensured. Furthermore, by combing various cost items, optimal maneuvers can be solved for different scenarios [24]. However, for complex maneuvers, it is always hard to model the optimization problem, which has a great impact on the calculation efficiency and trajectory quality. In visual tracking [25,26], maneuvering is also required, vision-and-language navigation [27–30] provides another novel perspective. This method combines vision, language, and action which can turn relatively general natural-language instructions into robot agent actions. It is generally used in indoor complex environments and also has certain lightening significance for fixed-wing tracking of dynamic targets or maneuvering in complex outdoor environments.

For the representation of complex maneuvers, one recent algorithm that has been successfully proposed is the idea of key-frames [31], which selects several specific points in 3D space as the key-frames and various tasks can be accomplished. In [32], both the position and yaw angle are taken into account in the key-frame and the minimum snap trajectory is generated using polynomials. Furthermore, the author formulates the trajectory optimization problem as a quadratic programming issue and then enables the quadrotors to pass through multiple circular hoops quickly. Ref. [33] emphasizes the Bang-Bang characteristics of the minimum time trajectory and compares the existing time-optimal approaches, and through analysis, it is concluded that the polynomial method can only obtain non-optimal trajectories. On this basis, Foehn proposes complementary constraints [34] and solved both the time allocation and time-optimal trajectory planning problem elaborately. Through comparison and verification, it can be concluded that the trajectory designed by the algorithm is faster than that of professional pilots. For fixed-wing, a model is required, which can be obtained by identifying methods [35]. In addition, it focuses on the minimum time problem through setting a series of waypoints and utilizing the flight corridors and B-spline curves. Using numerical optimization methods to solve non-convex problems and cleverly designed initial guesses, the optimal trajectory is obtained. As for the fixed-wing UAV maneuver, it is hard to extract the trajectory features merely from position information. It is necessary to comprehensively consider the position and attitude features and design the corresponding key-frames for various maneuvers. Motion primitives also have a satisfactory effect on the decomposition of complex maneuvers [36]. Mueller and D'Andrea [15] proposed a framework for the efficient calculation of motion primitives. McGill University has presented a series of representative works based on the idea of motion primitives and maneuver optimization [37–39]. Dynamic motion primitives (DMPs) is a typical primitive which has been successfully utilized in the design of car driving motion libraries [40]. Inspired by the former works, we studied the dual quaternion-based dynamic motion primitives (DQ-DMPs) [41], which have the ability to learn and generalize maneuvers in SE(3) space. Nevertheless, it is hard for the dynamic motion primitive algorithm to ensure the kinodynamic feasibility.

Inspired by the above discussion, we propose a data-model-driven framework, which is shown in Figure 1, for fixed-wing UAV maneuvering optimization. The framework is based on a global model identified by teaching data, and uses an optimization method based on positional, attitude, and compound key-frames. Through numerical optimization, the time-optimal maneuver is finally obtained. Through comparison, the actions generated by this algorithm take less time than professional pilots. In complex actions, different primitives can be flexibly concatenated, which simplifies the generation of complex actions. The main contribution of this paper is listed as follows:

(1) We proposed a novel data-driven approach for model identification and key-frames extraction using the learning from demonstration principles. Then, complex maneuvers are decomposed into multiple motion primitives;

(2) Based on the motion primitives, the optimal maneuver generation issue is formulated into a time-optimal problem considering key-frames which the UAV must pass by. The connection method of different primitives was also considered in this paper for practicability;

(3) The proposed framework was verified thoroughly in simulation experiments, and it was possible to deduce that this framework is applicable for flight maneuvers in reality.

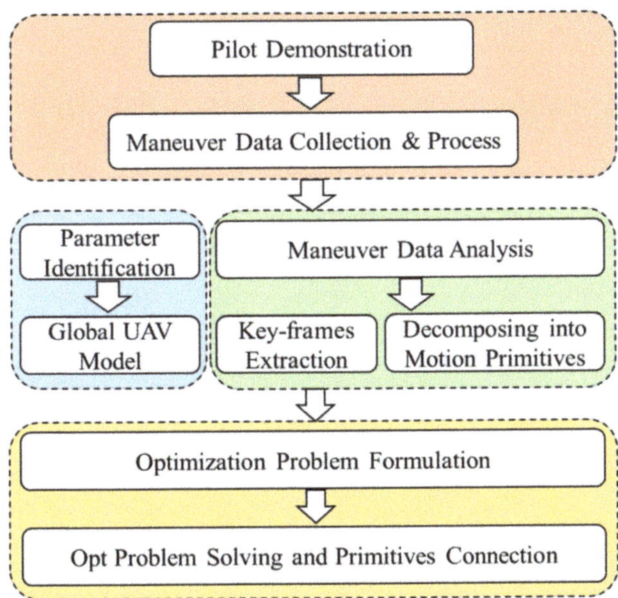

Figure 1. Hybrid-driven optimal maneuvering flight planning framework. We start with pilot demonstration and maneuver data collection and then perform model identification and maneuver key-frame and motion primitive analysis. Finally, based on the global model and key-frames, the optimal primitives are generated, and the corresponding primitives are concatenated into a complete maneuver.

In the next section, we discuss the basic knowledge. Sections 3 and 4 introduce data collection and acrobatic maneuver optimization, respectively. The experimental results are introduced in Section 5, and the conclusion in Section 6.

2. Preliminaries

2.1. Global Fixed-Wing Model

The aircraft model plays an important role in our algorithm framework. In order to obtain feasible maneuvers, we need to establish a relatively accurate rigid-body model. Due to the singular problem of Euler angle during the big maneuver, this paper adopted the quaternion model [42] to calculate:

$$
\begin{aligned}
\dot{\mathbf{p}}^i &= \mathbf{R}(\mathbf{q}) \cdot \mathbf{v}^b \\
\dot{\mathbf{q}} &= \tfrac{1}{2}\left[\boldsymbol{\omega}^b \otimes\right]\mathbf{q} \\
\dot{\mathbf{v}}^b &= m^{-1}(\mathbf{F}+\mathbf{T}) - \boldsymbol{\omega}^b \times \mathbf{v}^b \\
\dot{\boldsymbol{\omega}}^b &= \mathbf{J}^{-1}\left(\boldsymbol{\tau} - \boldsymbol{\omega}^b \times \mathbf{J}\boldsymbol{\omega}^b\right)
\end{aligned}
\tag{1}
$$

For translational kinematics equation, the $\mathbf{p}^i = [x,y,z] \in \mathbb{R}^3$ is the position of the inertial coordinate system, and $\mathbf{v}^b = [u,v,w]$ is the speed of the body coordinate system. The two state quantities are connected by a conversion matrix, which is composed of $\mathbf{q}$, where $\mathbf{q} = [q_0, q_1, q_2, q_3]^T \in SO(3)$ is a unit quaternion given $\|\mathbf{q}\| = 1$. The rotation matrix is:

$$
\mathbf{R}(\mathbf{q}) = \begin{bmatrix}
(q_0{}^2 + q_1{}^2 - q_2{}^2 - q_3{}^2) & 2(q_1 q_2 - q_0 q_3) & 2(q_1 q_3 + q_0 q_2) \\
2(q_1 q_2 + q_0 q_3) & (q_0{}^2 - q_1{}^2 + q_2{}^2 - q_3{}^2) & 2(q_2 q_3 - q_0 q_1) \\
2(q_1 q_3 - q_0 q_2) & 2(q_2 q_3 + q_0 q_1) & (q_0{}^2 - q_1{}^2 - q_2{}^2 + q_3{}^2)
\end{bmatrix}
\tag{2}
$$

In rotational kinematics equation, $\boldsymbol{\omega}^b = [p^b, q^b, r^b]$ where p^b, q^b, r^b is angular rate, we can write:

$$
\left[\boldsymbol{\omega}^b \otimes\right] = \begin{bmatrix}
0 & -p^b & -q^b & -r^b \\
p^b & 0 & r^b & -q^b \\
q^b & -r^b & 0 & p^b \\
r^b & q^b & -p^b & 0
\end{bmatrix}
\tag{3}
$$

In the translational dynamic equation, $\mathbf{T} = [F_T, 0, 0]$, F_T is the thrust, and

$$
\mathbf{F} = \mathbf{F}_A + \mathbf{F}_g = \begin{bmatrix}
F_x + 2(q_1 q_3 - q_0 q_2)mg \\
F_y + 2(q_2 q_3 + q_0 q_1)mg \\
F_z + 2\left(q_0{}^2 - q_1{}^2 - q_2{}^2 + q_3{}^2\right)mg
\end{bmatrix}
\tag{4}
$$

where $\mathbf{F}_A$ represents the term of aerodynamic force and $\mathbf{F}_g$ represents gravity-related items, F_x, F_y, F_z are the aerodynamic forces in the x-, y-, and z-axes, respectively.

The last one is the rotational dynamics equation, where $\boldsymbol{\tau} = [M_x, M_y, M_z]^T$ is the aerodynamic moment, and $\mathbf{J}$ is the moment of inertia which is composed by:

$$
\mathbf{J} = \begin{bmatrix}
J_{xx} & -J_{xy} & -J_{xz} \\
-J_{xy} & J_{yy} & -J_{yz} \\
-J_{xz} & -J_{yz} & J_{zz}
\end{bmatrix}
\tag{5}
$$

There are many unknown coefficients in this model as well as the margin of state quantities. These quantities have a great impact on the performance of a UAV and the completion of the maneuver. In Section 3, we identify these unknown quantities in a data-driven way.

2.2. Acrobatic Maneuver

Acrobatic maneuvers are generally summarized based on the pilot's actual flight experience and have strong practical significance. Acrobatic flights are a competitive sport or also a performance event. Therefore, remote control flight is also a way to achieve it. Figure 2 shows a typical fixed-wing maneuvering process. The International Federation of Aeronautics (FAI) is the main maker of acrobatics rules. FAI also provides a number of basic maneuvers for acrobatic events such as the Cuban eight [43]. This article mainly focuses on the realization of these basic maneuvers on UAVs.

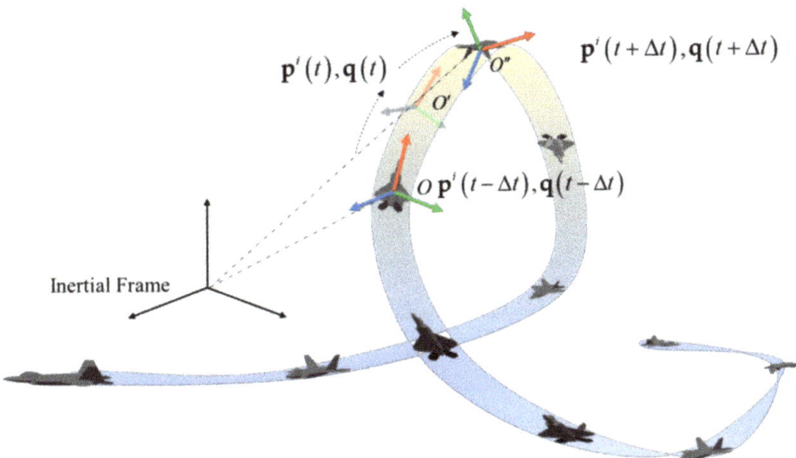

Figure 2. A typical maneuvering flight. The UAV has experienced multiple different position and attitude changes in a continuous period of time.

In contrast, manned pilots are restricted by physiological limits, remote control flight is safer and more flexible, but there are communication delays and the impact of visual distance. For autonomous drones, it removes the limitations of visual range and physiological limits and has the potential to achieve acrobatics.

3. Data Collection and Model Identification

3.1. Acrobatic Maneuver Data Collection

Maneuver attitude and other information can be obtained by collecting flight data. A maneuver has high requirements for data accuracy and frequency. With the improvement in sensor accuracy and miniaturization, a large amount of manual flight experience can be accurately recorded through data. UAV design, modeling, and flight testing can be realized through data collection. In actual flight, accelerometers, magnetometers, and other sensors can be used to record the aircraft's position, attitude, control inputs, and other data [43]. On the other hand, flight simulation technology is also an idea to solve the problem of flight maneuvers. We built a flight simulation system that can be used to collect flight data. As shown in Figure 3, the simulation system was built by the open-source flight control px4 and the flight simulation software X-Plane. The remote controller sends the control instructions to the flight controller. The control signal is processed by the internal program and then sent to the X-Plane simulator; this is a typical hardware-in-the-loop simulation system. For different maneuvers, we collected the position, attitude, and other information of the aircraft through expert teaching.

Figure 3. Flight maneuver demonstration and learning system including (1) remote controller, (2) signal receiver, (3) Pixhawk-v5, (4) flight joystick, (5) ground control station, and (6) flight simulation software X-Plane.

3.2. Model Parameter Identification

As mentioned in Section 1, the global fixed-wing model contains many unknown coefficients, mainly in aerodynamic forces and moments:

$$
\begin{aligned}
F_x &= \tfrac{1}{2}\rho V_a^2 S_{ref} C_x,\; F_y = \tfrac{1}{2}\rho V_a^2 S_{ref} C_y,\; F_z = \tfrac{1}{2}\rho V_a^2 S_{ref} C_z \\
M_x &= \tfrac{1}{2}\rho V_a^2 S_{ref} b_{ref} C_l,\; M_y = \tfrac{1}{2}\rho V_a^2 S_{ref} c_{ref} C_m,\; M_z = \tfrac{1}{2}\rho V_a^2 S_{ref} b_{ref} C_n
\end{aligned}
\tag{6}
$$

where ρ is the air density, $S_{ref}, b_{ref}, c_{ref}$ are the reference wing area, reference wing span, and average aerodynamic chord length, respectively, $C_x, C_y, C_z, C_l, C_m, C_n$ are the aerodynamic coefficient and moment coefficient, $\delta_e, \delta_a, \delta_r$ are the deflection of the elevator, aileron and rudder, respectively.

The calculation method of the angle of attack and the angle of sideslip are $\alpha = \arctan(w/u)$ and $\beta = \arcsin(v/V_a)$, where the airspeed is $V_a = \sqrt{u^2 + v^2 + w^2}$.

According to [44], the aerodynamic coefficients could be expressed using the global aerodynamic model in Equations (7) and (8).

$$
\begin{aligned}
C_x &= f_x(\alpha) = C_{x_0} + C_{x_\alpha}\alpha + C_{x_{\alpha^2}}\alpha^2 \\
C_y &= f_y\!\left(\beta, p^b, r^b, \delta_a, \delta_r\right) = C_{y_0} + C_{y_\beta}\beta + C_{y_p}\frac{p^b b_{ref}}{2V_a} + C_{y_r}\frac{r^b b_{ref}}{2V_a} + C_{y_{\delta a}}\delta_a + C_{y_{\delta r}}\delta_r \\
C_z &= f_z(\alpha, \delta_e) = C_{z_0} + C_{z_\alpha}\alpha + C_{z_{\delta e}}\delta_e + C_{z_{\alpha^2}}\alpha^2
\end{aligned}
\tag{7}
$$

$$
\begin{aligned}
C_l &= f_l\!\left(\beta, p^b, r^b, \delta_a, \delta_r\right) = C_{l_0} + C_{l_\beta}\beta + C_{l_p}\frac{p^b b_{ref}}{2V_a} + C_{l_r}\frac{r^b b_{ref}}{2V_a} + C_{l_{\delta a}}\delta_a + C_{l_{\delta r}}\delta_r \\
C_m &= f_m\!\left(\alpha, q^b, \delta_e\right) = C_{m_0} + C_{m_\alpha}\alpha + C_{m_q}\frac{q^b c_{ref}}{2V_a} + C_{m_{\delta e}}\delta_e + C_{m_{\alpha^2}}\alpha^2 \\
C_n &= f_n\!\left(\beta, p^b, r^b, \delta_a, \delta_r\right) = C_{n_0} + C_{n_\beta}\beta + C_{n_p}\frac{p^b b_{ref}}{2V_a} + C_{n_r}\frac{r^b b_{ref}}{2V_a} + C_{n_{\delta a}}\delta_a + C_{n_{\delta r}}\delta_r
\end{aligned}
\tag{8}
$$

where $C_{x_0}, C_{x_\alpha}, C_{x_{\alpha^2}}, C_{y_0}, C_{y_\beta}, C_{y_p}, C_{y_r}, C_{y_{\delta a}}, C_{y_{\delta r}}, C_{z_0}, C_{z_\alpha}, C_{z_{\delta e}}$ and $C_{z_{\alpha^2}}$ represent the coefficients related to the aerodynamic force, and $C_{l_0}, C_{l_\beta}, C_{l_p}, C_{l_r}, C_{l_{\delta a}}, C_{l_{\delta r}}, C_{m_0}, C_{m_\alpha}, C_{m_q}, C_{m_{\delta e}}, C_{m_{\alpha^2}}, C_{m_0}, C_{m_\alpha}, C_{m_q}, C_{m_{\delta e}}, C_{m_{\alpha^2}}, C_{n_0}, C_{n_\beta}, C_{n_p}, C_{n_r}, C_{n_{\delta a}}, C_{n_{\delta r}}$ are the coefficients related to the aerodynamic moments.

A least squares method was carried out to estimate the unknown coefficients in Equation (1). The optimization object was set as follows:

$$\min_{\substack{C_x, C_y, C_z \\ C_l, C_m, C_n}} \sum_{k=1}^{N} \left\| \eta^k \right\|_2^2 + \left\| \varepsilon^{(k)} \right\|_2^2 \tag{9}$$

subject to:

$$\varepsilon^{(k)} = \omega_{k+1}^b - \omega_k^b - \omega_k^b \times \mathbf{J}\omega_k^b \Delta t - \frac{1}{2}\rho S_{ref} V_a^2 \mathbf{J}^{-1} [\hat{C}_l, \hat{C}_m, \hat{C}_n] \Delta t \tag{10}$$

$$\eta^{(k)} = \mathbf{v}_{k+1}^b - \mathbf{v}_k^b - \frac{1}{2}m^{-1}\rho S_{ref} V_a^2 [\hat{C}_x, \hat{C}_y, \hat{C}_z] \Delta t - \omega_k^b \times \mathbf{v}_k^b \tag{11}$$

$$\hat{C}_l = b_{ref} f_l\left(\beta, p^b, r^b, \delta_a, \delta_r\right), \hat{C}_m = c_{ref} f_m\left(\alpha, q^b, \delta_e\right), \hat{C}_n = b_{ref} f_n\left(\beta, p^b, r^b, \delta_a, \delta_r\right) \tag{12}$$

It is worth noting that the optimization problem in Equation (9) could be solved utilizing the nonlinear least squares optimization methods such as the Levenberg–Marquardt algorithms. However, the divergence of the optimization problem or the convergence to the suboptimal solution might occur. Moreover, in order to diminish the side effects of over-fitting and the errors during the integration, we chose 10 s of flight data for the identification by trial and error. The results of system identification are listed in Section 5.

4. Optimal Acrobatic Maneuver Design and Generation

In this section, we propose a maneuver optimization algorithm based on teaching data and accurate models. We first introduce the two basic concepts of the algorithm in maneuver design: key-frames and motion primitives. Then different kinds of key-frames are listed and the entire optimization problem for maneuver is formulated. Finally, we conducted a detailed analysis of the solution of the proposed optimization problem.

4.1. Key-Frames and Motion Primitives

The concept of key-frames is often used in computer animation and simultaneous localization and mapping (SLAM) to represent frames that are decisive over a period of time. As mentioned in Section 1, this concept is also used to represent the necessary waypoints in motion planning. We introduced it into maneuvers.

As we can see in Figure 4, the same maneuver has the same typical characteristics of position and attitude changes. The key-frames are used to indicate the position and attitude that play a key role in a maneuver. The momentary state is a short-term key-frame, and the continuous state is a long-term key-frame.

Motion primitives explain the execution of complex motions based on action units. In maneuvers, simple maneuvers can be regarded as a single motion primitive without segmentation. For complex maneuvers, since it contains a variety of different pose change segments, it will be difficult to calculate if viewed as a whole, and different optimization goals should be considered for different segments, so it is necessary to divide it into multiple motion primitives.

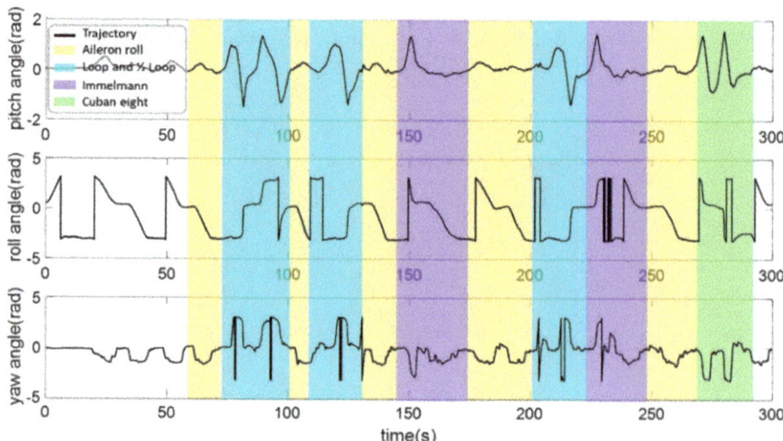

Figure 4. Some typical maneuver teaching data. The pilot performed different maneuvers continuously over a period of time, which is marked with different colors.

4.2. Maneuver Optimization

In maneuver or maneuver primitives, a drone is required to complete a certain position and attitude change in the shortest time; inspired by [24,31–35], we formulated this problem as a keyframe-based time optimization problem. A schematic diagram of the problem is shown in Figure 5.

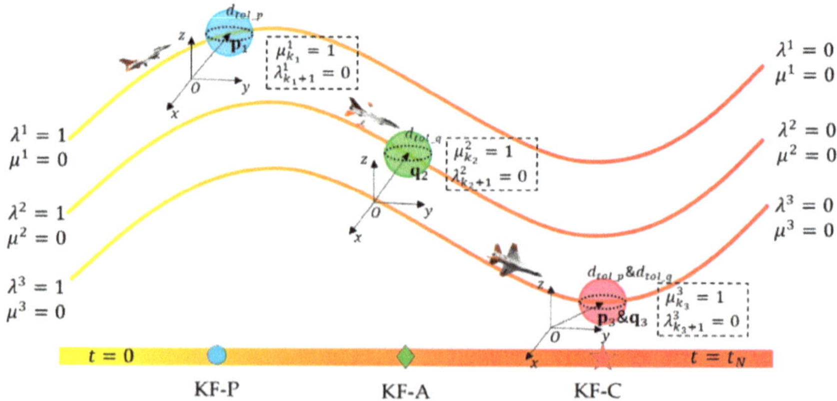

Figure 5. These figures show the key-frame-based optimization problem. The drone is required to sequentially pass through several different types of key-frames within a period of time, marked with different colors. We show the changes in the process state variables and process change variables in the process.

First, we set the state quantity to $\mathbf{x}_{dyn} = [\mathbf{p}, \mathbf{q}, \mathbf{v}, \omega]$, input $\mathbf{u} = [\delta_e, \delta_a, \delta_r, F_T]^T$ which needs to meet the constraints of the kinematics and dynamics model in Equation (1). The state quantity must first satisfy the start and end constraints and must be given or set free. The control input and some state quantities need to meet the upper and lower limits of UAV performance such as $\mathbf{u}_{min} \leq \mathbf{u}_k \leq \mathbf{u}_{max}$ and $\omega_{min} \leq \omega_k \leq \omega_{max}$. We selected the direct multiple shooting method to solve the optimization problem. Suppose the total time is t_N, discretize it into N segments, $dt = t_N/N$, and the index is k. Meanwhile, the state quantity needs to be discretized. In order to reduce the calculation error, the 4th order Runge–Kutta was used for numerical integration.

$$\mathbf{x}_{dyn,k+1} - \mathbf{x}_{dyn,k} - dt \cdot \mathbf{f}_{RK4}\left(\mathbf{x}_{dyn,k}, \mathbf{u}_k\right) = 0$$
$$\mathbf{f}_{RK4}\left(\mathbf{x}_{dyn,k}, \mathbf{u}_k\right) = 1/6 \cdot (\mathbf{k}_1 + 2\mathbf{k}_2 + 2\mathbf{k}_3 + \mathbf{k}_4) \tag{13}$$

where $\mathbf{k}_1, \mathbf{k}_2, \mathbf{k}_3, \mathbf{k}_4$ are integral terms composed of $\mathbf{x}_{dyn,k}$, $\mathbf{u}_k$, and dt.

Suppose the number of key-frames is set to M, indexed by i, and M-dimensional process state variables λ and M-dimensional process change variables μ are introduced to record the completion of key-frames, which meet the constraints:

$$\lambda_{k+1} - \lambda_k + \mu_k = 0, \lambda_1 - 1 = 0, \lambda_{N+1} = 0 \tag{14}$$

$$\mu_k^i \cdot \left(\left\|\mathbf{x}_{dyn\_k} - \mathbf{x}_{dyn\_i}\right\|_2\right) = 0, i \in [1, M] \tag{15}$$

As is shown in Figure 5, process state variables λ saves the state of event completion. λ is 1 at the beginning and 0 at the end. μ is a process change variable, which can be used to record the occurrence of instantaneous events. Through the constraint that μ^i multiplied by the state quantity part is equal to 0, it is required that when the event does not occur, the state quantity part is not 0 and μ^i is 0; when the event occurs, the state quantity part is 0 and μ^i is 1, then the corresponding λ^i changes from 1 to 0 permanently. Under this framework, λ^i can be used to represent the time period scale, which is used to record the requirements that need to be met during the process of two instantaneous events.

Corresponding to the sphere in Figure 5, the slack variable needs to be added due to the influence of discrete calculation, and different types of key-frames are introduced below:

(1) Positional key-frame (KF-P)

Short-term position constraints:

$$\mu_k^i \cdot \left(\left\|\mathbf{P}_k - \mathbf{p}_i\right\|_2 - dp_k^i\right) = 0$$
$$-dp_k^i \le 0 \; dp_k^i - d_{tol\_p} \le 0 \tag{16}$$

where $i \in [1, M], k \in [1, N+1]$, and in incomplete position constraints, only a part of the position variables are constrained; take flight altitude as an example:

$$\mu_k^i \cdot \left(|z_k - z_i| - dp_k^i\right) = 0 \tag{17}$$

During the flight, some state variables will be constrained for a long time such as altitude maintenance. Process state variables λ^i divides t_N into M + 1 fragments; take altitude maintenance as an example, we can set long-term position constraints as:

$$\lambda_k^1 \cdot |z_k - z_i| = 0 \tag{18}$$

$$\left(\lambda_k^{i+1} - \lambda_k^i\right) \cdot |z_k - z_i| = 0 \tag{19}$$

$$\left(1 - \lambda_k^M\right) \cdot |z_k - z_i| = 0 \tag{20}$$

(2) Attitude key-frame (KF-A)

Short-term attitude constraints:

$$\mu_k^i \cdot \left(\left\|\mathbf{q}_k - \mathbf{q}_i\right\|_2 - dq_k^i\right) = 0$$
$$-dq_k^i \le 0 \; dq_k^i - d_{tol\_q} \le 0 \tag{21}$$

If the requirements for posture are not so strict, we can set incomplete attitude constraint by Appendix A, for example, only constrain a certain angle:

$$\mu_k^i \cdot \left(\frac{2(q_{2k}q_{3k} + q_{0k}q_{1k})}{1 - 2(q_{1k}{}^2 + q_{2k}{}^2)} - \tan\phi_i - dq_k^i \right) = 0 \tag{22}$$

$$\mu_k^i \cdot \left((2q_{0k}q_{2k} - 2q_{1k}q_{3k}) - \sin\theta_i - dq_k^i \right) = 0 \tag{23}$$

$$\mu_k^i \cdot \left(\frac{2(q_{1k}q_{2k} + q_{0k}q_{3k})}{1 - 2(q_{2k}{}^2 + q_{3k}{}^2)} - \tan\psi_i - dq_k^i \right) = 0 \tag{24}$$

In the calculation process, the range of pitch angle is generally $[-\pi/2, \pi/2]$, and the range of roll and yaw is $[-\pi, \pi]$. The above key-frames of roll angle and yaw angle are not one-to-one correspondence. The calculation will produce singularities, the constraint can be changed to the following equation, which will increase a certain amount of calculation.

$$\mu_k^i \cdot \left(\arctan2\left(\frac{2(q_{2k}q_{3k} + q_{0k}q_{1k})}{1 - 2(q_{1k}{}^2 + q_{2k}{}^2)} \right) - \phi_i - dq_k^i \right) = 0 \tag{25}$$

(3) Compound key-frame (KF-C)

In some special scenarios, there are requirements for the position and attitude of the drone. We proposed a compound key-frame, if the same μ^i is used, the position and attitude can be constrained at the same time.

$$\begin{aligned} \mu_k^i \cdot \left(\|\mathbf{p}_k - \mathbf{p}_i\|_2 - dp_k^i \right) &= 0 \\ \mu_k^i \cdot \left(\|\mathbf{q}_k - \mathbf{q}_i\|_2 - dq_k^i \right) &= 0 \end{aligned} \tag{26}$$

In addition to these constraints, if we want to ensure the order of key-frames, we need to set the following constraints.

$$\lambda_k^i - \lambda_k^{i+1} \le 0 \quad \forall\, i \in [1, M-1] \tag{27}$$

Based on the above statements, our optimization variables are integrated as $x_{opt} = [t_N, x_0, \ldots, x_N]$, where:

$$\mathbf{x}_{opt,k} = \begin{cases} \left[\mathbf{x}_{dyn,k}, \mathbf{u}_k, \lambda_k, \mu_k, \mathbf{dp}_k, \mathbf{dq}_k \right] & for\ k \in [1, N] \\ \left[\mathbf{x}_{dyn,N+1}, \lambda_{N+1} \right] & for\ k = N+1 \end{cases} \tag{28}$$

The most basic goal of the designed maneuver is to minimize the time and require the maneuver to be completed in the shortest time. In order to improve the execution effect, we added the minimum energy term, and every motion primitive need to be adjusted according to its characteristics.

$$x_{opt}^* = \min L(\mathbf{x}_{opt}) = \min_{\mathbf{x}} \left(\sigma \cdot t_N + \tau \cdot \sum_{1}^{N} \left(\delta_e(k)^2 + \delta_a(k)^2 + \delta_r(k)^2 \right) \right) \tag{29}$$

Algorithm 1 flow is summarized as follows:

Algorithm 1. Fixed-wing UAV optimal maneuver generation.

1:    Input: L pieces of maneuver trajectory obtained by demonstration
2:    Output: optimal trajectory of single maneuver ξ (divided into primitives ξ^j)
3:       for $\forall l \in [1, L]$ do
4:          split the teaching trajectory into multiple maneuvers ξ
5:          extract the key-frames $\mathbf{p}_i, \mathbf{q}_i$ of each maneuver
6:          decompose ξ into motion primitives ξ^j according to certain principles
7:       end for
8:       for $\forall j \in [1, J]$ do
9:          if ξ^j exists, or similar ξ^j exists
10:          use the related ξ^j directly
11:          else
12:          constructing ξ^j as keyframe-based optimization problems
13:
$$\mu_k^i \cdot \left(\|\mathbf{p}_k - \mathbf{p}_i\|_2 - dp_k^i \right) = 0$$
14:
$$\mu_k^i \cdot \left(\|\mathbf{q}_k - \mathbf{q}_i\|_2 - dq_k^i \right) = 0 \text{ etc.}$$
15:          set N, d_{tol_p}, d_{tol_q} and initial guess
16:          solve optimization with Ipopt
17:          if infeasible
18:             goto 15 or 5
19:          else
20:             set $\xi_{x\_dyn(1)}^{j+1} = \xi_{x\_dyn(N+1)}^{j}$
21:          end if
22:        end if
23:       end for
24:    process ξ^j and concatenate ξ^j to ξ
25:    return primitives ξ^j concatenated maneuver data ξ

---

5. Experiments and Discussion

In this section, the identification results of the UAV model were obtained based on the flight data and three types of maneuvers are studied in this section. As mentioned in Section 4, the corresponding key-frames and motion primitives could be extracted for a specific maneuver. Therefore, we propose a technique to construct different minimum-time maneuver problem through adjusting the parameters of the boundary and key-frames. This algorithm could help to find the optimal and physically realizable flight maneuvers.

5.1. Model Identification Results

The main parameters of the UAV are: mass $m = 3.24$ kg, air density $\rho = 1.225$ kg $\cdot$ m^{-3}, reference area $S_{ref} = 0.56$ m^2, wing span $b_{ref} = 1.83$ m, mean chord length $c_{ref} = 0.30$ m, moment of inertia $J_{xx} = 0.22$ kg $\cdot$ m^2, $J_{yy} = 0.31$ kg $\cdot$ m^2, $J_{zz} = 0.48$ kg $\cdot$ m^2, $J_{xy} = J_{xz} = J_{yz} = 0$ kg $\cdot$ m^2, and the gravity acceleration was assumed to be $g = 9.8$ m $\cdot$ s^{-2}.

According to Section 3, the structure of the aerodynamic coefficients were defined in Equations (7) and (8). The estimation of the unknown parameters was performed from the flight using the least square approach. During the flight demonstration, the pilot performed several types of maneuvers. Both of the control surfaces' deflections and the output of the system are recorded in time domain. Furthermore, to satisfy the kinematic and dynamic constraints, the upper and lower limits of the control inputs and system states are listed in Table 1.

Table 1. Some important properties of UAVs.

Property	Value	Property	Value
$m\,[\text{kg}]$	3.24	$J_{xx}[\text{kg}\cdot\text{m}^2]$	0.22
$\rho\,[\text{kg}\cdot\text{m}^{-3}]$	1.225	$J_{yy}[\text{kg}\cdot\text{m}^2]$	0.31
$S\,[\text{m}^2]$	0.56	$J_{zz}[\text{kg}\cdot\text{m}^2]$	0.48
$b\,[\text{m}]$	1.83	$J_{xy}[\text{kg}\cdot\text{m}^2]$	0
$\bar{c}\,[\text{m}]$	0.30	$J_{xz}[\text{kg}\cdot\text{m}^2]$	0
$g\,[\text{m}\cdot\text{s}^{-2}]$	9.8	$J_{yz}[\text{kg}\cdot\text{m}^2]$	0

Property	Value
C_x	$-0.1004 - 0.0928\alpha + 1.7729\alpha^2$
C_y	$0.0446 - 0.5724\beta + 0.1203\frac{p^b b_{ref}}{2V_a} + 0.1181\frac{r^b b_{ref}}{2V_a} - 0.0276\delta_a + 0.1584\delta_r$
C_z	$-0.4522 - 5.3550\alpha + 0.7406\delta_e - 4.3813\alpha^2$
C_l	$0.0128 - 0.0579\beta - 0.3590\frac{p^b b_{ref}}{2V_a} + 0.1420\frac{r^b b_{ref}}{2V_a} + 0.1519\delta_a + 0.0042\delta_r$
C_m	$-0.0057 - 0.2402\alpha - 10.6607\frac{q^b c_{ref}}{2V_a} + 0.5737\delta_e - 0.0750\alpha^2$
C_n	$-0.0068 + 0.0538\beta - 0.0489\frac{p^b b_{ref}}{2V_a} - 0.0831\frac{r^b b_{ref}}{2V_a} - 0.0139\delta_a - 0.0465\delta_r$

Input	Range	State quantity	Range
δ_e	$[-0.3, 0.3]$	$p^b[\text{rad}\cdot\text{m}^{-1}]$	$[-2\pi, 2\pi]$
δ_a	$[-0.3, 0.3]$	$q^b[\text{rad}\cdot\text{m}^{-1}]$	$[-2, 2]$
δ_r	$[-0.3, 0.3]$	$r^b[\text{rad}\cdot\text{m}^{-1}]$	$[-2, 2]$
$F_T[\text{N}]$	$[0, 65]$	$\alpha[\text{rad}]$	$[-\pi/18, \pi/4]$

5.2. Optimization Simulation Setup

In this section, we investigate several types of maneuvers and conducted simulation experiments separately. First, we evaluate the loop maneuver which contains a single primitive. Meanwhile, only the positional key-frames were employed for this motion. Then, another classical maneuver named the Immelmann turn was taken into consideration. In this motion, two parts of the primitives, which contains the positional and attitude key-frames, respectively, are concatenated. Thirdly, the half Cuban eight, which was similar to the former one, was also evaluated, and the compound key-frames were proposed. Furthermore, the concatenation of two half Cuban eight results in a whole Cuban eight, which is a sophisticated motion containing multiple positional key-frames and compound key-frames.

As is shown in Figure 1, maneuver optimization is an important part of the framework, which can be referred to in detail in Algorithm 1. In this paper, all the maneuver optimization problems were carried out on CasADi [45] with Ipopt [46] optimization approach. The initial flight status and parameters of the optimization are listed prior to the results. For ease of presentation, we chose the original point at the initial position before the aerobatic flight.

5.2.1. Loop Maneuver

The Loop is a maneuver mainly performed in the vertical plane. At the beginning of the motion, the UAV keeps trim flight and starts to pitch up. After the pitch angle finishes a 360-degree turn, the UAV returns to the beginning position and maintains the trim condition.

It is worthy to mention that, even though the human pilot is well trained, the geometrical size and shape of different demonstrated trajectories are not completely consistent. Therefore, it is almost impossible for the human pilot to realize an optimal maneuver. However, the non-optimal trajectories still share similar features. Through sufficient trial-and-error, we found that the dominant feature of the loop was the position of key-frames. Nevertheless, the number/values of the positional key-frames are essential to obtain a reasonable circular trajectory. The set of the positional key-frames and the initial flight conditions are listed in Table 2. As for the optimization, we set the number of interval points $N = 210$, the positional tolerance $d_{tol\_p} = 0.4$ m, the parameters of the Equation (29) are set as $\sigma = 1, \tau = 0.1$, and the simulation results are illustrated in Figure 6.

Table 2. Flight status and key-frames of the loop maneuver.

Start	Value	End	Value
Position (m)	$[-2,0,0]$	Position (m)	$[0,0,0]$
Attitude (quaternion)	$[1,0,0,0]$	Pitch (deg)	0
Velocity (m/s)	$[15,0,0]$	Others	free
Angular rate (rad/s)	$[0,0,0]$		
Key-frame	**Value**		
Short-term positional key-frame	$[7.07,0,2.93]$, $[10,0,10]$, $[0,0,20]$ $[-7.07,0,16.57]$, $[-10,0,10]$, $[-7.07,0,2.43]$		

As shown in Figure 6, the drone passed through all the positional key-frames in a short period of time ($t_N = 3.22$ s). Once the drone meets the tolerance of each key-frames, the corresponding λ^i changes from 1 to 0. Furthermore, the UAV returns to the initial position after finishing the whole maneuver, which is difficult for a human pilot. Even though only one primitive is considered in this scenario, the half loop maneuver can be extracted from the results naturally.

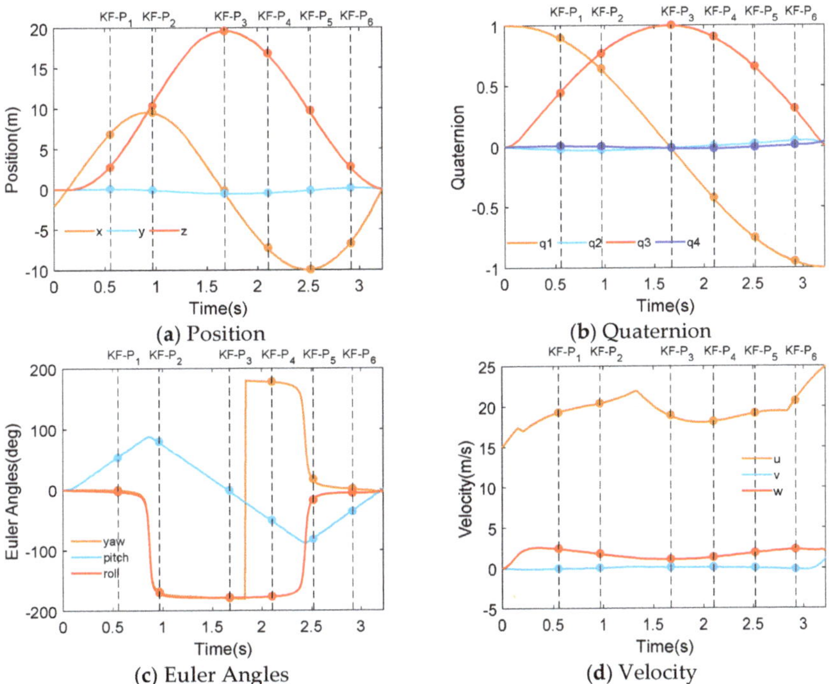

(a) Position

(b) Quaternion

(c) Euler Angles

(d) Velocity

Figure 6. *Cont.*

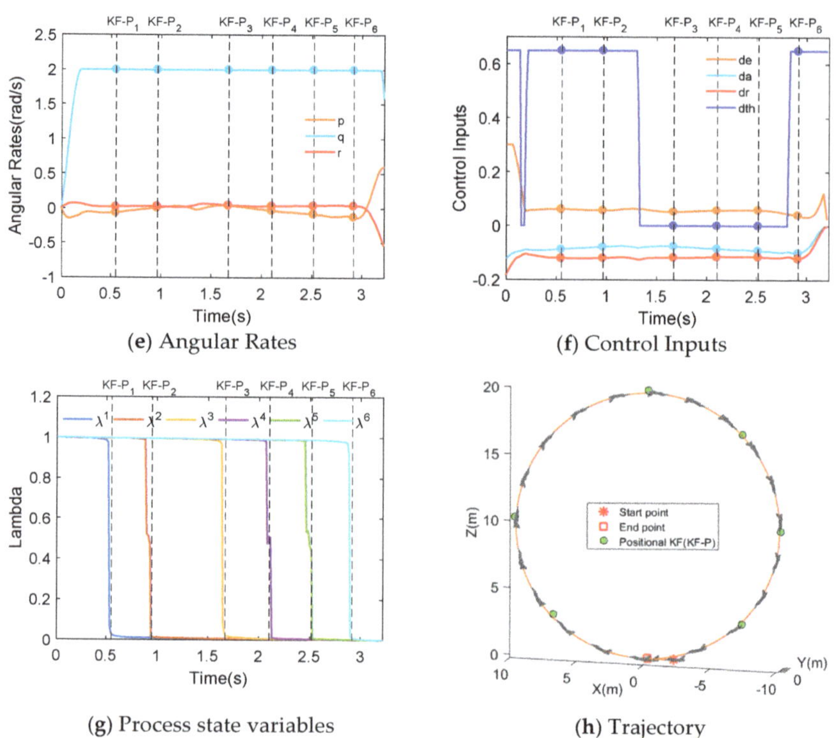

(**e**) Angular Rates

(**f**) Control Inputs

(**g**) Process state variables

(**h**) Trajectory

Figure 6. The experiment results for the loop maneuver including state variables, control inputs, process state variables, and maneuvering trajectory. The thrust shrunk by a factor of 100, and the state at the positional key-frame is marked with a circle. In some cases, the roll angle and yaw angle jumped between 180 and −180. (**h**) is the trajectory of UAV loop maneuver, including UAVs with different attitudes, the characteristic points are marked with different shapes.

5.2.2. The Immelmann Maneuver

The Immelmann maneuver is also known as upside-down half-roll. During the maneuver, the UAV first performs a half loop from trim flight. As soon as the aircraft reaches the top of the circular trajectory, it spins around the x-axis and executes a 180° roll. Finally, the UAV resumes to trim flight with an opposite direction compared to the initial condition.

We decompose the Immelmann into two concatenated primitives: half Loop and 180° roll for a better description. For the first primitive, there are two ways to obtain its near-optimal form: extracting from the Loop maneuver directly and modeling this motion into a two-point boundary value problem (BVP).

Modeling the optimization problem of 180° roll is significantly different from the former cases. Rolling with high angular rates will not only result in a large displacement in the forward direction, but also lead to deviations both in height and heading angle. Firstly, we set a short-term attitude key-frame (KF-A) which contained a 90° roll to ensure the motion completion. Then, in addition to the time factor, the displacement in the forward direction is also considered. The optimal solution is expressed as follows:

$$x^* = \min L(x) = \min_x \left(T + 0.1 \cdot \sum_1^N \left(\delta_e(k)^2 + \delta_a(k)^2 + \delta_r(k)^2 \right) + 0.1 \cdot |x_{N+1} - x_1| \right) \quad (30)$$

Meanwhile, the final states (position, attitude, velocities, and angular rates) of the half loop maneuver was set as the initial condition of the rolling primitive. Furthermore, the y-axis and z-axis constraints were also added to the final states of the rolling primitive. The initial condition of the second primitive is listed in Table 3. We set $N = 100$, $d_{tol\_q} = 0.04$, the results can be obtained by calculation and the result of concatenated primitives for the whole Immelman maneuver is shown in Figure 7.

Table 3. Flight status and key-frame of $\frac{1}{2}$ roll maneuver primitive.

Start	Value	End	Value
Position (m)	[0,0,20]	x (m)	free
Attitude (quaternion)	[0,0,1,0]	y (m)	0
Velocity (m/s)	[21.284,0.748,1.014]	z (m)	20
Angular rate (rad/s)	[0.226,1.799,−0.326]	Attitude (quaternion)	[0,0,0,−1]
Control input	[0.0171,−0.0318,−0.0023,65]		

Key-Frame	Value
Short-term angle key-frame	$\theta = -\pi/2$

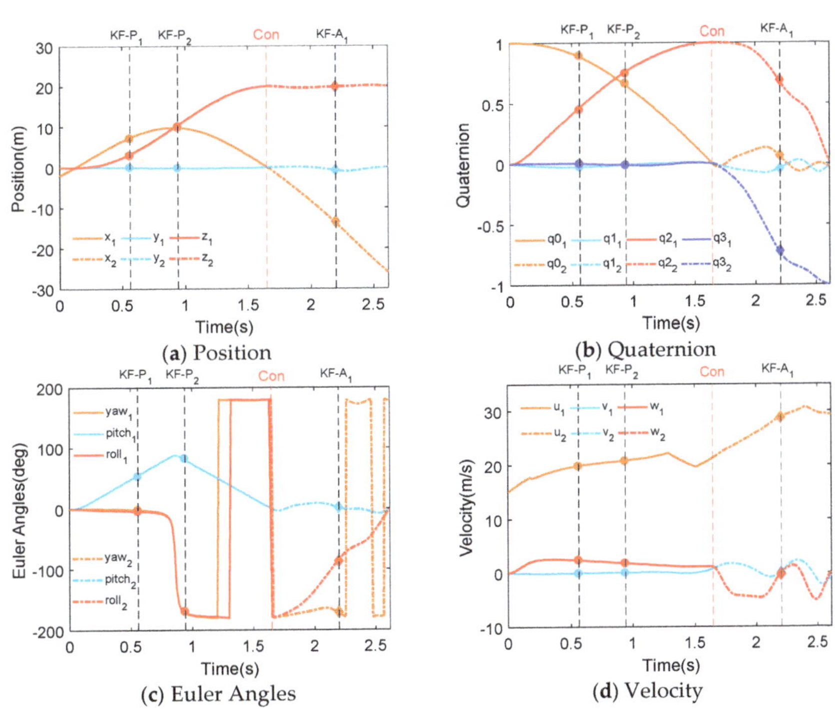

(**a**) Position

(**b**) Quaternion

(**c**) Euler Angles

(**d**) Velocity

Figure 7. *Cont.*

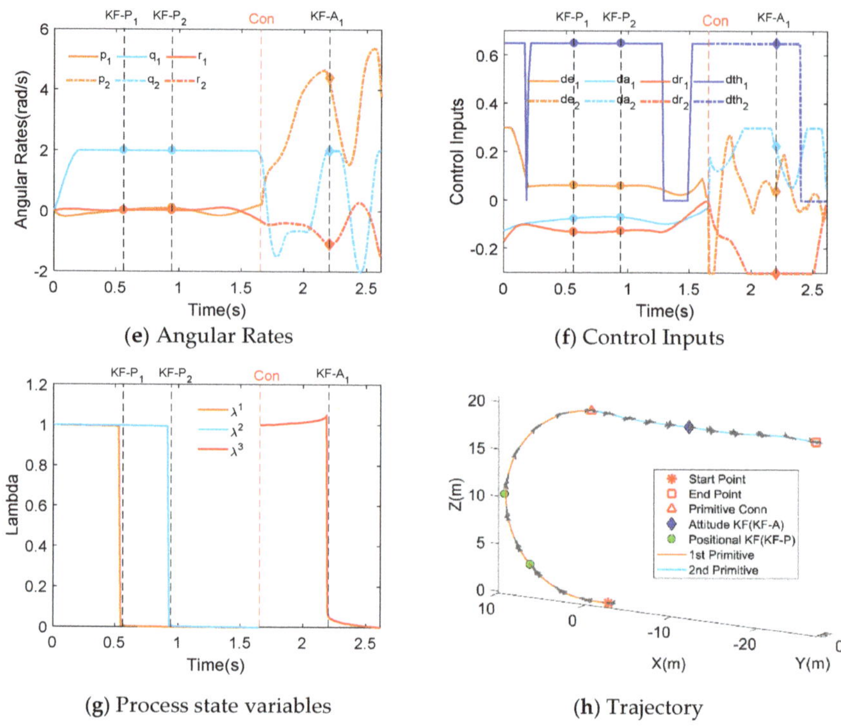

(e) Angular Rates

(f) Control Inputs

(g) Process state variables

(h) Trajectory

Figure 7. The experimental results for the Immelmann maneuver including state variables, control inputs, process state variables, and maneuvering trajectory. The thrust shrunk by a factor of 100, and the half loop primitive is represented by a solid line, the roll primitive is represented by a dashed line, the state at the attitude key-frame is marked by a diamond, and the red, dashed line represented the connection of the primitive. It should be noted that different primitives have different process change variables. (**h**) is the trajectory of UAV Immelmann maneuver, including UAVs with different attitudes, the characteristic points are marked with different shapes.

As is shown in Figure 7, the entire Immelmann maneuver takes only 2.6155 s, and each state quantity is well connected. It can be seen that the entire roll primitive takes only 0.9631 s, the drone smoothly passes the 90 degree roll angle key-frame and flies forward 26.1 m, finally flies out horizontally. The entire Immelmann maneuver takes only 2.6155 s, and each state quantity is well connected.

It is worth noting that the concatenated primitives might be intrinsically sub-optimal. However, the sub-optimal solutions are sufficient for practical applications and capable of generalizing more maneuvers that never demonstrated.

5.2.3. Half Cuban Eight and Cuban Eight Maneuver

The Cuban eight is a sophisticate maneuver consisting of two 3/4 Loops followed by $180°$ rolling. From the view of the ground, the UAV's trajectory is a vertical figure of eight. Due to the symmetric property, the left/right part of the motion, named half Cuban eight, can be evaluated first. To facilitate the calculation, the half Cuban eight was decomposed into a half loop (the same as the Immelmann turn's) and a special rolling primitive.

For the rolling primitive, we set a compound key-frame (KF-C), which requires the UAV to pass through the center of the Cuban eight with a $90°$ roll angle. Due to the inertia of rotation, the UAV will continue to rotate around the x-axis. Furthermore, the angular limitations are considered to avoid large angle deviations. The initial flight conditions are listed in Table 4.

Table 4. Flight status and key-frames of special rolling primitive.

Start	Value	End	Value
Position (m)	$[-2,0,0]$	Position (m)	$[-30,0,0]$
Attitude (quaternion)	$[0,0,1,0]$	Pitch (deg)	0
Velocity (m/s)	$[21.284,0.748,1.014]$	Roll (deg)	0
Angular rate (rad/s)	$[0.226,1.799,-0.326]$	Yaw (deg)	0
Control input	$[0.0171,-0.0318,-0.0023,65]$	Angular rate (rad/s)	$[0,0,0]$

Key-Frame	Value
Short-term Compound key-frame	$P = [-15,0,10]\ \theta = -\pi/2$
Long-term attitude key-frame	$\theta \leq 5\pi/180$

The two key-frames can be formulated as follows:

$$\mu_k^1 \cdot \left(\|\mathbf{p}_k - \mathbf{p}_1\|_2 - dp_k^1 \right) = 0, \ \mu_k^1 \cdot \left(\arctan2\left(\frac{2(q_{2k}q_{3k} + q_{0k}q_{1k})}{1 - 2(q_{1k}^2 + q_{2k}^2)} \right) + \frac{\pi}{2} - dq_k^i \right) = 0 \quad (31)$$

$$\left(1 - \lambda_k^1 \right) \cdot \left(\arctan2\left(\frac{2(q_{2k}q_{3k} + q_{0k}q_{1k})}{1 - 2(q_{1k}^2 + q_{2k}^2)} \right) - 5\pi/180 \right) \leq 0 \quad (32)$$

In the rolling primitive, we set $N = 100$, $d_{tol\_p} = 0.1m$, $d_{tol\_q} = 0.04$, the coefficients in Equation (29) are set as $\sigma = 1, \tau = 0.1$ for both primitives, and we can connect the two primitives into a half Cuban eight, and the results are shown in Figure 8.

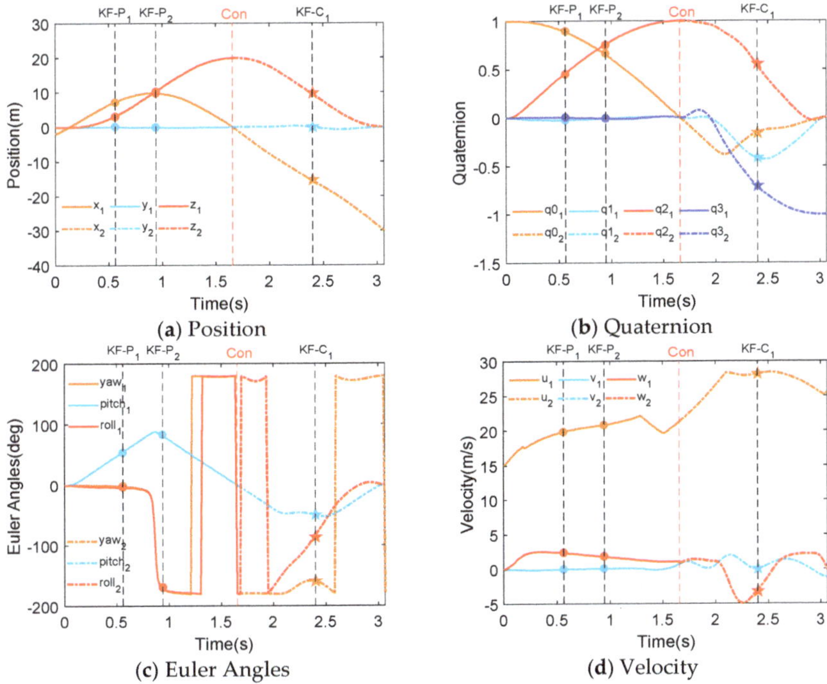

Figure 8. *Cont.*

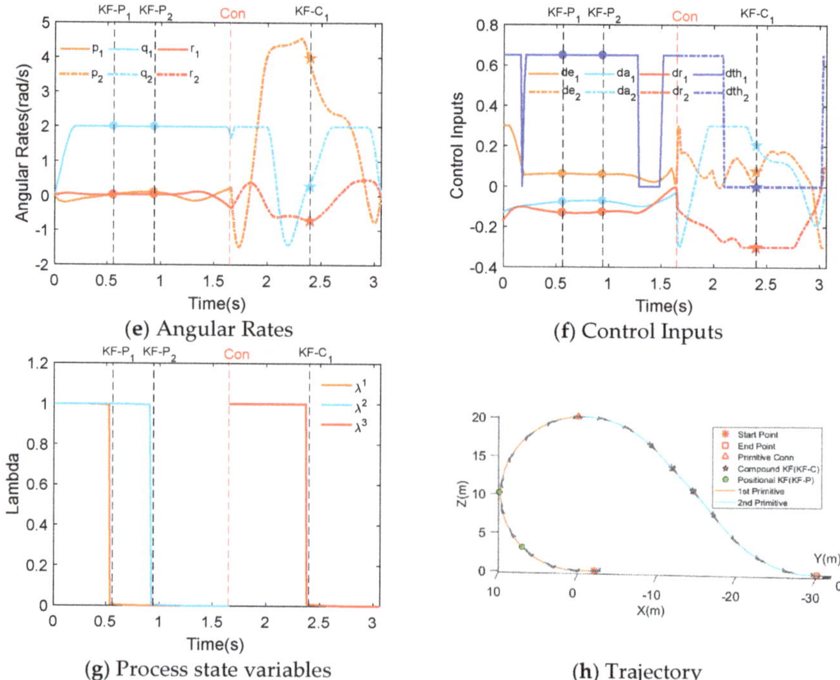

(e) Angular Rates

(f) Control Inputs

(g) Process state variables

(h) Trajectory

Figure 8. The experiment results for the half Cuban eight maneuver including state variables, control inputs, process state variables, and maneuvering trajectory. The thrust shrunk by a factor of 100, and the half loop primitive is represented by a solid line, the special rolling primitive is represented by a dashed line, the compound key-frame is marked with a pentacle. (**h**) is the trajectory of UAV half Cuban eight maneuver, including UAVs with different attitudes, the characteristic points are marked with different shapes.

The results of the optimization of the half Cuban eight are illustrated in Figure 9. It takes 1.4118 s for the rolling primitive and 3.0642 s for the half Cuban eight. It can be seen that the UAV accurately crossed the center point with a 90° roll angle, and the roll angle was always less than 5° in the subsequent primitive. Nevertheless, the velocity at the end of the half Cuban eight maneuver is much larger than the one at the start point. This phenomenon will hinder the completion of the concatenate loop. Therefore, we considered limiting the speed to $[20, 0, 0]$ of the drone at the end of the first rolling primitive, although this will increase the entire maneuver time to some extent.

Even though the two identical half-Cuban eights cannot be joined directly, there are still many similarities between them. Modeling a mirroring problem from the original problem and using data symmetry facilitate to obtain the optimization results. As shown in Figure 9, with the limitation on the terminal velocity, the duration time of the rolling primitive increased by 0.3898 s, and the engine thrust maintained at 0 N. Basically, the sub-optimal Cuban eight maneuver is completed, which takes 6.4856 s in total, and the primitives are well connected.

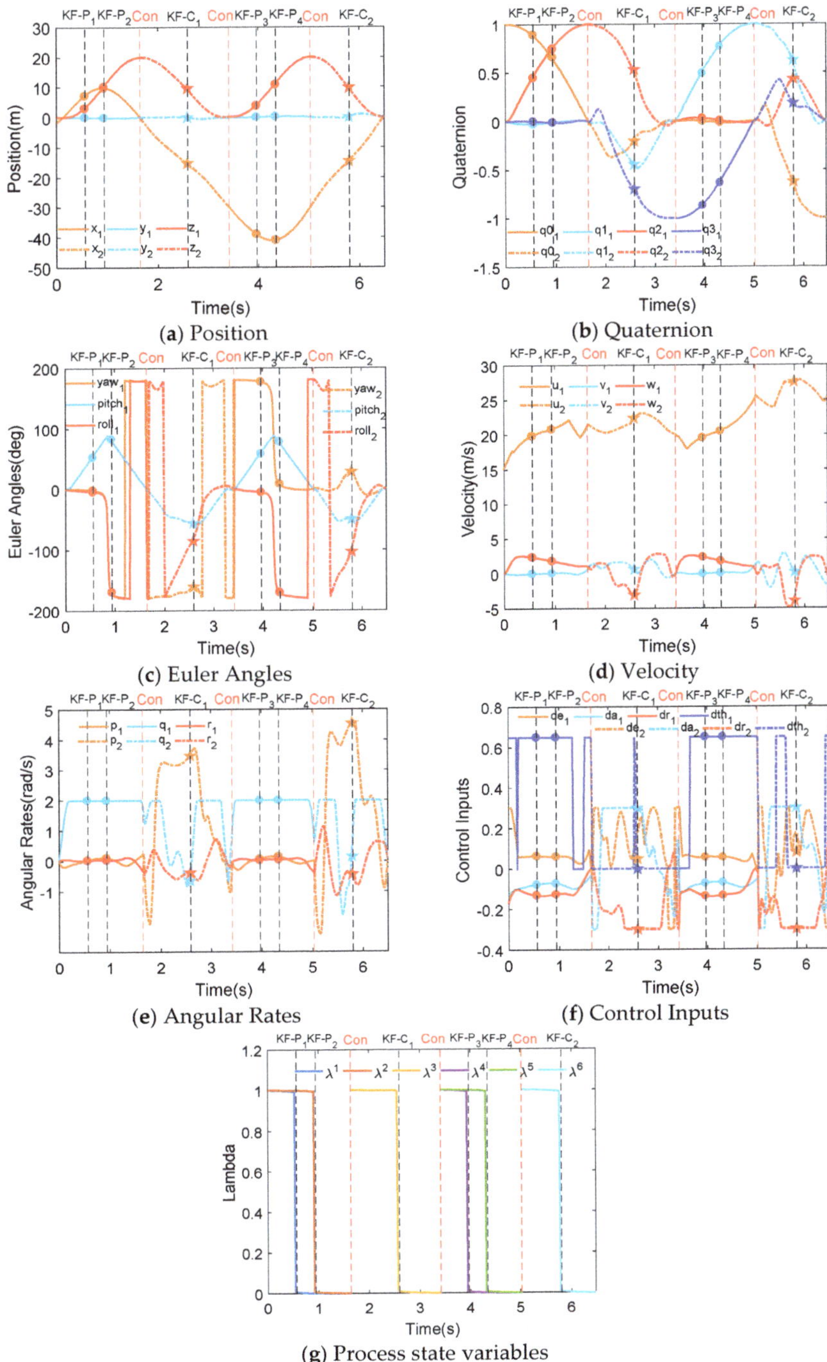

Figure 9. *Cont.*

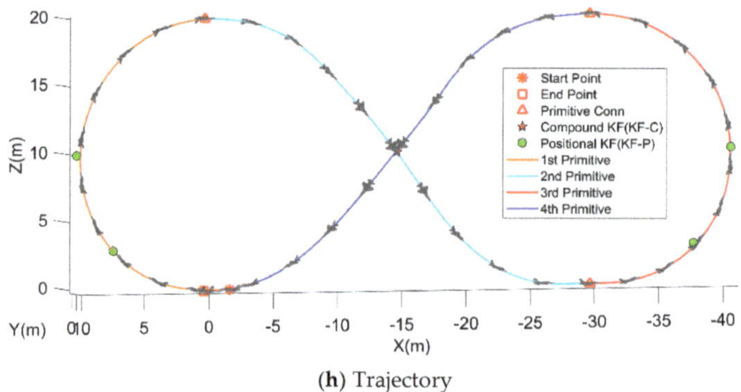

(**h**) Trajectory

Figure 9. The experiment results for the Cuban eight maneuver including state variables, control inputs, process state variables, and maneuvering trajectory. The whole maneuver contains four primitives, the two loop parts are represented by solid lines, and the rolling part is represented by dashed lines. (**h**) is the trajectory of UAV Cuban eight maneuver, including UAVs with different attitudes, the characteristic points are marked with different shapes.

5.3. Benchmark Comparisons

To further illustrate the superiority of our approach, a comparison with three other methods is conducted. The first strategy is through collecting the maneuvering trajectories performed by the human pilot. The second one is the Gaussian pseudo-spectral method proposed in Morales's work [24]. Moreover, the 3-DOF model introduced [23] was also utilized in our approach, formulated in Appendix B. The simulation results are depicted in the Figure 10.

It can be seen from Figure 10; Figure 11 that the trajectories obtained from manual flights are inferior to the other methods. Even though Morales's method's performance is slightly better than the manual flight, it cannot solve the period constraints. The flight time is significantly longer than those of our proposed method and the 3-DOF model. The method proposed in this paper was based on the global model, and motion segmentation is needed increasing computational costs. It is worth noting that our approach can flexibly connect those segments, resulting in various maneuvers. In the result of the mass point model, since our model was simplified, and there was no need to segment the trajectories, the maneuvering time optimized was slightly shorter than our method. Therefore, the optimization results based on the 3-DOF model can be utilized as an initial guess in our approach or for benchmark comparison. Since the 3-DOF model was simplified by ignoring the kinodynamic constraints, it cannot be directly used for maneuvering planning.

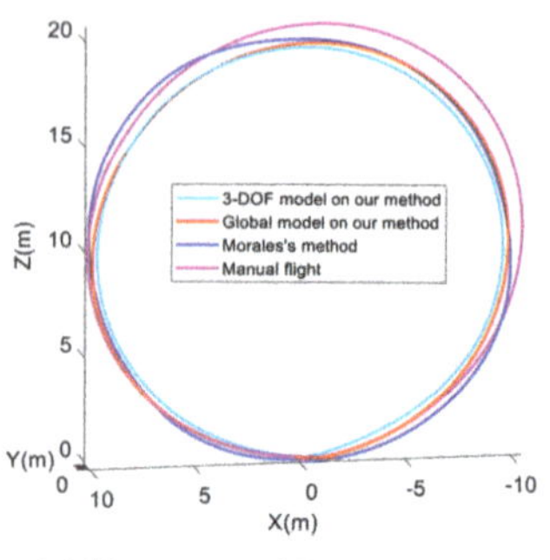

(**a**) Trajectory of Loop maneuver

(**b**) Trajectory of Immelmann maneuver

Figure 10. *Cont.*

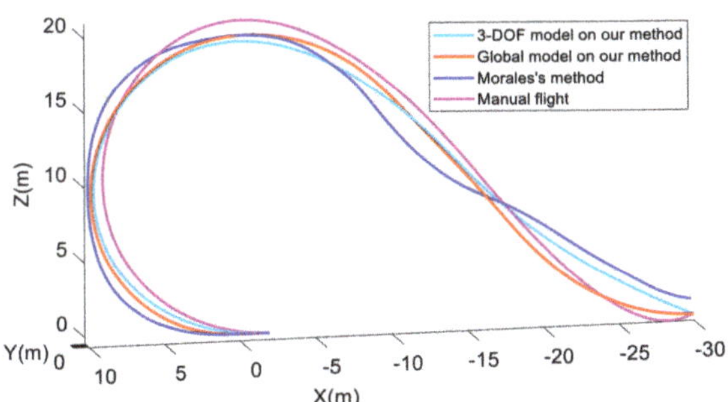

(**c**) Trajectory of Half Cuba eight maneuver

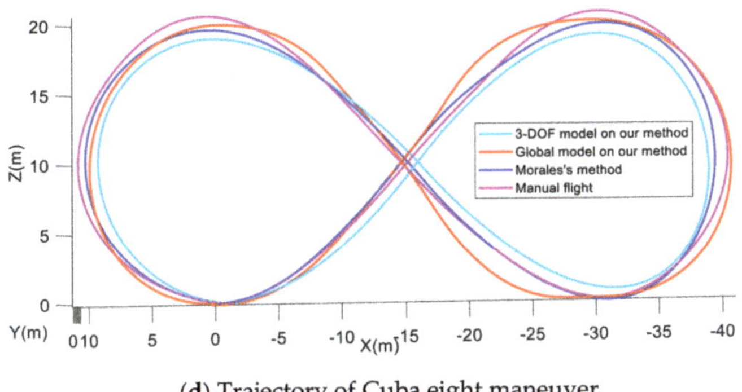

(**d**) Trajectory of Cuba eight maneuver

Figure 10. Comparison of maneuvering trajectories optimized by four methods. The magenta, red, sky blue, and blue curves represent manual flight, proposed approach, 3-DOF model, and Morales's method, respectively.

On this basis, we list the statistical flight time data in the Figure 11 for quantitative comparison of each method.

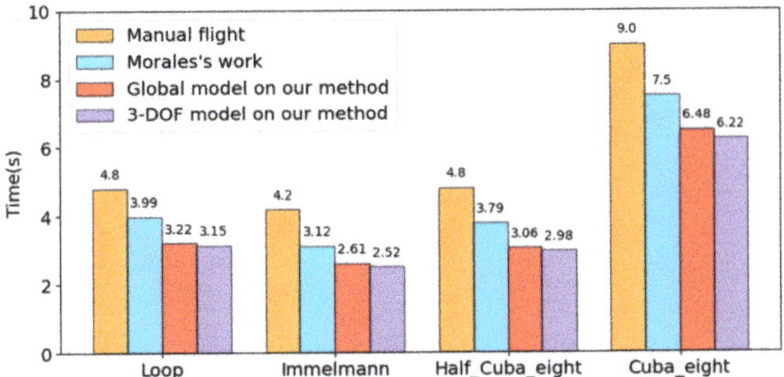

Figure 11. The time characteristics of different methods in different maneuvers, the orange, red, purple, and blue box represent manual flight, proposed approach, 3-DOF model, and Morales's method, respectively.

In order to conduct a more thorough analysis, we performed statistics on the running time of several methods. All the experiments were executed on an Intel Core i7-4710MQ CPU with 8 GB RAM that runs at 2.50 GHz.

It can be seen from Table 5 that the minimal-time maneuver optimization is time-consuming. Our work decomposes the optimization into the calculation of the 3-DOF model (initial guess) and the re-optimization of the global model. Compared with Morales's work, the optimized solution can be obtained faster.

Table 5. Calculation efficiency comparison.

Acrobatic Maneuver and Methods	Morales's Work	Proposed-Global Model	Proposed-3-DOF Model
Loop	20 min	10 min	45 s
Immelmann	6 min	2 min	48 s
Half Cuban eight	8 min	3 min	42 s
Cuban eight	20 min	8 min	2 min

5.4. Analysis of Maneuver Optimization Algorithm

It worth noting that the optimization process of maneuvers is quite time-consuming due to the inherent highly non-convex property. There are several methods that can accelerate the calculations and avoid the infeasible problem in Ipopt solver.

(1)   Global Fixed-wing model

Our optimization algorithm requires an accurate model in order to obtain the extreme motion of the UAV. Although the simplification of the model was conducive to simplifying the calculation, it was difficult to get the most realistic movement conditions. In our multiple maneuvering experiments, the performance of the drone was pushed to the limit, which was in line with the minimum time movement characteristics.

(2)   Key-frames and primitives design

To the best knowledge of the authors, it's unfeasible to propose a unified approach of key-frame extraction and motion decomposition for various maneuvers. Choosing the number of the key-frames and primitives requires a balance between computational complexity and geometric accuracy. Furthermore, most of the maneuvers can be decomposed into several longitudinal and lateral primitives.

(3)   Initial guess

The initial guess is essential for the nonlinear programming problem. In this paper, a reasonable initial guess was obtained either from the optimization results based on the point-mass model ($\ddot{\mathbf{p}}^i = \mathbf{a}$, with input $\mathbf{u} = \mathbf{a}$, $\|\mathbf{a}\| \leq a_{\max}$) or simply from the demonstration data using interpolation.

(4)   Optimization constraints

The small slacking variables can be added into the nonlinear constraints to accelerate the calculation and avoid the infeasible solutions. For instance, the complimentary constraint can be adjusted into:

$$0 \leq \mu_k^i \cdot \left(\|\mathbf{p}_k - \mathbf{p}_i\|_2 - dp_k^i \right) \leq 10^{-3}$$

In addition, excessive slacking variables can be added into nonlinear constraints to obtain an initial solution, which can be considered as the initial guess in the next step of optimization iteratively.

6. Conclusions

In this paper, a hybrid-driven flight maneuver optimization framework was proposed. Based on the demonstrated maneuvers performed by an experienced pilot, the parameters of the UAV model along with the features of maneuvers were extracted. Then, the idea of key-frames and maneuver primitives were proposed which can design and concatenate different maneuvers more flexibly. Based on the above work, we formulated maneuver planning into a time-optimal problem, the feasibility of the framework was fully verified through designing the positional, attitude, and compound key-frames in the loop, Immelmann, half Cuban eight, and Cuban eight maneuvers. Through comparison, it can be concluded that the designed action was faster than professional pilots, and the results provide verifications on the optimality of the solutions and the effectiveness of the proposed approach.

Future works will concentrate on realizing our algorithm on a UAV platform experimentally.

Author Contributions: Conceptualization, investigation, methodology, writing—original draft preparation, R.Z.; software, resources, data curation, R.Z., S.C., and K.Z.; writing—review and editing, S.C. and H.Y.; supervision, project administration, K.Z. and Y.H.; funding acquisition, Y.H. All authors have read and agreed to the published version of the manuscript.

Funding: This research received no external funding.

Institutional Review Board Statement: Not applicable.

Informed Consent Statement: Not applicable.

Data Availability Statement: All data and programs were collected and created by us. Please contact the first author of this article for relevant information.

Acknowledgments: Thanks to the research environment provided by the Institute of Unmanned Systems, the National University of Defense Technology, and the Nanjing Institute of Telecommunications Technology, National University of Defense Technology, for its administrative support.

Conflicts of Interest: The authors declare no conflict of interest. The funders had no role in the design of the study; in the collection, analyses, or interpretation of data; in the writing of the manuscript, or in the decision to publish the results.

Appendix A

Euler angle can be obtained by quaternion as follow:

$$\theta = \arcsin(2q_0q_2 - 2q_1q_3)$$
$$\phi = \arctan2\left(\frac{2(q_2q_3+q_0q_1)}{1-2(q_1^2+q_2^2)}\right)$$
$$\psi = \arctan2\left(\frac{2(q_1q_2+q_0q_3)}{1-2(q_2^2+q_3^2)}\right) \tag{A1}$$

$$\arctan2(y,x) = \begin{cases} \arctan(y/x), & x > 0 \\ \arctan(y/x) + \pi, & y \geq 0, x < 0 \\ \arctan(y/x) - \pi, & y < 0, x < 0 \\ +\pi/2, & y > 0, x = 0 \\ -\pi/2, & y < 0, x = 0 \\ undefined, & y = 0, x = 0 \end{cases} \tag{A2}$$

Appendix B

Aircraft Point-mass model:

$$\dot{x} = V \cos\gamma \cos\chi$$
$$\dot{y} = V \cos\gamma \sin\chi$$
$$\dot{z} = V \sin\gamma \tag{A3}$$

where γ is the angle of trajectory, χ is the course angle.

$$\dot{V} = \frac{T\cos\alpha\cos\beta - D'}{m} - g\sin\gamma$$
$$\dot{\gamma} = \frac{T(\cos\kappa\sin\alpha + \sin\kappa\cos\alpha\sin\beta) + L'\cos\kappa + C'\sin\kappa}{mV} - \frac{g\cos\gamma}{V}$$
$$\dot{\chi} = \frac{T(\sin\kappa\sin\alpha - \cos\kappa\cos\alpha\sin\beta) + L'\sin\kappa - C'\cos\kappa}{mV\cos\gamma} \tag{A4}$$

where κ is the aerodynamic roll angle, D, C, L are the components of the aerodynamic forces, described in the aerodynamic reference system, calculated as follows:

$$
\begin{bmatrix} D' \\ C' \\ L' \end{bmatrix} = \begin{bmatrix} \cos(\alpha)\cos(\beta) & \sin(\beta) & \sin(\alpha)\cos(\beta) \\ -\cos(\alpha)\sin(\beta) & \cos(\beta) & -\sin(\alpha)\sin(\beta) \\ -\sin(\alpha) & 0 & \cos(\alpha) \end{bmatrix} \times \begin{bmatrix} -1/2 \cdot \rho V_a^2 S_{ref} C_x \\ -1/2 \cdot \rho V_a^2 S_{ref} C_y \\ -1/2 \cdot \rho V_a^2 S_{ref} C_z \end{bmatrix} \tag{A5}
$$

References

1. Ju, C.; Son, H.I. Multiple UAV systems for agricultural applications: Control, implementation, and evaluation. *Electronics* **2018**, *7*, 162. [CrossRef]
2. Tang, T.; Hong, T.; Hong, H.; Ji, S.; Mumtaz, S.; Cheriet, M. An improved UAV-PHD filter-based trajectory tracking algorithm for multi-UAVs in future 5G IoT scenarios. *Electronics* **2019**, *8*, 1188. [CrossRef]
3. Zhou, Y.; Wu, C.; Wu, Q.; Eli, Z.M.; Xiong, N.; Zhang, S. Design and analysis of refined inspection of field conditions of oilfield pumping wells based on rotorcraft uav technology. *Electronics* **2019**, *8*, 1504. [CrossRef]
4. Chen, C.L.; Deng, Y.Y.; Weng, W.; Chen, C.H.; Chiu, Y.J.; Wu, C.M. A traceable and privacy-preserving authentication for UAV communication control system. *Electronics* **2020**, *9*, 62. [CrossRef]
5. Stellin, M.; Sabino, S.; Grilo, A. LoRaWAN Networking in Mobile Scenarios Using a WiFi Mesh of UAV Gateways. *Electronics* **2020**, *9*, 630. [CrossRef]
6. Mellinger, D.; Michael, N.; Kumar, V. Trajectory generation and control for precise aggressive maneuvers with quadrotors. *Int. J. Robot. Res.* **2012**, 1–11. [CrossRef]
7. Hehn, M.; Ritz, R.; DAndrea, R. Performance benchmarking of quadrotor systems using time-optimal control. *Auton. Robot.* **2012**, *33*, 69–88. [CrossRef]
8. Kaufmann, E.; Loquercio, A.; Ranftl, R.; Muller, M.; Koltun, V.; Scaramuzza, D. Deep drone acrobatics. RSS: Robotics, Science, and Systems. *arXiv* **2021**, arXiv:2006.05768.
9. Gilhyun, R.; Tal, E.; Karaman, S. Multi-fidelity black-box optimization for time-optimal quadrotor maneuvers. RSS: Robotics, Science, and Systems. *arXiv* **2020**, arXiv:2006.02513.
10. Bry, A.; Richter, C.; Bachrach, A.; Roy, N. Aggressive flight of fixed-wing and quadrotor aircraft in dense indoor environments. *Int. J. Robot. Res.* **2015**, *34*, 969–1002. [CrossRef]
11. Levin, M. Maneuver Design and Motion Planning for Agile Fixed-Wing UAVs. Ph.D. Thesis, McGill University, Montreal, Canada, April 2019.
12. Tang, J.; Singh, A.; Goehausen, N. Parameterized Maneuver Learning for Autonomous Helicopter Flight. In Proceedings of the IEEE International Conference on Robotics and Automation, Anchorage, AK, USA, 4–8 May 2010.
13. Abbeel, P. Apprenticeship Learning and Reinforcement Learning with Application to Robotic Control. Ph.D. Thesis, Stanford University, Stanford, CA, USA, August 2008.
14. Abbeel, P.; Coates, A.; Ng, A.Y. Autonomous Helicopter Aerobatics through Apprenticeship Learning. *Int. J. Robot. Res* **2010**, *29*, 1608–1639. [CrossRef]
15. Mueller, M.W.; Hehn, M.; D'Andrea, R. A computationally efficient motion primitive for quadrocopter trajectory generation. *IEEE Trans. Robot.* **2015**, *31*, 1294–1310. [CrossRef]
16. Quan, L.; Han, L.; Zhou, B. Survey of UAV motion planning. *IET Cyber-Syst. Robot.* **2020**, *2*, 14–21. [CrossRef]
17. Liu, S.; Mohta, K.; Atanasov, N.; Kumar, V. Search-based motion planning for aggressive flight in se (3). *IEEE Robot. Autom. Lett.* **2018**, *3*, 2439–2446. [CrossRef]
18. Watterson, M.; Liu, S.; Sun, K.; Smith, T.; Kumar, V. Trajectory optimization on manifolds with applications to quadrotor systems. *Int. J. Robot. Res.* **2020**, *39*, 303–320. [CrossRef]
19. Wang, Z.; Zhou, X.; Xu, C.; Gao, F. Geometrically Constrained Trajectory Optimization for Multicopters. *arXiv* **2021**, arXiv:2103.00190.
20. Basescu, M.; Moore, J. Direct nmpc for Post-Stall Motion Planning with Fixed-Wing uavs. In Proceedings of the 2020 IEEE International Conference on Robotics and Automation (ICRA), Paris, France, 31 May–31 August 2020.
21. Fan, Y.; Lutze, F.H.; Cliff, E.M. Time-optimal lateral maneuvers of an aircraft. *J. Guid. Control. Dyn.* **1995**, *18*, 1106–1112. [CrossRef]
22. Komduur, H.J.; Visser, H.G. Optimization of vertical plane cobralike pitch reversal maneuvers. *J. Guid. Control. Dyn.* **2002**, *25*, 693–702. [CrossRef]
23. Imado, F.; Heike, Y.; Kinoshita, T. Research on a new aircraft point-mass model. *J. Aircr.* **2011**, *48*, 1121–1130. [CrossRef]
24. Morales, M.A.; Silvestre, F.J.; Neto, A.B.G. Equations of motion for optimal maneuvering with global aerodynamic model. *Aerosp. Sci. Technol.* **2018**, *77*, 206–216. [CrossRef]
25. Du, D.; Qi, Y.; Yu, H. The Unmanned Aerial Vehicle Benchmark: Object Detection and Tracking. In Proceedings of the European Conference on Computer Vision (ECCV), Munich, Germany, 8–14 September 2018.
26. Wen, L.; Zhu, P.; Du, D. Visdrone-sot2018: The Vision Meets Drone Single-Object Tracking Challenge Results. In Proceedings of the European Conference on Computer Vision (ECCV), Munich, Germany, 8–14 September 2018.

27. Qi, Y.; Wu, Q.; Anderson, P. Reverie: Remote Embodied Visual Referring Expression in Real Indoor Environments. In Proceedings of the IEEE/CVF Conference on Computer Vision and Pattern Recognition, Seattle, WA, USA, 14–19 June 2020.
28. Hong, Y.; Rodriguez-Opazo, C.; Qi, Y. Language and visual entity relationship graph for agent navigation. *arXiv* **2020**, arXiv:2010.09304.
29. Qi, Y.; Pan, Z.; Hong, Y.; Yang, M.H.; Hengel, A.; Wu, Q. Know What and Know Where: An Object-and-Room Informed Sequential BERT for Indoor Vision-Language Navigation. *arXiv* **2021**, arXiv:2104.04167.
30. Qi, Y.; Pan, Z.; Zhang, S. Object-and-action aware model for visual language navigation. In Proceedings of the Computer Vision–ECCV 2020: 16th European Conference, Glasgow, UK, 23–28 August 2020.
31. Gebhardt, C.; Hepp, B.; Nägeli, T. Airways: Optimization-based Planning of Quadrotor Trajectories According to High-Level User Goals. In Proceedings of the 2016 CHI Conference on Human Factors in Computing Systems, San Jose, CA, USA, 7–12 May 2016.
32. Mellinger, D.; Kumar, V. Minimum Snap Trajectory Generation and Control for Quadrotors. In Proceedings of the 2011 IEEE International Conference on Robotics and Automation, Shanghai, China, 9–13 May 2011.
33. Foehn, P.; Davide, S. CPC: Complementary progress constraints for time-optimal quadrotor trajectories. *arXiv* **2020**, arXiv:2007.06255.
34. Foehn, P.; Romero, A.; Scaramuzza, D. Time-optimal planning for quadrotor waypoint flight. *Sci. Robot.* **2021**, *6*, eabh1221. [CrossRef]
35. Rousseau, G.; Maniu, C.S.; Tebbani, S.; Babel, M.; Martin, N. Minimum-time B-spline trajectories with corridor constraints. Application to cinematographic quadrotor flight plans. *Control. Eng. Pract.* **2019**, *89*, 190–203. [CrossRef]
36. Frazzoli, E.; Dahleh, M.A.; Feron, E. Maneuver-Based Motion Planning for Nonlinear Systems with Symmetries. *IEEE Trans. Robot.* **2005**, *21*, 1077–1091. [CrossRef]
37. Bulka, E.; Nahon, M. Autonomous Fixed-Wing Aerobatics: From Theory to Flight. In Proceedings of the 2018 IEEE International Conference on Robotics and Automation (ICRA), Brisbane, Australia, 21–25 May 2018.
38. Bulka, E.; Nahon, M. Automatic Control for Aerobatic Maneuvering of Agile Fixed-Wing UAVs. *J. Intell. Robot. Syst.* **2019**, *93*, 85–100. [CrossRef]
39. Levin, J.M.; Nahon, M.; Paranjape, A.A. Real-time motion planning with a fixed-wing UAV using an agile maneuver space. *Auton. Robot.* **2019**, *43*, 2111–2130. [CrossRef]
40. Wang, B.; Gong, J.; Chen, H. Motion Primitives Representation, Extraction and Connection for Automated Vehicle Motion Planning Applications. *IEEE Trans. Intell. Transp. Syst.* **2020**, *21*, 3931–3945. [CrossRef]
41. Zhang, R.; Hu, Y.; Zhao, K. A Novel Dual Quaternion Based Dynamic Motion Primitives for Acrobatic Flight. In Proceedings of the 2021 5nd International Conference on Robotics and Automation Sciences (ICRAS), Wuhan, China, 11–13 June 2021.
42. Sonneveldt, L. *"Nonlinear F-16 Model Description"*,*Control and Simulation Div. Faculty of Aerospace Engineering*; Delft University of Technology: Delft, The Netherlands, 2006.
43. Zhang, R.; Zhang, J.; Yu, H. Review of Modeling and Control in UAV Autonomous Maneuvering Flight. In Proceedings of the 2018 IEEE International Conference on Mechatronics and Automation (ICMA), Changchun, China, 5–8 August 2018.
44. Dutra, D.A. Collocation-Based Output-Error Method for Aircraft System Identification. In Proceedings of the AIAA Aviation 2019 Forum, Dallas, TX, USA, 17–21 June 2019.
45. Andersson, J.; Joris, G.; Horn, G.; Rawlings, J.B.; Diehl, M. CasADi: A software framework for nonlinear optimization and optimal control. *Math. Program. Comput.* **2018**, *11*, 1–36. [CrossRef]
46. Wächter, A.; Biegler, L.T. On the implementation of an interior-point filter line-search algorithm for large-scale nonlinear programming. *Math. Program.* **2006**, *106*, 25–57. [CrossRef]

 *electronics*

Article

Estimation of Airspeed, Angle of Attack, and Sideslip for Small Unmanned Aerial Vehicles (UAVs) Using a Micro-Pitot Tube

Gennaro Ariante [1,*], Salvatore Ponte [2], Umberto Papa [1] and Giuseppe Del Core [1]

[1] Department of Science and Technology, University of Naples "Parthenope", 80133 Naples, Italy; umberto.papa@collaboratore.uniparthenope.it (U.P.); giuseppe.delcore@uniparthenope.it (G.D.C.)

[2] Department of Engineering, University of Campania "L. Vanvitelli", 81031 Aversa, CE, Italy; salvatore.ponte@unicampania.it

* Correspondence: gennaro.ariante@studenti.uniparthenope.it

Abstract: Fixed and rotary-wing unmanned aircraft systems (UASs), originally developed for military purposes, have widely spread in scientific, civilian, commercial, and recreational applications. Among the most interesting and challenging aspects of small UAS technology are endurance enhancement and autonomous flight; i.e., mission management and control. This paper proposes a practical method for estimation of true and calibrated airspeed, Angle of Attack (AOA), and Angle of Sideslip (AOS) for small unmanned aerial vehicles (UAVs, up to 20 kg mass, 1200 ft altitude above ground level, and airspeed of up to 100 knots) or light aircraft, for which weight, size, cost, and power-consumption requirements do not allow solutions used in large airplanes (typically, arrays of multi-hole Pitot probes). The sensors used in this research were a static and dynamic pressure sensor ("micro-Pitot tube" MPX2010DP differential pressure sensor) and a 10 degrees of freedom (DoF) inertial measurement unit (IMU) for attitude determination. Kalman and complementary filtering were applied for measurement noise removal and data fusion, respectively, achieving global exponential stability of the estimation error. The methodology was tested using experimental data from a prototype of the devised sensor suite, in various indoor-acquisition campaigns and laboratory tests under controlled conditions. AOA and AOS estimates were validated via correlation between the AOA measured by the micro-Pitot and vertical accelerometer measurements, since lift force can be modeled as a linear function of AOA in normal flight. The results confirmed the validity of the proposed approach, which could have interesting applications in energy-harvesting techniques.

Keywords: unmanned aircraft systems (UASs); IMU; pressure sensors; angle-of-attack estimation; autonomous flight; flight mechanics

Citation: Ariante, G.; Ponte, S.; Papa, U.; Del Core, G. Estimation of Airspeed, Angle of Attack, and Sideslip for Small Unmanned Aerial Vehicles (UAVs) Using a Micro-Pitot Tube. *Electronics* **2021**, *10*, 2325. https://doi.org/10.3390/electronics10192325

Academic Editors: Esteban Tlelo-Cuautle and Maysam Abbod

Received: 26 July 2021
Accepted: 16 September 2021
Published: 22 September 2021

Publisher's Note: MDPI stays neutral with regard to jurisdictional claims in published maps and institutional affiliations.

1. Introduction

Small unmanned aerial vehicles (UAVs), with maximum gross takeoff mass <10 kg, normal operating altitude <1200 ft above ground level (AGL) and airspeed <100 knots according to the U.S. Department of Defense (DoD) classification [1] (p. 12), are an easy-to-use and economical way to perform tasks that can be fulfilled without human involvement, or for flights in unconventional missions or constrained space. UAVs can also be an optimal solution as a test bench for new sensor systems or embedded flight management systems. When subsystems are integrated to improve characteristics such as estimation of the vehicle's state vector, autonomy, and guidance, navigation, and control (GNC) capabilities, we categorize unmanned aircraft systems (UASs) as semiautonomous, remotely operated, and fully autonomous [2–5]. A UAS comprises several subsystems that include the aircraft, its payload, the control station(s) (and, often, other remote stations known as ground stations (GSs)), aircraft launch and recovery subsystems where applicable, support subsystems, communication subsystems, and transport subsystems. Recent improvements of technologies such as global navigation satellite systems (GNSSs), inertial measurement units (IMUs), light detection and ranging systems (LiDAR), microcontrollers, imaging sensors, and micro-electromechanical systems (MEMS) significantly contributed to the progress of UAS technology and functionalities. UASs must also be considered as part of a local or global air transport/aviation environment with its rules, regulations, and disciplines [6].

Typically, for real-time trajectory and speed estimation, IMUs are used. Much work has been published in the field of extrapolation, filtering, and fusion of IMUs and other sensor data (GPS/GNSS, LiDAR, sonar, etc.) for attitude and velocity estimation [7–11], and for meteorology and wind estimation for atmospheric energy harvesting [12–17]. For small, low-cost UAVs, minimization of sensing requirements and expansion of flight envelope, endurance, and mission capability are priorities. Accurate measurement of airspeed could be critical to a UAV mission, affecting the safety of the aircraft, and allowing performance enhancement by specific flight strategies and energy-saving maneuvers (for example, gust soaring) [18]. Usually, large airplanes are equipped with arrays of multihole Pitot probes, properly calibrated along the fuselage, that measure wind velocity, angle of attack (AOA, α), and angle of sideslip (AOS, β), which are variables that contain useful information about performance and safety in normal and abnormal flight conditions. However, few papers have considered practical methods for AOA and AOS estimation for small UAVs or light aircraft, due to the large weight, size, cost, and power consumption of the measuring devices. Estimates of these parameters can be used for fault detection and changes due to structural damage or adverse flight conditions that could alter the aerodynamic coefficients of a UAV, and could also allow precise control of small UAVs in particular maneuvers, such as vertical landing [19] or high dynamic flight [20]. A comprehensive review of some methodologies and estimation approaches for typical fixed-wing and rotary-wing UAVs can be found in [21], in which the authors point out that multihole probes based on pressure measurements are difficult to calibrate for low-speed, low-altitude flights; optical methods are unreliable for small UAVs; and potentiometer-based vane probes could be suitable for applications in mini flyers. In the literature, attempts at estimation without direct measurements of AOA and AOS in small UAVs, using algebraic methods or multistage fusion algorithms with linear time-varying models, can also be found [22,23].

The advent of miniaturization has permitted the use of small, low-weight, low-power sensors, allowing the possibility of installing a Pitot tube on a UAV. This paper, based on a preliminary design presented by the authors in [24], focuses on trajectory measurement and control of a mini-UAV using a micro-Pitot tube for speed (true air speed, TAS) and AOA and AOS estimation. The approach aims to enhance the performance of small UAVs, constrained by onboard energy due to the aircraft's limited size, by following specific flight strategies defined by particular atmospheric formations. In low-level flight, typical for small UAVs, the turbulence intensity is significantly increased, due to the ground proximity [12], and wind estimation plays a crucial role in optimizing the onboard energy consumption, both for fixed-wing and rotary-wing aircraft. Onboard estimation of AOA and AOS by differential pressure measurements could improve control of critical roll motions, potentially allowing active control methodologies for stabilization or energy-harvesting strategies. The sensors used in this research are a differential pressure sensor and a 10 degrees of freedom (DoF) inertial measurement unit (IMU), both managed by a microcontroller Arduino for data acquisition and TAS, AOA, and AOS estimation. We used 1D Kalman filtering to estimate and remove measurement noise, and complementary filtering was used to provide an estimate of the attitude from IMU acceleration and angular rate data [25,26]. Indoor sessions were needed for system setup and calibration. Experimental data were collected using a prototype of the system in different test conditions by the Parthenope Flight Dynamics Laboratory (PFDL) team of the University of Naples "Parthenope" (Italy). After the characterization of the system, it will be tested in real flight campaigns using a small UAV (a quadcopter) with a maximum take-off mass (MTOM) of 2 kg/visual line of sight (VLOS). These limitations, considered in [6], guaranteed safety of flight, and all the related operations were considered not critical.

The paper is structured as follows: after establishing the theoretical framework in Section 2 (kinematic model, error analysis, and Kalman-filter-based estimation technique), in Section 3 the sensors used are characterized. Section 4 reports the experimental activity (pressure-sensor calibration, estimation of velocity, angle of attack, angle of sideslip, and attitude), describing the tests performed and analyzing the numerical results. Final considerations and future work (Section 5) conclude the paper.

2. Theoretical Framework

2.1. Kinematic Model

We begin with the aircraft kinematics [27–29], assuming a rigid-body model, and referencing Figure 1. Let $\mathbf{v}_{ac}^{B} = \begin{pmatrix} u & v & w \end{pmatrix}^{T}$ denote the velocity vector (ground speed, relative to Earth) in the aircraft's body coordinate frame (CF), with T denoting transpose. Let $\mathbf{v}_{ac}^{N} = \begin{pmatrix} u_g & v_g & w_g \end{pmatrix}^{T}$ denote the velocity vector with components referred to an Earth-fixed north–east–down (NED) CF. The UAV kinematics are:

$$\begin{aligned} \dot{u} - rv + qw &= a_x \\ \dot{v} - pw + ru &= a_y \\ \dot{w} - qu + pv &= a_z \end{aligned} \tag{1}$$

where $\mathbf{a} = \begin{pmatrix} a_x & a_y & a_z \end{pmatrix}^{T}$ is the acceleration vector, decomposed in the body CF, and p, q, r are the body-frame angular rates.

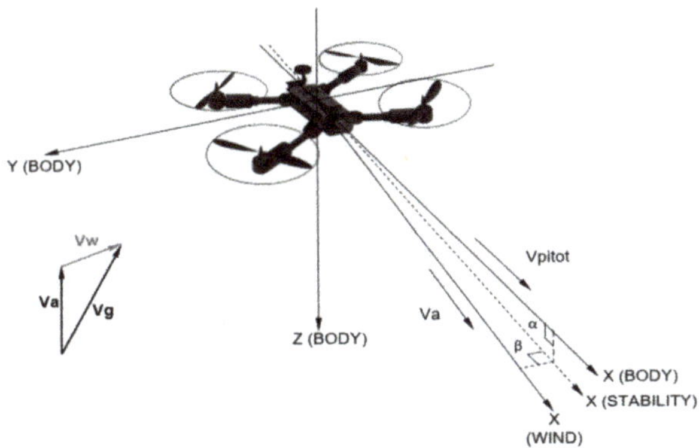

Figure 1. Airspeed definition and wind triangle.

The wind velocity vector relative to the Earth, decomposed in the NED CF, is $\mathbf{v}_{w}^{N} = \begin{pmatrix} u_w & v_w & w_w \end{pmatrix}^{T}$. The relation among the airspeed (velocity with respect to the surrounding air), the ground speed (velocity with respect to the Earth frame), and the wind velocity (with respect to the Earth frame) is:

$$\mathbf{v}_r = \mathbf{v}_g - \mathbf{v}_w \tag{2}$$

The rotation matrix for moving from the vehicle-carried NED frame to the body frame, $\mathbf{R}_{N}^{b}$, is defined by roll (ϕ, positive up), pitch (θ, positive right), and yaw (ψ, positive clockwise) angles [27] (Chap. 2):

$$\mathbf{R}_{N}^{b} = \begin{bmatrix} \cos\theta\cos\psi & \cos\phi\sin\psi & -\sin\theta \\ \sin\phi\sin\theta\cos\psi - \cos\phi\sin\psi & \sin\phi\sin\theta\sin\psi + \cos\phi\cos\psi & \sin\phi\cos\theta \\ \cos\phi\sin\theta\cos\psi + \sin\phi\sin\psi & \cos\phi\sin\theta\sin\psi - \sin\phi\cos\psi & \cos\phi\cos\theta \end{bmatrix} \tag{3}$$

Therefore, the relative velocity (airspeed) vector $\mathbf{v_r} = (\; u_r \quad v_r \quad w_r \;)^T$ in the body CF is:

$$\begin{bmatrix} u_r \\ v_r \\ w_r \end{bmatrix} = \begin{bmatrix} u \\ v \\ w \end{bmatrix} - \mathbf{R_N^b} \begin{bmatrix} u_w \\ v_w \\ w_w \end{bmatrix} \tag{4}$$

According to [27], the airspeed vector body-frame components can be expressed in terms of the airspeed magnitude, angle of attack, and sideslip angle as:

$$\begin{bmatrix} u_r \\ v_r \\ w_r \end{bmatrix} = V_{TAS} \begin{bmatrix} \cos\alpha\cos\beta \\ \sin\beta \\ \sin\alpha\cos\beta \end{bmatrix} \tag{5}$$

The AOA, α, is defined as the angle between the longitudinal (X) axis of the airframe and the freestream velocity (or relative wind), measured in the Y–Z plane of the body CF, and is positive when there is uplift (pitch-up), whereas the AOS, β, is measured between the X-body axis of the airframe and the relative wind velocity vector, and is positive for wind coming from starboard (right side). Inverting Equation (5), the angles α, β and the true airspeed V_{TAS} are given by:

$$\alpha = \tan^{-1}\left(\frac{w_r}{u_r}\right) \tag{6}$$

$$\beta = \sin^{-1}\left(\frac{v_r}{V_{TAS}}\right) \tag{7}$$

$$V_{TAS} = \sqrt{u_r^2 + v_r^2 + w_r^2} \tag{8}$$

The calibrated airspeed V_{CAS} is derived from [30] (Chap. 3):

$$V_{CAS} = \sqrt{\frac{1}{\rho_{sl}} 7 P_{sl} \left(\frac{\Delta P}{P_{sl}} + 1\right)^{\frac{2}{7}} - 1} \tag{9}$$

where ρ_{sl} and P_{sl} are the sea-level standard atmospheric values of air density and pressure, respectively ($P_{sl} = 101.325$ kPa, $\rho_{sl} = 1.225$ kg/m^3 at 15 °C, or 288.15 K [31]), and ΔP is the measured differential pressure. When Mach numbers are small (less than 0.3), Equation (8) is related to Equation (9) by:

$$V_{CAS} = V_{TAS} \sqrt{\sigma} \tag{10}$$

where σ is the relative density; i.e., the ratio between the actual air density and the standard air density at sea level. For low-level flights and small velocities (typical of small UAV mission scenarios), V_{CAS}. can be assumed as equal to V_{TAS}.

2.2. Error Analysis

Assuming statistically independent observations, the variance of the calculated value of α (Equation (6)), σ_α^2, can be evaluated using the special law of propagation of variances (SLOPOV) [32] (Chapter 6):

$$\sigma_\alpha^2 = \sigma_{w_r}^2 \left(\frac{u_r}{u_r^2 + w_r^2}\right)^2 + \sigma_{u_r}^2 \left(\frac{-w_r}{u_r^2 + w_r^2}\right)^2 \tag{11}$$

where $\sigma_{w_r}^2$ and $\sigma_{u_r}^2$ are the variance of the measured quantities w_r and u_r, respectively Equation (11) can be easily rearranged as follows:

$$\sigma_\alpha^2 = \left(\frac{1}{\frac{u_r}{w_r} + \frac{w_r}{u_r}}\right)^2 \left[\left(\frac{\sigma_{u_r}}{u_r}\right)^2 + \left(\frac{\sigma_{w_r}}{w_r}\right)^2\right] \tag{12}$$

As far as σ_β^2 is concerned, using Equation (7) and the SLOPOV, it can be shown that:

$$\sigma_\beta^2 = \left(\frac{u_r^2 v_r}{AB}\right)^2 \left(\frac{\sigma_{u_r}}{u_r}\right)^2 + \left(\frac{v_r A}{B}\right)^2 \left(\frac{\sigma_{v_r}}{v_r}\right)^2 + \left(\frac{v_r w_r^2}{AB}\right)^2 \left(\frac{\sigma_{w_r}}{w_r}\right)^2 \tag{13}$$

where $A = \sqrt{u_r^2 + w_r^2}$, and $B = u_r^2 + v_r^2 + w_r^2 = V_{TAS}^2$. Assuming that the vertical winds are zero (usually correct for a nonturbulent atmosphere), and measuring the sideslip angle in the X–Y plane of the body CF to maintain independence (i.e., decoupling) between α and β, then $A = u_r$, $B = u_r^2 + v_r^2$, and Equation (13) can be expressed in a form similar to Equation (12):

$$\sigma_\beta^2 = \left(\frac{1}{\frac{u_r}{v_r} + \frac{v_r}{u_r}}\right)^2 \left[\left(\frac{\sigma_{u_r}}{u_r}\right)^2 + \left(\frac{\sigma_{v_r}}{v_r}\right)^2\right] \tag{14}$$

From Equations (12) and (14), it can be seen, by finding the maxima of the functions $\frac{1}{\frac{u_r}{w_r} + \frac{w_r}{u_r}}$ and $\frac{1}{\frac{u_r}{v_r} + \frac{v_r}{u_r}}$, that the errors in α and β are maximized when the velocity ratios $\frac{u_r}{w_r}$ or $\frac{w_r}{u_r}$ (for β, $\frac{u_r}{v_r}$ or $\frac{v_r}{u_r}$) are equal to 1, and σ_α^2 and σ_β^2 increase as the velocity components become small; i.e., the vehicle approaches a hovering flight (in which knowledge of AOA and AOS becomes less important to the control strategy). Figure 2a, as an example, shows σ_α in degrees as a function of u_r and w_r, assuming a typical value of 0.2 m/s for the standard deviations of the measured velocity components (σ_{u_r} and σ_{w_r}).

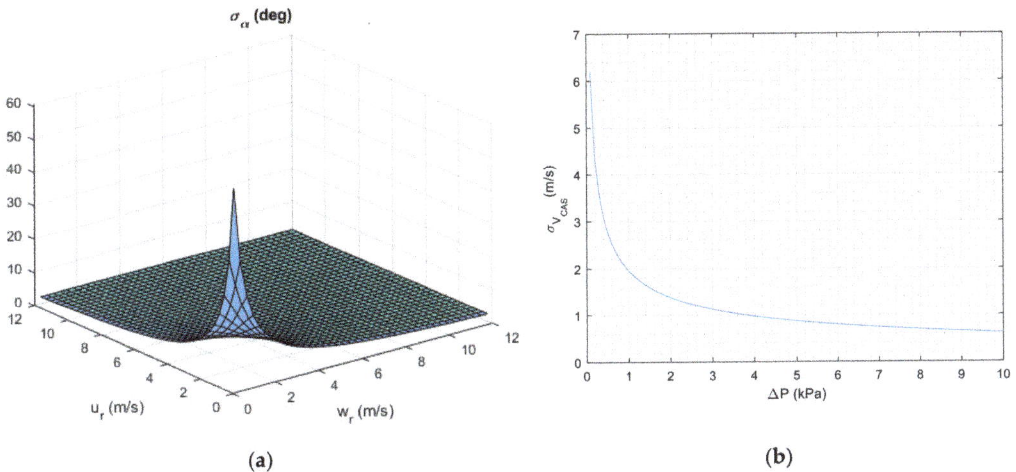

Figure 2. (**a**) Standard deviation (in degrees) of the AOA, α, as a function of the measured velocity components. (**b**) Standard deviation of the calibrated airspeed.

Finally, using Equation (9), the variance of the calibrated airspeed is found to be:

$$\sigma_{V_{CAS}}^2 = \left(\frac{1}{\rho_{sl}} \frac{\left(\frac{\Delta P}{P_{sl}} + 1\right)^{-\frac{5}{7}}}{V_{CAS}}\right)^2 \sigma_{\Delta P}^2 \tag{15}$$

where $\sigma_{\Delta P}^2$ is the variance of the differential pressure measurements ($\sigma_{\Delta P}$ set to 0.1 kPa; see Table 2, Section 3). Figure 2b shows the propagated error on V_{CAS} (i.e., $\sigma_{V_{CAS}}$) as a function of ΔP, with operating pressure range of the sensor from 0 to 10 kPa, as from Table 2.

2.3. Measurement Noise Estimation via Kalman Filtering

We applied Kalman filtering [33] for measurement noise estimation and removal, using a simple 1D formulation. The measurement process was modeled as:

$$\begin{cases} x_{k+1} = Ax_k + w_k \\ y_k = Cx_k + v_k \end{cases} \tag{16}$$

where x_k is the k-th measurement; w_k is the (Gaussian, zero-mean) model noise with variance Q (initialized to 0); v_k is the (Gaussian, zero-mean) measurement noise, with variance R, derived from the sensors' technical datasheet (see Section 3); and A and C are scalar quantities equal to 1. The (scalar) variance of the estimation error is P.

The predictor–corrector sequence (Kalman filtering) used in our work was implemented in the Arduino Integrated Development Environment (IDE) as a function (`KF_nr`, where "nr" stands for "noise removal") called whenever a new measurement (`Data`) was available from the sensor. The equations of the a priori estimation error, the Kalman gain K_k, the updated estimate $\hat{x}_{k+1|k+1}$ (with current data y_{k+1}), and the minimum-square a posteriori estimation error $P_{k+1|k+1}$, are, respectively:

$$P_{k+1|k} = P_{k|k} + Q \tag{17}$$

$$K_{k+1} = \frac{P_{k+1|k}}{P_{k+1|k} + R} \tag{18}$$

$$\hat{x}_{k+1|k+1} = \hat{x}_{k+1|k} + K_{k+1}\left(y_{k+1} - \hat{x}_{k+1|k}\right) \tag{19}$$

$$P_{k+1|k+1} = P_{k+1|k}(1 - K_{k+1}) \tag{20}$$

The Arduino code translated these equations (using the compound addition "+=") as shown in Table 1.

Table 1. The 1D Kalman filtering code implemented for measurement noise reduction.

```
float KF_nr(float Data)
{
P += Q;
K = P/(P+R);
X += K*(Data - X);
P = (1-K)*P;
return X;
}
```

3. Sensor System

The system used for velocity, AOA, AOS, and attitude estimation is composed of the following:

- Differential pressure sensor (Freescale Semiconductor MPX2010DP micro-Pitot);
- Static pressure sensor (Bosch Sensortec BMP180);
- IMU (DFRobot 10 DOF MEMS IMU sensor);
- Microcontroller (Arduino Mega 2560).

3.1. Differential Pressure Sensor (MPX2010DP)

Estimating the CAS (Equation (9)) requires evaluation of the dynamic pressure. The Freescale Semiconductor, Inc. MPX2010DP (Figure 3) silicon piezoresistive pressure sensor [34] provides a very accurate and linear voltage output directly proportional to the applied differential pressure, in the range of 0–10 kPa.

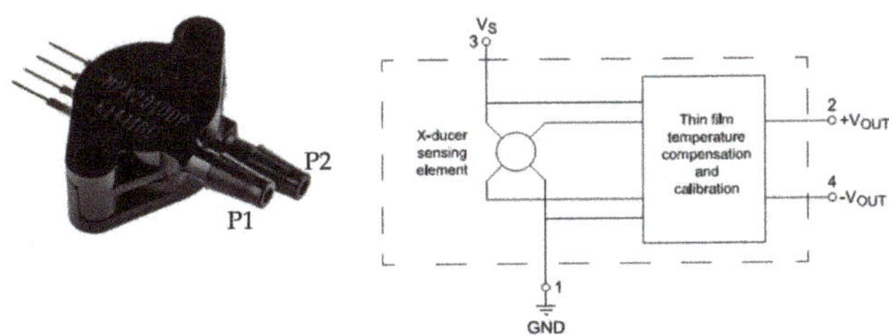

Figure 3. The MPX2010DP differential pressure sensor and its schematics. P1 and P2 are the pressure side and the vacuum side, respectively.

The output voltage of the differential gauge sensor increases with increasing pressure applied to the positive pressure side (port P1 in Figure 3) relative to the vacuum side (port P2). The sensor is designed to operate with positive differential pressure applied (P1 > P2). Table 2 shows some technical specifications at 25 °C.

Table 2. The main operating characteristics of the MPX2010DP.

Dimensions, mass	$29 \times 29 \times 18$ mm, <20 g
Pressure range, operating temperature	0–10 kPa, -40 °C–125 °C
Supply voltage, current, power consumption	10 VDC, 6 mA, 60 mW
Temperature compensation	0 °C to +85 °C
Full-scale span (VFSS)	25 mV
Offset, sensitivity	Max $\pm$ 1 mV, 2.5 mV/kPa
Offset stability, linearity	$\pm$0.5% VFSS, max $\pm$ 1% VFSS
Response time (10% to 90%)	1 ms
Warm-up time	20 ms

The sensor housed a single monolithic silicon die with the strain gauge and thin film resistor network integrated. The chip was laser-trimmed for precise span, offset calibration, and temperature compensation, allowing optimal linearity, low pressure and temperature hysteresis ($\pm$0.1%VFSS from 0 to 10 kPa and $\pm$0.5%VFSS from -40 °C to 125 °C, respectively), and an excellent response time. It fulfilled the in-flight requirements.

3.2. Digital Pressure Sensor (BMP180)

The static pressure also was acquired by the BMP180 digital pressure sensor (Figure 4) in order to obtain redundant measurements and more accurate estimates. The BMP180, produced by Bosch Sensortec, consists of a piezoresistive sensor, an analog-to-digital converter, a control unit with E^2PROM, and a serial I^2C interface, which allowed for easy system integration by direct connection to commercial microcontrollers [35]. The 16-bit (or 19-bit in high-resolution mode) pressure data, in the range of 300–1100 hPa (from +9000 m to -500 m related to sea level) and 16-bit temperature data were compensated by the calibration data stored in the embedded E^2PROM. Detailed sensor features are reported in Table 3.

Figure 4. The BMP180 digital pressure sensor.

Table 3. Key features of the BMP180.

Dimensions, mass	20 × 25 mm, < 5 g
Supply voltage, power consumption	2.5 VDC (typical), < 2 mW
Supply current @ 1 sample/s, 25 °C	5 μA (standard mode), 7 μA (hi-res mode)
Standby current @ 25 °C	0.1 μA
Operating temperature	−40 °C to +85 °C
Pressure-sensing range (altitude)	300–1100 hPa (9000 m to −500 m ASL)
Pressure (altitude) resolution	Up to 0.03 hPa (0.25 m)
Operating temperature/resolution	−40 °C to 85 °C/0.1 °C
RMS noise	3 Pa
Long-term stability (12 months)	±1.0 hPa

3.3. 10 DOF IMU

The DFRobot 10 DOF IMU sensor [36] (Figure 5) is a low-power (10 mW), compact-size (26x18mm) board, fully compatible with the Arduino microcontroller family, and integrates an Analog Device 10-bit ADXL345 accelerometer with up to ±16 g dynamic range, 0.312×10^{-5}-g sensitivity, and ±40-mg drift [37]; a Honeywell HMC5883L magnetometer [38]; a 16-bit ITG-3205 gyro with ±2000°/s full-scale range, 0.014°/s sensitivity, and ±1°/s drift [39]; and a Bosch BMP085 pressure sensor. The IMU is a polysilicon surface micromachined structure built on top of a silicon wafer. Polysilicon springs suspend the structure over the surface of the wafer and provide resistance against acceleration forces.

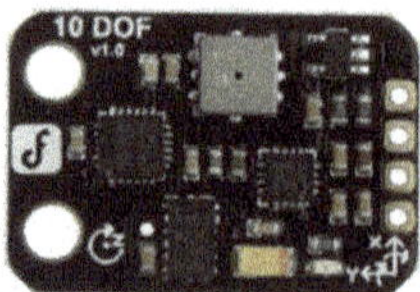

Figure 5. The 10 DOF MEMS IMU.

IMU measurements were used in this research to estimate the acceleration components and the angular rates in Equation (1), and to check the effectiveness of the micro-Pitot approach for velocity and attitude determination.

3.4. Microcontroller

The Arduino Mega 2560 board (102 × 53 mm, weight 37 g) is a microcontroller board based on the ATmega2560. It has 54 digital input/output pins (of which 14 can be used as PWM outputs), 16 analog inputs, 4 UARTs (hardware serial ports), a 16-MHz crystal oscillator, a USB connection, a power jack, an ICSP header, and a 5 V voltage supply. The board has 250-mA current absorption (1.25-W power consumption) when running code and providing power to external sensors. The board contains everything needed to support the microcontroller; simply connecting it to a computer with a USB cable or powering it with an AC-to-DC adapter or battery allows the board be used and work in an integrated development environment (IDE) based on the Processing language project.

4. Simulations, Results, and Performance Evaluation

Schematics and a prototype of the acquisition system are shown in Figures 6 and 7, respectively. The indoor experimental campaigns were performed in the PFDL of the University of Naples "Parthenope", Naples, Italy.

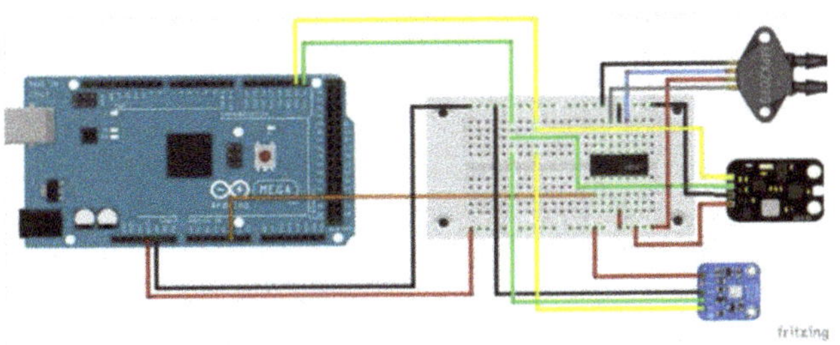

Figure 6. Electric scheme of the developed system.

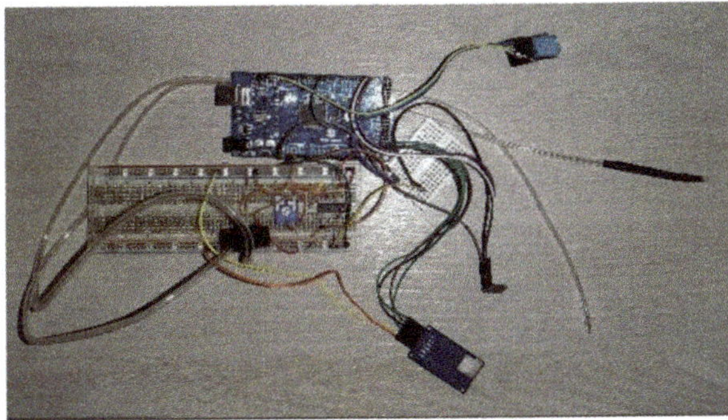

Figure 7. Prototype of the data-acquisition system.

The system was tested using a fan as airflow generator while acquiring the differential pressures from the MPX2010DP sensor. The sensor's measurement range is 0–1023 (10 bits). Data were collected in 3-min acquisitions at a sampling frequency of 10 Hz. A digital anemometer (Proster PST-TL017 Handheld Anemometer [40]) was used for reference measurement of the airstream. Table 4 reports some technical specifications of the PST-TL017 device, shown in Figure 8.

Table 4. Technical data for the PST-TL017.

Power supply, current consumption	3 V CR2032 battery, 3 mA
Operating humidity	≤90% RH
Storage temperature	−40 to 60 °C (−40 to 140 °F)
Weight	50 g
Air velocity resolution and range	0.1 m/s, 0–30 m/s
Air temperature range	−10 to +45 °C (14–113 °F)
Air temperature resolution	0.2 °C, 0.36 °F
Air temperature accuracy	±2 °C, ±3.6 °F

Figure 8. The PST-TL017 anemometer used for the airflow speed check.

4.1. Pressure-Sensor Calibration

To reduce measurement noise, 1D Kalman filtering was applied to the raw data, postprocessed in the Matlab® software environment; the results are shown in Figure 9. Preliminarily, for sensor bias estimation, digital pressure data (10-bit strings) were collected in quiet airflow ($v_w = 0$).

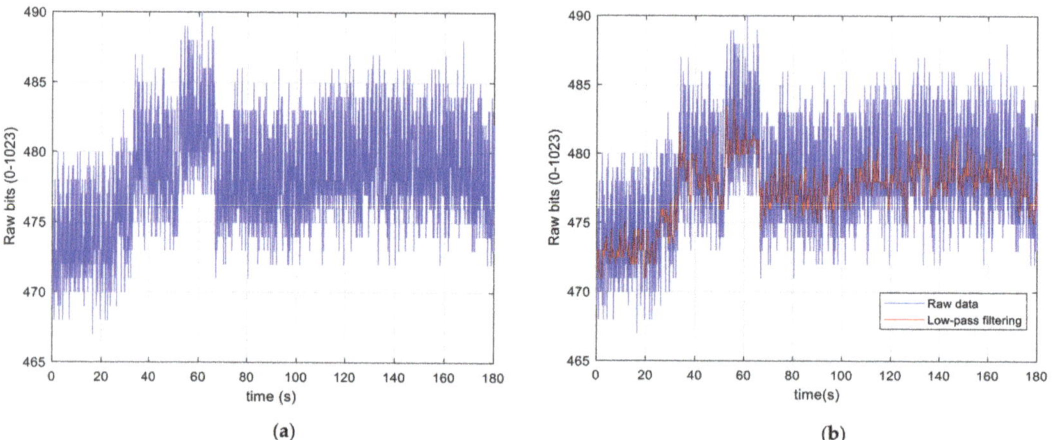

Figure 9. (a) Raw data from the MPX2010DP, with $v_w = 0$. (b) Comparison between the raw data and the low-pass-filtered data.

The MPX2010DP sensor used the raw Arduino 5 V voltage source, and the output voltage V_{out} was amplified by a two-stage differential op-amp circuit with default gains of 101 (first stage) and 6 (second stage), to obtain a signal in the range of 0–5 V. Inherent in the MPX2010 family of pressure sensors is a zero-pressure offset voltage, typically up to 1 mV. At a 5 V supply voltage, the zero-pressure offset was 0.5 mV, which corresponded to a 0.39 V offset voltage in the first op-amp stage. This offset was amplified by the second stage and appeared as a DC offset at V_{out} with no differential pressure applied. Using the design considerations described in [41], at zero pressure we expected a theoretical voltage value after the second gain stage of 2.34 V, and a pressure range of 0–2 kPa. Since the supply voltage was 5 V, the available signal for the actual pressure was $5 - 2.34 = 2.66$ V.

When $v_w = 0$, the sensor's DN (digital number) average output was 477, which corresponded to $V_{out} = V_{bias} = 2.33$ V (very close to the expected value), according to:

$$V_{out} = \left(\frac{5}{1023}\right) DN \qquad (21)$$

The average bias voltage acquired at $v_w = 0$ (Figure 10), V_{bias}, was subtracted from measurements with values of $v_w \neq 0$ to estimate the differential pressure and velocity.

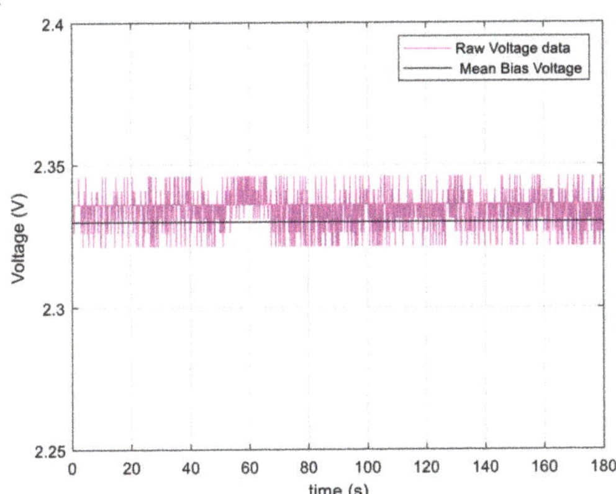

Figure 10. The bias voltage (output at $v_w = 0$).

To convert voltage into differential pressure (ΔP, in kPa) and evaluate the velocity from Equation (9), the following linear relation was applied:

$$\Delta P \, (kPa) = \frac{2 \, (V_{out} - V_{bias})}{V_{max} - V_{bias}} \tag{22}$$

where $V_{max} = 5$ V, $V_{bias} = 2.33$ V, and 2 is the full scale of the sensor (2 kPa).

4.2. Indoor Tests—Velocity Estimation

The indoor experiments were performed by mounting the equipment on a test bench on which the airflow was generated by a domestic fan, placed at 0.5 m from the micro-Pitot tube. The environmental conditions were: 33% relative humidity at 27 °C, collected by the DHT11 sensor, a low-cost, small-size (12 mm × 15.5 mm, mass < 5 g), low-power (12.5-mW power consumption when operating at 5 V, 2.5 mA) temperature and humidity sensor with a calibrated 16-bit digital signal output on a single-wire serial interface, with a 20–80% relative humidity range (accuracy 5%) and 0–50 °C temperature range with ± 2 °C accuracy [42].

Three airflow conditions were set as follows:

- Speed 0: 0 m/s (for system calibration);
- Speed 1: 2.5 0.1 m/s (measured by the anemometer);

- Speed 2: randomly varying airflow.

Data were acquired at a 10 Hz sampling rate. Bias estimation and sensor calibration (Section 4.1) were performed in the "Speed 0" condition. Figures 11 and 12 show the acquired raw (unfiltered) and filtered data in the "Speed 1" and "Speed 2" (random wind speed) conditions, respectively. As a check of the effectiveness of the calibration strategy, the average estimated velocity value in the "Speed 1" test condition (Figure 11) was found to be 2.56 m/s; this corresponded to a relative error equal to $(2.56 - 2.5)/2.56 = 2.3\%$. Raw data smoothing was performed as shown in Figures 11a and 12a with a moving average filter (the Matlab function `smoothdata`) in the time domain, comparable (but not as effective) to low-pass filtering in the frequency domain, while Figures 11b and 12b show the effect of 1D KF on the processed data.

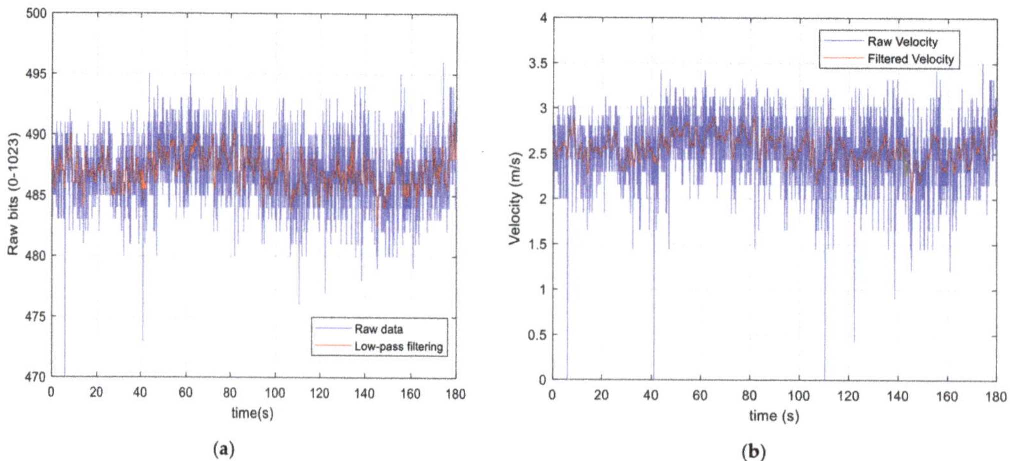

Figure 11. (**a**) Raw data collection during the "Speed 1" test. (**b**) Velocity acquisition (raw data vs. KF data) calculated by Equation (9).

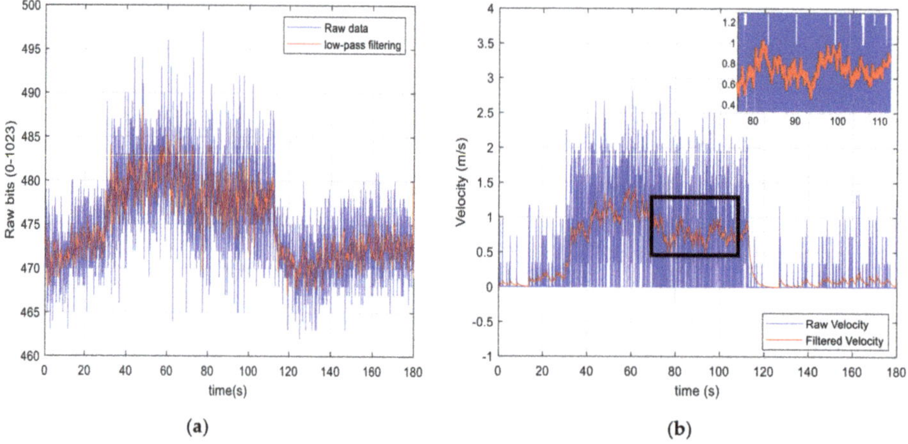

Figure 12. (**a**) Raw data vs. low-pass-filtered data (DN of sensor measurement). (**b**) Raw data vs. KF data (calculated velocity) in test condition "Speed 2".

4.3. Indoor Tests—AOA, AOS, and Attitude Estimation

AOA and AOS were calculated according to Equations (5) and (6). The entire system was successively mounted on a movable structure (Figure 13) to simulate various attitude changes of the UAV during the indoor tests. The following assumptions were made:

- Static data acquisition: the system was fixed and simply surrounded by the constant airflow coming from the fan;
- Pitot tube aligned with the airframe longitudinal axis: the relative velocity was equal to the airflow velocity, and from Equation (4):

$$\begin{bmatrix} u_r \\ v_r \\ w_r \end{bmatrix} = \begin{bmatrix} 0 \\ 0 \\ 0 \end{bmatrix} - \mathbf{R}_N^b \begin{bmatrix} u_w \\ 0 \\ 0 \end{bmatrix} \tag{23}$$

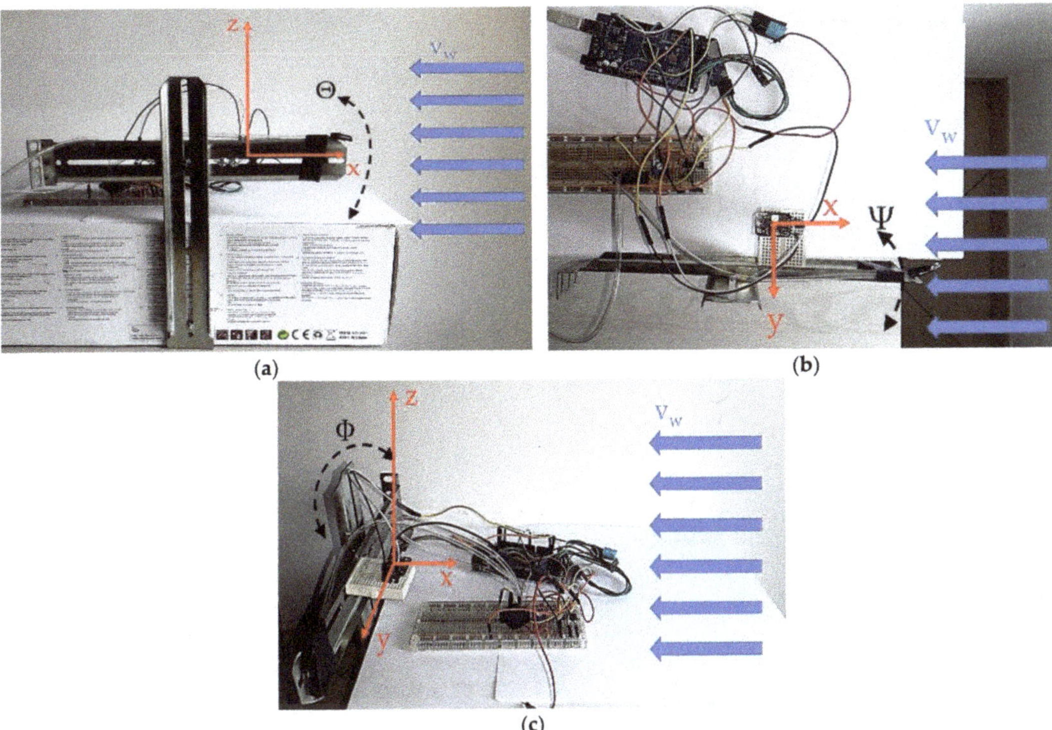

Figure 13. The prototype mounted on a movable structure to simulate pitch (**a**), yaw (**b**), and roll (**c**) variations.

Several test cases were set up, as shown in Table 5, in which the actual wind velocity ($|\mathbf{v_w}|$) was measured by the digital anemometer, and the controlled ("true") attitude angles were selected by using a graduated scale mounted on the structure. For example, in test case 1, the system was set in a horizontal position, with the X–Y plane parallel to the ground (yaw, roll, and pitch angles equal to 0°, calculated by the IMU), and the "true" wind velocity was set to 7 m/s.

Table 5. The test cases and parameter setup.

Test Case	$|\mathbf{v_w}|$ (m/s)	θ (deg)	ϕ (deg)	ψ (deg)
1	7	0	0	0
2	7	0	45	0
3	7	0	90	0
4	7	45	0	0
5	7	0	0	45
6	11.5	0	0	0
7	7	Varying	Varying	Varying

To estimate roll and pitch angles from the IMU, a complementary filter was applied, fusing accelerometer and gyroscope data, to reduce drift and noise errors [43,44]:

$$\begin{bmatrix} \hat{\phi}_k \\ \hat{\theta}_k \\ \hat{\psi}_k \end{bmatrix} = \gamma \begin{bmatrix} \hat{\phi}_{k-1} \\ \hat{\theta}_{k-1} \\ \hat{\psi}_{k-1} \end{bmatrix} + \begin{bmatrix} \dot{\phi}_{k-1} \\ \dot{\theta}_{k-1} \\ \dot{\psi}_{k-1} \end{bmatrix} \Delta t + (1-\gamma) \begin{bmatrix} \phi_{acc,k-1} \\ \theta_{acc,k-1} \\ 0 \end{bmatrix} \tag{24}$$

where $\hat{\phi}_k, \hat{\theta}_k, \hat{\psi}_k$ are the estimates of roll, pitch, and yaw angles at the instant k; γ is the filter coefficient; $\omega = \begin{bmatrix} \dot{\phi} \ \dot{\theta} \ \dot{\psi} \end{bmatrix}^T$ is the angular rate vector derived from gyroscopic measurements [45]; and ϕ_{acc}, θ_{acc}, are the angles derived from the accelerometer data vector **a**. These can be derived from triple-axis tilt calculation, which evaluates the angles ϕ' between the gravity vector and the accelerometer's z-axis (with positive direction opposite to gravity); θ' between the horizon (initially coincident with the accelerometer xy-plane) and the x-axis, coincident with the body longitudinal axis; and ψ' between the horizon and the y-axis of the accelerometer [46]:

$$
\left.
\begin{aligned}
\phi' &= \tan^{-1}\left(\frac{\sqrt{a_x^2 + a_y^2}}{a_z} \right) \\[2mm]
\theta' &= \tan^{-1}\left(\frac{a_x}{\sqrt{a_y^2 + a_z^2}} \right) \\[2mm]
\psi' &= \tan^{-1}\left(\frac{a_y}{\sqrt{a_x^2 + a_z^2}} \right)
\end{aligned}
\right\}
\tag{25}
$$

The roll angle ($a_x = 0$) is given by $\tan^{-1} a_y/a_z$, and pitch ($a_y = 0$) is given by $\tan^{-1} a_x/a_z$. The yaw angle was estimated by simply integrating the gyroscopic yaw rate. The filter coefficient was determined by:

$$
\gamma = \frac{\tau}{\tau + \Delta t}
\tag{26}
$$

where τ is the time constant of the filter. Figure 14 shows the KF velocity estimation and the Euler angles with the complementary filter during data collection in static conditions (test case 1).

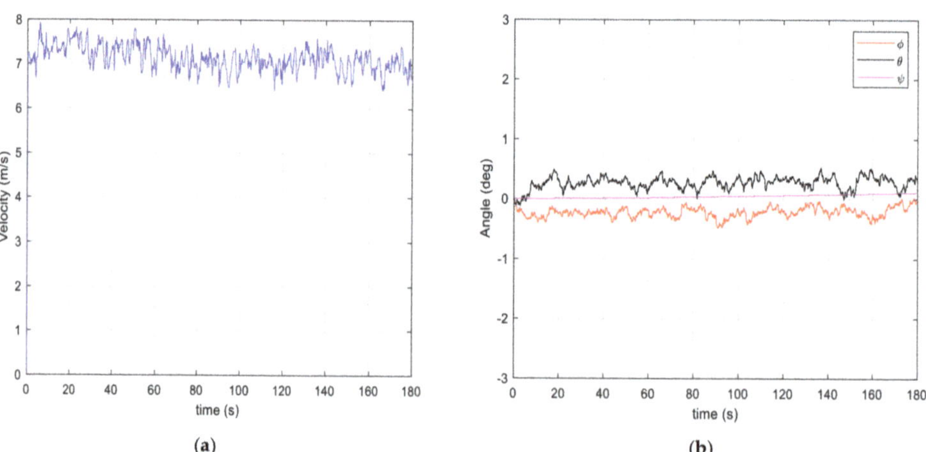

Figure 14. (**a**) The filtered velocity measured by the differential pressure sensor in test case 1. (**b**) The Euler angles after complementary filtering (fusion of gyroscope and accelerometer data from the IMU).

Following Equation (4), the relative velocity $\mathbf{v}_r$ and its components $\begin{pmatrix} u_r & v_r & w_r \end{pmatrix}^T$ were evaluated, whereas Equations (6) and (7) were used to estimate the AOA (α) and AOS (β). Table 6 shows the average values of the estimates and the experimental results for the seven test cases devised, and Figures 15–17 show the collected data and the estimates during test conditions 3, 6, and 7, respectively.

Table 6. Estimates of v_r, AOA, and AOS in the seven experimental setups (filtered data).

Test Case	v_r^T (m/s)	Relative Error	α (deg)	β (deg)
1	(−7.14 0.00 0.03)	2.0%	−0.26	−0.05
2	(−6.66 0.01 0.42)	4.9%	−3.67	−0.08
3	(−6.93 −0.13 0.08)	1.0%	−0.55	−0.86
4	(−3.51 −0.21 2.90)	8.5%	−39.5	−2.72
5	(−3.17 −4.10 0.03)	4.6%	−0.51	−47.1
6	(−10.5 −0.05 0.01)	4.8%	−0.06	−0.23
7	(−6.77 −0.01 0.26)	3.4%	−1.74	−0.07

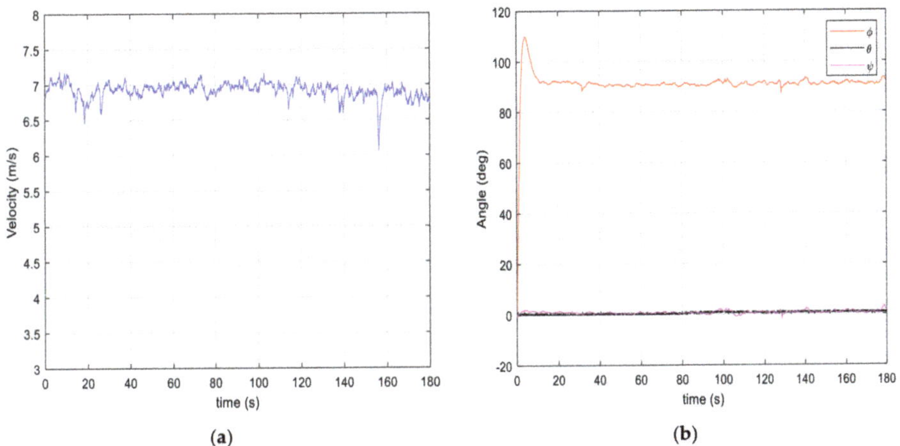

Figure 15. The filtered velocity (**a**) and yaw, pitch, and roll angles (**b**) during test case 3.

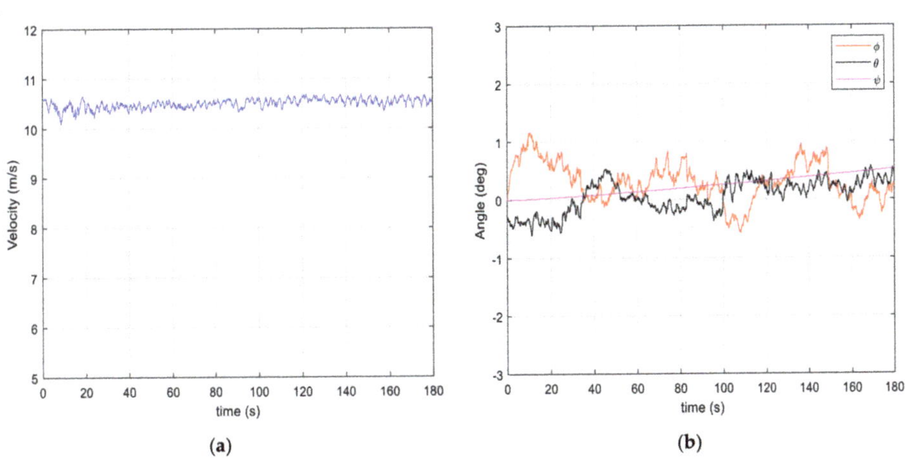

Figure 16. The filtered velocity (**a**) and filtered Euler angles (**b**) during test case 6.

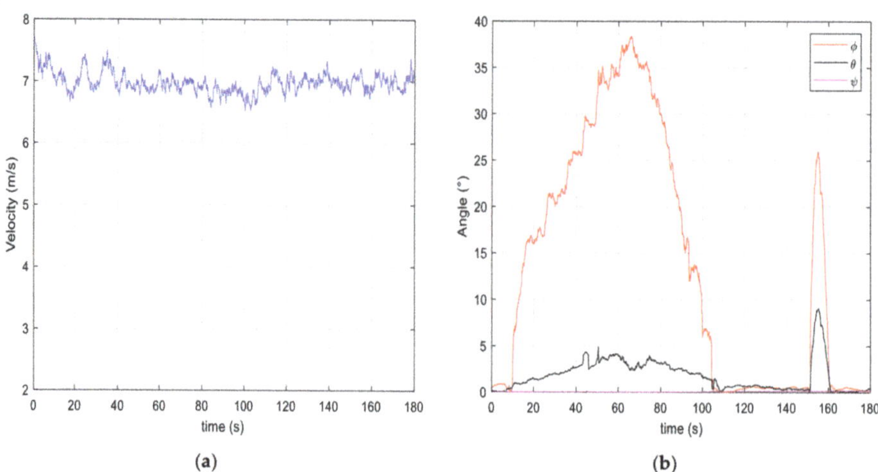

Figure 17. Data collection in dynamic conditions (test case 7, simulation of a roll variation from 0–40°). (**a**) filtered velocity, (**b**) filtered Euler angles.

A simple statistical analysis was conducted on the estimates of airspeed, AOA, and AOS to evaluate the system's performance. Table 7 shows the standard deviations $\sigma_{V_{CAS}}$, σ_α, σ_β of the estimated values (σ_α and σ_β before and after filtering), while Figure 18 shows a plot of σ_α and σ_β as a function of the velocity magnitude, together with the relative least-square fits. The experimental data were in agreement with the growth in errors as the velocities approached zero, according to the developed error analysis (Section 2.2).

Table 7. Standard deviations of estimated CAS, AOA, and AOS (raw and filtered data).

Test Case	$\sigma_{V_{CAS}}$ (m/s)	σ_α (deg)		σ_β (deg)	
		Raw	Filtered	Raw	Filtered
1	0.32	3.2	0.10	0.10	0.03
2	0.29	5.3	0.32	0.42	0.04
3	0.27	4.0	0.39	0.78	0.50
4	1.41	13	6.1	7.7	0.47
5	0.60	3.3	0.15	1.2	1.1
6	0.35	2.2	0.14	1.1	0.16
7	0.76	1.7	1.66	11	0.06

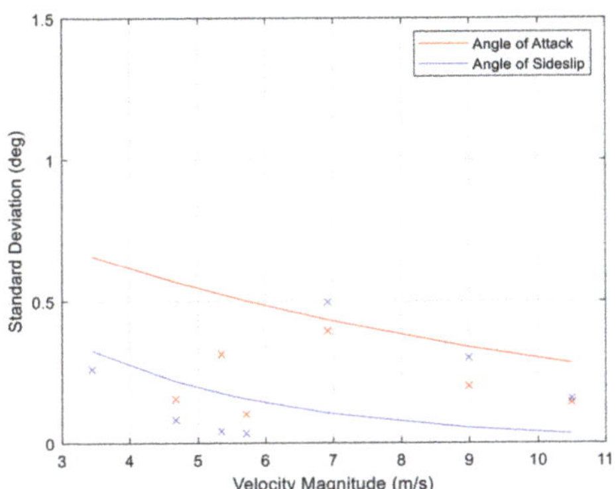

Figure 18. The standard deviation of the angle of attack and sideslip vs. magnitude of the relative velocities: numerical values and least-square fit.

5. Conclusions and Further Work

This paper presented a kinematic model for estimation of airspeed, angle of attack α, and sideslip angle β for small UAVs equipped with low-cost, off-the-shelf commercial navigation sensors. The devised system used a miniaturized differential pressure sensor (micro-Pitot tube) and an IMU for attitude determination, managed by a microcontroller. The calibration technique used for the pressure sensor returned estimation errors of less than 3%. The system performance was evaluated using experimental results from indoor benchmarking tests that emulated some basic dynamics of typical UAV missions. As shown in Table 6, the relative error that affected airspeed measurements was found to be less than 9% in all the test scenarios considered, or less than 5% if we excluded test case 4 shown in Table 6, relative to a nonrealistic value of the AOA (which was typically in the range of $-2°$ to $15°$). The estimation errors of α, β, and V_{CAS} were found to be within $1.7°$ (excluding the nonrealistic case $\alpha = 45°$), $0.5°$, and 1.4 m/s, respectively, in good agreement with the theoretical values derived from the law of error propagation, and consistent with other authors' work [15,19,20,26]. The proposed approach showed a promising potentiality for implementation of real-time control laws to increase the flight envelope by exploiting attitude measurements and direct knowledge of α and β. Using AOA and AOS estimates as in-flight feedback inputs to the autopilot control loop also could help to improve the endurance of small aircraft (typically in the range 15–45 min) by implementing specific flight strategies according to the wind conditions, or even optimizing the trajectory by gaining power in energy-harvesting techniques. There are, however, some limitations of the proposed methodology:

- The alignment of the micro-Pitot tube (differential pressure sensor) to the longitudinal axis of the UAV must be performed very precisely in order to avoid biases in the estimations of the AOA and AOS. Moreover, the sensor must be located reasonably far from the rotary wings (considering a quadcopter or a multirotor VTOL UAV) to avoid turbulent airflow added by the rotors.
- High velocities (>20 m/s) create differential pressure values out of the available sensor range (0.10 kPa at 10 V supply, 0–2 kPa at 5 V supply). This was not an issue for the micro-UAV applications devised by the authors (the system will be installed on a quadcopter with maximum velocity on the order of 10 m/s), but could be a problem for larger aircraft.
- The mass and size requirements of our system (<150 g, typical dimensions of the boxed prototype of $120 \times 60 \times 30$ mm) fit typical mini- and micro-UAV payload constraints, but the power consumption of the system (in the range 1–2W) could significantly reduce the aircraft's endurance (which was in the order of 15–30 min for typical small

UAVs). Therefore, careful engineering considerations must be devised to reduce the impact of the system in terms of flight mission duration.

Future work will involve the realization of a minimum-size, minimum-weight version of the sensor suite, to be strapdown-mounted on a small UAV with a 2 kg maximum take-off weight (MTOW); the implementation of outdoor tests in typical flight and atmospheric conditions; and further developments of the error models and the attitude kinematics, to improve the accuracy of airflow angle and sideslip estimates in various flight and wind conditions.

Author Contributions: Conceptualization, methodology, G.A., S.P., and U.P.; software and validation, G.A., S.P., and U.P.; formal analysis, investigation, data curation, G.A., S.P., and U.P.; resources, G.D.C.; writing—original draft preparation, G.A., U.P., and S.P.; writing—review and editing, S.P.; supervision, funding acquisition, G.D.C. All authors have read and agreed to the published version of the manuscript.

Funding: This work was supported by the University of Naples "Parthenope" (Italy) Internal Research Project DIC (Drone Innovative Configurations, CUP I62F17000060005-DSTE 338).

Acknowledgments: The authors wish to thank Alberto Greco for his invaluable support during the laboratory tests, and the four anonymous reviewers who helped them to enhance the scientific and technical content of the paper with their enlightening observations.

Conflicts of Interest: The authors declare no conflict of interest.

References

1. U.S. Army UAS Center of Excellence (ATZQ-CDI-C). *Eyes of the Army"—U.S. Army Roadmap for Unmanned Aircraft Systems 2010–2035*; Army UAS CoE Staff: Fort Rucker, AL, USA, 2010.
2. Papa, U. Embedded Platforms for UAS Landing Path and Obstacle Detection. In *Studies in Systems, Decision and Control*; Springer: New York, NY, USA, 2018; Volume 136.
3. Nawaz, H.; Ali, H.M.; Massan, S.-U. Applications of unmanned aerial vehicles: A review. *3c Tecnol.* **2019**. [CrossRef]
4. González-Jorge, H.; Martínez-Sánchez, J.; Bueno, M.; Arias, A.P. Unmanned aerial systems for civil applications: A review. *Drones* **2017**, *1*, 2. [CrossRef]
5. Shakhatreh, H.; Sawalmeh, A.H.; Al-Fuqaha, A.; Dou, Z.; Almaita, E.; Khalil, I.; Othman, N.S.; Khreishah, A.; Guizani, M. Unmanned Aerial Vehicles (UAVs): A survey on civil applications and key research challenges. *IEEE Access* **2019**, *7*, 48572–48634. [CrossRef]
6. ENAC (Ente Nazionale per l'Aviazione Civile). *Regolamento Mezzi aerei a pilotaggio remoto. Amendment 1, 14/07/2020*; Enac: Rome, Italy, 2020.
7. Abdelkrim, N.; Aouf, N. Robust INS/GPS sensor fusion for UAV localization using SDRE nonlinear filtering. *IEEE Sens. J.* **2010**, *10*, 789–798.
8. Gross, J.N.; Gu, Y.; Rhudy, M.B.; Gururajan, S.; Napolitano, M.R. Flight-test evaluation of sensor fusion algorithms for attitude estimation. *IEEE Trans. Aerosp. Electron. Syst.* **2012**, *48*, 2128–2139. [CrossRef]
9. Yang, X.; Mejias Alvarez, L.; Garratt, M. Multi-sensor data fusion for UAV navigation during landing operations. In Proceedings of the Australasian Conference on Robotics and Automation (ACRA 2011), Australian Robotics and Automation Association, Melbourne, Australia, 7–9 December 2011.
10. Del Pizzo, S.; Papa, U.; Gaglione, S.; Troisi, S.; Del Core, G. A vision-based navigation system for landing procedure. *Acta IMEKO* **2018**, *7*, 102–109. [CrossRef]
11. Ariante, G.; Papa, U.; Ponte, S.; Del Core, G. UAS for positioning and field mapping using LIDAR and IMU sensors data: Kalman filtering and integration. In Proceedings of the IEEE 5th International Workshop on Metrology for AeroSpace (MetroAeroSpace), Turin, Italy, 19–21 June 2019.
12. Wang, B.H.; Wang, D.B.; Ali, Z.A.; Ting, B.T.; Wang, H. An overview of various kinds of wind effects on unmanned aerial vehicle. *Meas. Control.* **2019**, *52*, 731–739. [CrossRef]
13. Tian, P.; Chao, H.; Rhudy, M.; Gross, J.; Wu, H. Wind sensing and estimation using small fixed-wing unmanned aerial vehicles: A survey. *J. Aerosp. Inf. Syst.* **2021**, *18*, 132–143. [CrossRef]
14. Shimura, T.; Inoue, M.; Tsujimoto, H.; Sasaki, K.; Iguchi, M. Estimation of wind vector profile using a hexarotor unmanned aerial vehicle and its application to meteorological observation up to 1000 m above surface. *J. Atmos. Ocean. Technol.* **2018**, *35*, 1621–1631. [CrossRef]
15. Petrich, J.; Subbarao, K. On-board wind speed estimation for Uavs. In Proceedings of the AIAA Guidance, Navigation, and Control Conference, Portland, OR, USA, 8–11 August 2011.

16. Watkins, S.; Milbank, J.; Loxton, B.J.; Melbourne, W.H. Atmospheric winds and their implications for microair vehicles. *AIAA J.* **2006**, *44*, 2591–2600. [CrossRef]

17. Langelaan, J.W.; Alley, N.; Neidhoefer, J. Wind field estimation for small unmanned aerial vehicles. *J. Guid. Control. Dyn.* **2011**, *34*, 1016–1030. [CrossRef]

18. Gavrilovic, N.; Benard, E.; Pastor, P.; Moschetta, J.-M. Performance improvement of small UAVs through energy-harvesting within atmospheric gusts. In Proceedings of the AIAA SciTech Forum, Grapevine, TX, USA, 9–13 January 2017.

19. Johansen, T.A.; Cristofaro, A.; Sorensen, K.; Hansen, J.M.; Fossen, T.I. On estimation of wind velocity, angle-of-attack and sideslip angle of small UAVs using standard sensors. In Proceedings of the International Conference on Unmanned Aircraft Systems (ICUAS), Denver, CO, USA, 9–12 June 2015.

20. Perry, J.; Mohamed, A.; Johnson, B.; Lind, R. Estimating angle of attack and sideslip under high dynamics on small UAVs. In Proceedings of the ION GNSS 21st International Technical Meeting of the Satellite Division, Savannah, GA, USA, 16–19 September 2008; pp. 1165–1173.

21. Sankaralingam, L.; Ramprasadh, C. A comprehensive survey on the methods of angle of attack measurement and estimation in UAVs. *Chin. J. Aeronaut.* **2019**, *33*, 749–770. [CrossRef]

22. Ramprasadh, C.; Prem, S.; Sankaralingam, L.; Deshpande, P.; Dodamani, R. A simple method for estimation of angle of attack. *IFAC-Pap.* **2018**, *51*, 353–358. [CrossRef]

23. Ramprasadh, C.; Arya, H. Multistage-fusion algorithm for estimation of aerodynamic angles in mini aerial vehicle. *J. Aircr.* **2012**, *49*, 93–100. [CrossRef]

24. Ariante, G.; Papa, U.; Ponte, S.; Del Core, G. Velocity and attitude estimation of a small unmanned aircraft with micro Pitot tube and Inertial Measurement Unit (IMU). In Proceedings of the IEEE 7th International Workshop on Metrology for Aerospace (MetroAer-oSpace), Pisa, Italy, 22–24 June 2020.

25. De Marina, H.G.; Pereda, F.J.; Giron-Sierra, J.M.; Espinosa, F. UAV attitude estimation using unscented Kalman filter and TRIAD. *IEEE Trans. Ind. Electron.* **2011**, *59*, 4465–4474. [CrossRef]

26. Zhang, Q.; Xu, Y.; Wang, X.; Yu, Z.; Deng, T. Real-time wind field estimation and pitot tube calibration using an extended Kalman filter. *Mathematics* **2021**, *9*, 646. [CrossRef]

27. Beard, R.W.; McLain, T.W. *Small Unmanned Aircraft: Theory and Practice*; Princeton University Press: Princeton, NJ, USA, 2012.

28. Cai, G.; Chen, B.M.; Lee, T.H. An overview on development of miniature unmanned rotorcraft systems. *Front. Electr. Electron. Eng. China* **2009**, *5*, 1–14. [CrossRef]

29. Quan, Q. *Introduction to Multicopter Design and Control*; Springer: Singapore, 2017.

30. Yechout, T.; Morris, S.L.; Bossert, D.E.; Hallgren, W.F. Introduction to aircraft flight mechanics: Performance, static stability, dynamic stability, and classical feedback control. In *AIAA Education Series*; American Institute of Aeronautics and Astronautics, Inc.: Reston, VA, USA, 2003.

31. International Organization for Standardization (ISO). *Standard Atmosphere*; ISO: London, UK, 1975.

32. Ghilani, C.D. *Adjustment Computations—Spatial Data Analysis*, 5th ed.; John Wiley and Sons, Inc.: Hoboken, NJ, USA, 2010.

33. Grewal, M.S.; Andrews, A.P. *Kalman Filtering—Theory and Practice Using MATLAB®*, 3rd ed.; John Wiley and Sons, Inc.: Hoboken, NJ, USA, 2008.

34. NXP B., V. MPX2010 Series—10 kPa Temperature Compensated Pressure Sensors—Product Datasheet. Rev. 14. 2021. Available online: https://www.nxp.com/docs/en/data-sheet/MPX2010.pdf (accessed on 1 July 2021).

35. Bosch Sensortec. BMP180 Digital Pressure Sensor, Rev. 2.8, Doc. No. BST-BMP180-DS000-12. 2015. Available online: https://git.wapakema.de/walter/Wetterstation/raw/commit/5d7e482b2baaa35a205b06021f282d10bd271cd3/python/BMP280.pdf (accessed on 1 July 2021).

36. DFRobot, 2018. 10-Dof MEMS IMU Sensor V2.0. Available online: https://www.dfrobot.com/wiki/index.php/10_DOF_Mems_IMU_Sensor_V2.0_SKU:_SEN0140 (accessed on 1 July 2021).

37. Analog Devices. *Small, Low Power, 3-Axis ±3g Accelerometer ADXL335-345. Rev. 0*; One Technology Way: Norwood, MA, USA, 2009. Available online: https://www.sparkfun.com/datasheets/Components/SMD/adxl335.pdf (accessed on 1 July 2021).

38. Honeywell. *3-Axis Digital Compass IC HMC5883L. Rev E.*; Honeywell: Plymouth, MN, USA, 2013.

39. InvenSense Inc. *ITG-3200 Product Specification—Revision 1.7*; InvenSense Inc.: Sunnyvale, CA, USA, 2011.

40. Proster®. Proster TL017 Handheld Anemometer Wind Speed Meter Scale Gauge. 2012. Available online: http://www.prostereu.com/index.php/2015/07/24/tl017/ (accessed on 1 July 2021).

41. Romero, M.; Figueroa, R. *Low-Pressure Sensing Using MPX2010 Series Pressure Sensors. Application Note AN4010, Rev. 1*; Freescale Semiconductors Inc.: Austin, TX, USA, 2005.

42. Adafruit Industries. DHT11, DHT22 and AM2302 Sensors. 2020. Available online: https://cdn-learn.adafruit.com/downloads/pdf/dht.pdf (accessed on 3 September 2021).

43. Merhav, S. *Aerospace Sensor Systems and Applications*; Springer: New York, NY, USA, 1996.

44. Gui, P.; Tang, L.; Mukhopadhyay, S. MEMS based IMU for tilting measurement: Comparison of complementary and Kalman filter based data fusion. In Proceedings of the IEEE 10th conference on Industrial Electronics and Applications (ICIEA), Auckland, New Zealand, 15–17 June 2015.

45. Stančin, S.; Tomazic, S. On the interpretation of 3D gyroscope measurements. *J. Sens.* **2018**, *2018*, 1–8. [CrossRef]
46. Fisher, C.J. *Using an Accelerometer for Inclination Sensing. AN-1057 Application Note*; Analog Devices Inc.: Norwood, MA, USA, 2010.

 *electronics*

Article

Design, Simulation, Analysis and Optimization of PID and Fuzzy Based Control Systems for a Quadcopter

Isaac S. Leal [1], Chamil Abeykoon [2,*] and Yasith S. Perera [3]

1 Department of Electronics Design, Mid Sweden University, Holmgatan 10, 851 70 Sundsvall, Sweden; isaac.sanchezleal@miun.se
2 Aerospace Research Institute and Northwest Composites Centre, Department of Materials, Faculty of Science and Engineering, The University of Manchester, Oxford Road, Manchester M13 9PL, UK
3 Department of Textile and Apparel Engineering, University of Moratuwa, Moratuwa 10400, Sri Lanka; yasiths@uom.lk
* Correspondence: chamil.abeykoon@manchester.ac.uk

Abstract: Unmanned aerial vehicles or drones are becoming one of the key machines/tools of the modern world, particularly in military applications. Numerous research works are underway to explore the possibility of using these machines in other applications such as parcel delivery, construction work, hurricane hunting, 3D mapping, protecting wildlife, agricultural activities, search and rescue, etc. Since these machines are unmanned vehicles, their functionality is completely dependent upon the performance of their control system. This paper presents a comprehensive approach for dynamic modeling, control system design, simulation and optimization of a quadcopter. The main objective is to study the behavior of different controllers when the model is working under linear and/or non-linear conditions, and therefore, to define the possible limitations of the controllers. Five different control systems are proposed to improve the control performance, mainly the stability of the system. Additionally, a path simulator was also developed with the intention of describing the vehicle's movements and hence to detect faults intuitively. The proposed PID and Fuzzy-PD control systems showed promising responses to the tests carried out. The results indicated the limits of the PID controller over non-linear conditions and the effectiveness of the controllers was enhanced by the implementation of a genetic algorithm to autotune the controllers in order to adapt to changing conditions.

Keywords: quadcopter; unnamed aerial vehicle; dynamic model; PID controller; fuzzy logic; genetic algorithm; intelligent control

Citation: Leal, I.S.; Abeykoon, C.; Perera, Y.S. Design, Simulation, Analysis and Optimization of PID and Fuzzy Based Control Systems for a Quadcopter. *Electronics* **2021**, *10*, 2218. https://doi.org/10.3390/electronics10182218

Academic Editor: Nurul I. Sarkar

Received: 27 July 2021
Accepted: 7 September 2021
Published: 10 September 2021

Publisher's Note: MDPI stays neutral with regard to jurisdictional claims in published maps and institutional affiliations.

1. Introduction

In recent years, Unmanned Aerial Vehicles (UAVs) have gained a massive popularity around the globe. The current technological advances have made it possible to equip these vehicles with a variety of state-of-the-art technologies such as machine vision, global position systems or a simple video camera, and so forth. Moreover, these machines are becoming widely popular over other platforms such as aircraft or helicopters due to their flight characteristics. Planes, despite being excellent flying platforms, are not effective for flights operated indoors as they are difficult to maneuver in confined spaces. On the other hand, helicopters offer great mobility in all the environments but certainly are quite difficult to control. Furthermore, helicopters may not be an appropriate mode of flying when considering indoor flights due to their fast-rotating propellers. These reasons make quadcopters an excellent alternative for indoor flights. However, the design and control of quadcopters pose a number of challenges: the multivariate system makes modelling a difficult task; the existence of coupled variables makes it difficult to control the plant; the drone must operate in a three-dimensional environment with unplanned disturbances; the need for achieving a stable system with a high degree of precision and accuracy to allow the implementation of different technologies such as artificial vision, and so forth.

A quadcopter is a machine that can fly without a pilot but in order to realize this, an implementation of a reliable control system is necessary. Basically, a control system should control the velocity of the rotors, enabling the vehicle to fly stably and safely. Usually, linear control techniques should enable these types of vehicles to fly stably. However, a quadcopter possesses a very complex dynamic model and is very susceptible to wind or other unforeseeable climatological adversities. For this reason, a quadcopter requires a non-linear control system which can be improved by implementing an algorithm that can help the controller in making decisions to adapt to the possible adverse conditions. Intelligent control techniques are gradually becoming more effective in helping conventional control techniques to tackle these issues with an elevated level of abstraction.

The growing popularity of quadcopters stimulates a wide variety of projects to be devoted to this theme [1–8]. Several previous works related to quadcopters on the development of dynamic models and the designing of linear control as well as non-linear control are reviewed in detail in the following section.

1.1. Dynamic Models

A dynamic model is the basis of electronic control and hence, a number of previous works (discussed below) focused on the development of equations describing the motion of a quadcopter. It is necessary to consider the disturbances on the system when a quadcopter is flying over adverse weather conditions such as windy or rainy conditions, where a non-linear model is required to accurately model such situations. However, the implementation of a dynamic model is quite complex and hence, the realization of linear control (i.e., simplification of the dynamic model) should be necessary.

In the work presented by Amir and Abbass [9], a mathematical model of a quadcopter was proposed with the aim of obtaining a simplified model which would be useful in designing a control system. The proposed model was based on several assumptions (i.e., the quadcopter body and blades are rigid; there is no friction on the body surface; the free stream air velocity is zero; and so forth). It can be observed that the axes connecting the two motors coincided with the axes of the coordinate system, and a "plus" (+) configuration was assumed for the frame. The authors further showed the relationships between the thrust force and the voltage as well as the angular velocity and the voltage. In this way, the system input would be the voltage and the outputs would be the accelerations of each of the four degrees of freedom to be controlled, and the acceleration on the vertical axis and the three angular axes. The development presented in this work shows a clear and concise process where it is possible to distinguish each step that resulted in the equations of the dynamic model.

Pounds et al. [10] presented the development of a mathematical model of a quadcopter called the "X4-flyer". The system was modeled using Newton's laws of motion and then the design of a pilot augmentation control was proposed. The authors developed a control system to manage the pitch, roll and yaw angles of the plant. A model with disturbance inputs was created to determine the behavior of the plant over unplanned circumstances. The simulations were performed in MATLAB Simulink and the results showed a satisfactory performance at rotor speeds below 450 rads^{-1}. As the rotor speed increased, the system became unstable and the authors attributed this instability to the high frequency noise from the rotors, which adversely affected the data collected by the accelerometer.

Morar and Nascu [11] attempted to obtain a physical model similar to that of the AR.Drone (i.e., a quadcopter designed by Parrot). Obtaining a model that behaves similar to that of a commercial vehicle should be a difficult task in a research setting, due to the lack of available information (i.e., the exact details that a manufacturer has considered in their design). In the development of the model, the authors assumed that the frame is symmetrical between the "Oxy" system, and that the frame configuration corresponded to the "plus" (+) configuration. This model was also based on the Newton–Euler formulation and it was implemented in Simulink where the test results were very close to those of the real vehicle.

Sá et al. [12] aimed to design a platform on which the testing and evaluation of new control systems for quadcopters would be possible and they also reported the stages for designing and manufacturing a quadcopter from scratch. The proposed dynamic model was based on the Newton–Euler formulation and the frame configuration was of the "plus" (+) type. Although the quadcopter design was completed, the controllers were not designed in this work. The quadcopter was equipped with all the required components (i.e., sensors, communication systems, a power system, etc.) and its functionality was tested by adding different loads. However, the quadcopter designed in this work could carry only small loads (i.e., up to 800 g).

Fernando et al. [13] reported a dynamic model of a quadcopter to create a platform to facilitate the implementation of future automatic navigation systems. Similar to the work by Amir and Abbass [9], several assumptions were made to develop the mathematical model, the main assumptions being that the center of mass of the quadcopter coincides with the origin of the fixed frame and that the frame axes coincide with the principal axes of inertia. Furthermore, the "plus" (+) configuration was used for the frame. The development of the model was based on the Newton–Euler method. The authors also developed a control system which was based on four cascaded Proportional-Integral (PI) controllers, and they argued that this type of a configuration would be effective in limiting the effects of disturbances. Finally, the authors highlighted that the vehicle exhibited better stability and maneuverability during indoor flights than during outdoor flights. Perhaps the use of a genetic algorithm (GA) or a controller based on fuzzy logic could have helped the system to cope with the conditions existing in outdoor flights.

Barve and Patel [14] reported the development of a mathematical model, its simulation and a novel study called the "Altitude-Range-Analysis" for a quadcopter. The mathematical model was developed by using the Newton–Euler formulation and transformation matrices, and the configuration of the frame was of the "plus" (+) type. The main contribution of this work lies in the study of the feasible altitude ranges, which shows the altitude limitations of the quadcopter. The authors have considered several parameters for carrying out this analysis, such as the payload weight, rotor thrust, reference altitude positions, and so forth. Moreover, the authors modified their mathematical model to account for the air density variations, as well. The MATLAB simulator was used in this work, and it was mentioned that an altitude control system should tackle the changes in the air density during the flight. This is because, within the Earth's atmosphere, the air pressure/density is inversely related to the altitude. Hence, at a higher altitude, the air pressure is lower, which makes the air thinner, and therefore, the flight conditions are not constant [15]. This was confirmed by the simulation results, which indicated a lower altitude limit at a low reference altitude than the altitude limit at a high reference altitude [14].

Alkamachi and Erçelebi [16] reported a mathematical model and the design of an intelligent control system for a quadcopter. The mathematical model chosen was of the "H" type (i.e., corresponding to the shape of its chassis) and its development was based on the Newton–Euler method. The Newton formulation was used to model the translation dynamics and the Euler formulation for the rotational dynamics. The type "H" is not a very common type of chassis in modern quadcopter design, but it is interesting to note that its double symmetry feature makes it completely suitable to fly with four engines.

1.2. Linear Controllers and Algorithms

Work carried out by Patrascu et al. [17] showed the development of a control system for a three-dimensional crane. According to the authors, a three-dimensional crane is a complex non-linear multiple-input multiple-output (MIMO) system that has five output and three input signals. The authors argued that the use of a GA would enhance the performance of traditional controllers. The control strategy followed was based on the implementation of a

GA module in a control system based on Proportional-Integral-Derivative (PID) controllers. The GA module aimed to minimize the performance index and the three outputs of the plant that served to tune the PID controller. The results showed that the GA module contributed to improving the performance of the plant. However, the authors warned that the initialization process of the GA module is important and may even influence the outcome of the optimization. In this work, it is possible to see that the three-dimensional crane model shows some similarity to that of a quadcopter since both are non-linear MIMO systems with inherent linearities. Therefore, a GA should help a quadcopter control system as it did with the 3D crane.

Lim et al. [18] compared several open-source projects on quadcopters, namely Arducopter, Openpilot, Paparazzi, Pixhawk, Mikrokopter, KKmulticopter, Multiwii and Aeroquad, as of 2012. It could be seen that all the projects used PID controllers with anti-windup, even though the structure of the controllers showed slight differences. The KKmulticopter and the Mikrokopter used proportional (P) and PI controllers, respectively, while the Pixhawk and the Aeroquad employed PID controllers. The PI+P configuration was the most dominant, which could be found in the Arducopter, Paparazzi and Multiwii, where the P was for the inner loop and the PI was for the forward attitude error compensation. The Openpilot used a PI + PI configuration.

Argentim et al. [19] discussed different types of control systems for quadcopters. The main objective of their work was to test and explore which type of control exhibited the best performance for a quadcopter. Three control systems—an integral time absolute error (ITAE) tuned PID (ITAE is a criterion for regulation of PIDs where transient responses are obtained with a small overshoot and well-damped oscillations), a classic linear quadratic regulator (LQR) and an LQR tuned PID—were implemented with a dynamic model, and then simulations were performed to determine the behavior of the controllers over 10 different attitudes. Only the simulation results for the vertical attitude test were presented as the rest were analogous. The LQR controller showed excellent performance and good robustness, but with a significant transition delay. The PID offered a good dynamic response but turned out to be less robust than the LQR controller. Even though the LQR tuned PID exhibited a delayed response in certain attitudes, the authors claimed that it did not affect the accurate operation of the quadcopter and they concluded that it was a robust, versatile and easy-to-implement controller in a practical setting.

Boudjit and Larbes [20] presented a control system for a quadcopter, where the main objective was to stabilize the attitude of the vehicle subjected to various disturbances. First, a dynamic model of the vehicle was presented and then a control system based on PID controllers was designed. During the simulation, the authors verified that the vehicle responded better when the integral gain was set to zero. Therefore, the designed controller was a proportional-derivative (PD) controller. However, the authors commented that they found certain difficulties in having the controller face the initial conditions due to the non-linearities of the system. The tests for the real system showed a similar result where the PD controllers required high gains that led the motors to saturate. The rest of the tests showed satisfactory results where the vehicle was stabilized in different flight modes. Finally, the technology of navigation and control of a commercial AR.Drone was also presented.

In the work by Alkamachi and Erçelebi [16], a trajectory tracking controller was designed with four PID controllers: one for the altitude and the other three for the angular movements, roll, pitch and yaw. Moreover, the authors used a GA to automatically fine-tune the parameters of the PID controllers. The GA minimized the absolute tracking error, peak overshoot and the settling time for step inputs. The results shown could be considered excellent as the control system made the quadcopter follow the path with a minimum error. Finally, the authors showed an analysis that determines the robustness of the control system while evaluating the noise suppression, the perturbation rejection capacity and the sensitivity to uncertainties of the model parameters. The control system successfully performed in all the tests carried out. Moreover, the importance of intelligent control systems as a support for the conventional techniques was appreciated by this work as well.

Thu and Gavrilov [21] introduced a novel method of control for a quadcopter based on an L1 adaptive feedback control system. The key feature of the L1 adaptive control is the decoupling of adaptation from robustness [22]. The separation between the adaptation and the robustness is achieved by explicitly constructing the robustness specification in the formulation of the problem, understanding that the uncertainties in any feedback loop can be compensated only within the bandwidth of the control channel. Moreover, the authors explained that this modification of the formulation of the problem leads to the insertion of a limited bandwidth filter into the feedback path to maintain the control signal within the desired frequency range [21]. The control parameters were systematically determined based on the desired performance and robustness metrics. On the one hand, the rapid adaptation makes it possible to compensate for the undesirable effects of rapidly changing uncertainties and significant changes in system dynamics. Rapid adaptation is also critical for achieving predictable transient performance of the inputs and outputs of the system, without imposing persistence of excitation or resorting to high gain feedback. On the other hand, the limited bandwidth filter keeps the limited robustness margins far from zero in the presence of fast adaptation. In this sense, the bandwidth and the structure of the filter define the compensation between the performance and robustness of the closed-loop adaptation system. The simulation results indicated that the L1 adaptive controller showed excellent performance in trajectory tracking.

Njinwoua and Wouwer [23] suggested a cascade control strategy for the attitude control of a quadcopter, to handle the disturbances created by asymmetric actuators in the UAV. The proposed control system involved a PD controller in the inner loop to achieve stabilization, while a PI controller was used in the outer loop to handle disturbances. The simulation results showed that the proposed control system was capable of quickly adjusting to the disturbances, and an overshoot of about 10% and a settling time of 5 s were observed for 5% asymmetry on the drag and lift factors of the actuators.

Cedro and Wieczorkowski [24] employed the Wolfram Mathematica modelling software to optimize PID controller gains of a quadcopter. It was found that unconstrained optimization procedures were not suitable as they resulted in negative gains. Constrained procedures delivered satisfactory gain values, but they required the constraints to be very precisely defined. The best results were reported by the "RandomSearch" constrained optimization method with a penalty coefficient value of 1. The penalty coefficient was used to reduce the input signal level to prevent overshoots and oscillations, but its value should be selected using a trial-and-error approach.

Paredes et al. [25] presented (in the form of a tutorial) the design, simulation, implementation and testing of quadcopter controllers, which were designed based on a linearized quadcopter model. The aim was to assess the performance of these controllers when implemented on computationally limited embedded systems. The authors developed and compared quadcopter controllers based on PD, PID, LQR, LQR with integrators (LQR-I) and explicit model predictive control algorithms. The controllers were tested using inclined, circular flight sequences and position tracking error, velocity tracking error and also the control effort of these controllers was assessed. Based on the results, the authors claimed that the PID controller exhibited the best tracking performance, while the LQR-I controller provided the best compromise between the tracking performance, maintaining consistent simulation and experimental performance. The LQR controller was the fastest in terms of the average computation time. Furthermore, the authors inferred that the controllers were capable of meeting the time constraints imposed by the computationally limited embedded systems. This study is quite useful for future researchers not only to be used as a step-by-step guide in designing quadcopter controllers, but also in choosing a more suitable control strategy for computationally constrained applications.

1.3. Non-Linear Controllers and Algorithms

Hoffman [26] presented an overview on the use of evolutionary algorithms for the design of control systems based on fuzzy logic controllers and their optimization. The author explained that an evolutionary algorithm (such as a GA) works by adjusting membership functions and/or scaling factors based on an index that specifies the performance or the behavior of the control system. This work presented two applications: a control system for a car–pole system whose aim was to stabilize the pole in a vertical position keeping the car within the limits of the track, and another control system for a mobile robot whose mission was to avoid obstacles. In both experiments, an excellent performance was verified, and in the case of the mobile robot, the system was transferred to a real robot that could perform its function even in environments that had never been simulated. Furthermore, the author argued that the robustness of fuzzy controllers manages to absorb the uncertainties and inaccuracies generated by the environment. Furthermore, the technique of adaptive adjustment of the gains and the membership functions by means of an algorithm has considerably reduced the computational cost. It is also interesting to note how the tuning of a fuzzy controller (which is a slow and tedious process) can be performed agilely by means of an evolutionary algorithm.

Work presented by Zulfatman and Rahmat [27] proposed the design of a self-tuned Fuzzy-PID control to improve the performance of an electrohydraulic actuator. In this work, an algorithm was not used to tune the gains but the fuzzy itself was used to supply the gains. Hence, the authors attempted to demonstrate how the performance of the PID controller could be improved. Moreover, they used two inputs to the fuzzy controller: the error between the input and the response of the plant to the control signal (i.e., Error), and the change of error (i.e., derivate of the Error). The outputs obtained were the three gains of the PID (K_p, K_d and K_i). It is quite interesting to note that the authors succeeded in supplying the integral gain without an input of the integral error. Moreover, the authors claimed that this expert system performed well in all tests carried out.

Santos et al. [28] implemented an intelligent control system for a quadcopter. The proposed strategy was based on fuzzy logic and the authors affirmed that this provided flexibility and better efficiency. The model within the controller requires an adaptive tuning of the parameters, due to the independence of the variables, to achieve the satisfactory performance. The results obtained were promising as the system was able to control all the input variables while maintaining a balance among stability, speed of response and precision.

Sydney et al. [29] proposed a non-linear controller for a quadcopter to deal with highly variable environmental conditions such as wind gusts. In the dynamic model developed, the aerodynamic effects that would affect the vehicle under such conditions were taken into account (i.e., drag force, rotor blade flapping and induced thrust). A dynamic input/output feedback linearization controller was designed with a recursive Bayesian filter to eliminate the effects of wind. The control system was simulated with MATLAB Simulink. A 3D simulator was developed to visualize the results (i.e., the trajectories of the aircraft and wind) where a good level of performance was observed. The Bayesian recursive filter was able to accurately predict the parameters related to wind turbulence. Moreover, the proposed control strategy was applied to an autonomous ship landing application and a proximity flight application.

Fu et al. [30] introduced a fuzzy logic controller based on a novel cross-entropy optimization for a quadcopter, which enables the vehicle to avoid obstacles. An algorithm that could detect objects that arise in the scene was also designed (i.e., a system based on artificial vision). A simulation was carried out using MATLAB Simulink. A large number of actual flight tests were also carried out and the controller showed satisfactory results by precisely navigating to avoid the obstacles in the path. Furthermore, the cross-entropy optimization reduced the number of rules in the fuzzy system by 64%.

Gautam and Ha [31] proposed the modeling, control system design and simulation of a quadcopter. An intelligent auto tuning PID controller was proposed where the gains were achieved through the implementation of a fuzzy controller which was responsible for offering the most suitable gains to respond to the demands of the control system. The auto tuning fuzzy controller was achieved based on an extended Kalman filter algorithm, which is an optimal algorithm for managing computational resources that allows the addition of external parameters (such as odometry or scan). This consists of two stages: prediction and correction. An algorithm called Dijkstra was also applied in order to plan the shorter paths to avoid the obstacles in the way. The results were satisfactory in obtaining an autotuned intelligent PID controller, which can adapt to changing situations.

Fatan et al. [32] reported an altitude control system for a quadcopter. The system was based on an adaptive neural PID which was initially tuned with an evolutionary algorithm and then a GA was implemented to tune the gains. The authors emphasized that the main advantage of this method was the ability to adapt to new variations that arise during execution. Moreover, they mentioned that the learning rate presented a challenge, as the learning speed decreased the mobility of the coefficients, which in turn reduced the adaptive capacity of the controller. In this study, this issue was considered when designing the GA, which was a part of the controller.

Benavidez et al. [33] developed two fuzzy logic control systems for a Parrot AR Drone 2.0, with the aim of achieving landing and altitude control. The altitude fuzzy controller worked directly with a sonar input and had five rules to control the vehicle relative to the "z" axis. Nineteen rules were defined for the landing controller to establish a logical linear behavior relative to the three axes of rotation. The simulation results showed satisfactory control performance in moderately adverse environmental conditions.

Larrazabal and Peñas [34] proposed a control mechanism for the rudder of a ship. According to the authors, such a system shows a strong non-linearity and usually operates in an unstable environment. They implemented two control systems: a fuzzy logic controller, which was responsible for solving the problem of non-linearity, and a PID controller with self-tuning gains adjusted by GAs which make the system adaptable and computationally efficient (i.e., a low computational cost). This work demonstrated that the combination of traditional and intelligent techniques is suitable to manage complex problems.

Garcia-Aunon et al. [35] presented a route tracking algorithm for a UAV. The proposed algorithm has the ability of adapting its parameters using a fuzzy logic based approach. In fact, the models generated with tuned parameters performed better than the models with constant parameters as the tuned models were able to adapt to the variations of speed and trajectories quite well. On the other hand, the authors proposed different laws of tuning by studying the equations of movement which were based on the theoretical model and the appropriate simulations. Although comparable results were achieved with the kinematic law and the diffuse law, it was stated that the diffuse law was faster and easier to develop. Moreover, this work highlights the importance of parameter tuning (rather than using constant values) as this enables achieving better performance even if the system becomes a little more complex.

Domingos et al. [36] proposed a solution to the problem of tuning fuzzy controllers. The authors stated that a quadcopter is a non-linear and complex system, and these vehicles face non-stationary environments where the perturbations occur randomly. It makes the design and adjustment of fuzzy controllers a task that requires a significant effort. Therefore, the authors justified the inefficiency of classic fuzzy controllers and the need for the implementation of autonomous fuzzy controllers. The design of the system was carried out using a self-evolving parameter-free fuzzy rule-based controller (SPARC), which is a real-time adaptive controller. The advantage of this controller lies in its configuration. According to the authors, this controller does not need rules or other parameters so only the supply of the input and output ranges is sufficient. Furthermore, this controller needed to be trained to reduce the values of error and overshoot, thus obtaining data clouds to be used as control signals would help to follow the entry references. The performance of the controllers was evaluated through the application of the mean square error (MSE) technique in each of the plants. The authors showed that the MSE returned the square of the difference between the output of the plant and the reference signal at each instant, and then, the result was divided by the total simulation time. Therefore, it was shown that a lower result corresponded to better performance. In the table of results shown by the authors, it was appreciated that the results obtained are slimier for the SPARC controller. The simulation results showed a clear advantage from the point of view of design and efficiency of the autonomous fuzzy controllers (SPARC) over the classic fuzzy controllers.

Yazid et al. [37] proposed a self-tuning controller for trajectory tracking control of a quadcopter. They presented a first-order Takagi-Sugeno-Kang fuzzy logic controller (FLC) tuned using an evolutionary algorithm. The authors compared the performance of the FLC optimized with three major evolutionary algorithms: GA, particle swarm optimization and artificial bee colony (ABC) algorithm. The performance of the FLC was evaluated under three different flight conditions: constant step, varying step and sinusoidal functions. The results indicated that the ABC-FLC outperformed the other two in terms of faster settling time in the absence of overshoots. The GA-FLC had faster rise time and consequently larger overshoots, which is undesirable. The proposed method was effective in optimizing the fuzzy parameters to obtain good performance and robustness, without having to go through the tedious manual tuning process.

Pham et al. [38] introduced a control strategy for tracking the trajectory of a quadcopter, the mass of which changes with time. Continuous reduction of the mass as well as a sudden change in the mass during the flight were considered in the study. The authors proposed a cascaded structure, with two linear parameter varying H_∞ controllers in the inner loop to control the attitude and a combination of backstepping and PD controllers in the outer loop for altitude and longitudinal control. The simulation results showed that the quadcopter was able to track the trajectory under both types of mass variation, even when external disturbances were present.

Ali et al. [39] designed a feedback linearization based non-linear controller for a quadcopter, cascaded with sliding mode and backstepping based control, which can deal with uncertain external disturbances. The authors argued that most of the existing controllers reported in the literature do not account for the Coriolis effect and that the models are simplified using small signal approximations, which make the controllers less robust even against small disturbances. As a solution to this, the authors derived a non-linear model for the quadcopter, with fewer assumptions compared to existing models in the literature. Based on the simulation results, the authors demonstrated that the proposed controller showed slightly better tracking performance than a conventional PID controller for small step signals, while for larger step signals, it clearly outperformed the PID controller. Furthermore, it was found that larger step signals caused the PID controller to become unstable, while the proposed controller showed excellent performance. A similar behavior was observed for sinusoidal inputs, as well. Moreover, it was clear that both the advanced control strategies used for dealing with uncertainties (i.e., sliding mode and backstepping based control) had better transient performance than the PID controller, while the sliding mode control surpassed the backstepping based control in terms of steady state performance.

Guerrero-Sánchez et al. [40] developed and compared two control strategies (i.e., a PD control law with non-linear coupled term and a non-linear control law), with a simple structure and low computational complexity, to stabilize a quadcopter designed to transport a payload attached by a cable. In order to stabilize the attitude dynamics, a state-dependent differential Riccati equation control was also developed. For the numerical experiments, the authors considered a commercial Parrot AR Drone attached with a hanging payload. Based on the numerical simulations, the authors validated the suitability of the proposed controllers for a quadcopter carrying a payload. Moreover, the simulation results indicated that the non-linear control law was superior in terms of payload oscillation angle suppression, lower settling time and lower maximum swing.

It can be seen that in the previous studies, dynamic models were formulated by taking two reference systems into account (i.e., fixed and mobile). Rotational 3D matrices were applied to represent the orientation of the mobile system with respect to the fixed earth system. With these matrices, it is possible to represent the mobile system vectors using the vectors defined in the fixed system. The first step in constructing a dynamic model is the formulation of a kinematic model. This means that it is necessary to establish the reference systems and then find the position vectors, velocity and acceleration. It is obvious that the most common frame configuration used in the past studies is the "plus" (+) configuration. The quadcopter's frame configuration (i.e., +, H, or ×) determines the capacity of performing complicated maneuvers. Each configuration presents a different motor organization having one or two motors in propulsion of the vehicle to make the vehicle more powerful and capable, but it also compromises the control system. In the present study, the "×" configuration was implemented. The aim was to obtain a configuration that allows a greater pushing force to the vehicle. This is possible as the "×" configuration allows the use of two motors for longitudinal and lateral movements.

Based on the review of the existing literature, it is possible to see a great diversity in the control systems used for quadcopters such as traditional PIDs, LQR controllers and controllers based on fuzzy logic. The non-linear controllers have generally outperformed linear controllers due to their ability to conform to varying conditions. Autotuning of controller parameters is necessary to ensure the reliable performance of the controller under varying conditions. Evolutionary algorithms have widely been used in this context.

This work aims to design an advanced control system for a quadcopter which can handle the possible non-linear behaviors in the worst possible conditions. The dynamic model for the controller is developed using the "×" configuration due to the benefits discussed above, as opposed to the "+" configuration commonly used in the past research works. The study investigates the limitations and potentials of traditional PID and fuzzy based controllers and investigates the potential of enhancing their performance by autotuning their parameters using a GA. Therefore, five different control systems that include a traditional PID controller, two Fuzzy-PD controllers and two controllers autotuned using a GA are designed based on the developed dynamic model, and their performance is compared.

The rest of the paper is structured as follows. In Section 2, a detailed description of the dynamic modelling, the design of the controllers, the implementation of the GA and controller tuning is presented. Section 3 illustrates the details of the 3D simulator development. The simulation results are discussed in Section 4. Finally, the conclusions drawn from the results of this study are presented with future recommendations in Section 5.

2. Control System Design

This section presents the procedure followed in dynamic modelling, the design of the controllers, the implementation of the GA and the procedure used in achieving an optimum tuning.

2.1. Dynamic Modeling

A quadcopter is a vehicle that can freely move in the space, as it is a system with six degrees of freedom. In order to develop a model for a quadcopter, two systems of reference should be considered, where one is the earth and the other is fixed to the body of the quadcopter, as explained by Equations (1) and (2), respectively.

$$q_e = (\varepsilon_e, \eta_e) = (x, y, z, \phi, \sigma, \psi) \in R^6 \tag{1}$$

$$q_b = (\varepsilon_b, \eta_b) = (u, v, w, p, q, r) \in R^6 \tag{2}$$

where $\varepsilon_e = x, y, z$: Position vector in system 1;
$\eta_e = \phi, \sigma, \psi$: Angular position vector in system 1;
$\varepsilon_b = u, v, w$: Position vector in system 2;
$\eta_b = p, q, r$: Angular position vector in system 2.

Initially, both systems of reference stay together, but when the quadcopter starts its movement, the second system is displaced from the first system. This change in coordinates involves a set of changes in the components (i.e., relevant parameters) of the second system. This means that if the quadcopter is able to move (depending on the type of movement), it is necessary to modify the components u, v and w for the linear displacement and p, q and r for the angular displacement. When the coordinates on the second system are modified, a new set of coordinates can be obtained, describing the position of the second system with respect to the first system. Figure 1 represents a linear displacement of w and angular displacements of u and v vectors as a result of the changes in p and q values. This motion causes the second system fixed on the quadcopter's body to move to U', V' and W'. Hence, it is possible to determine the position of the vehicle with respect to the system which is fixed to the earth (i.e., system 1).

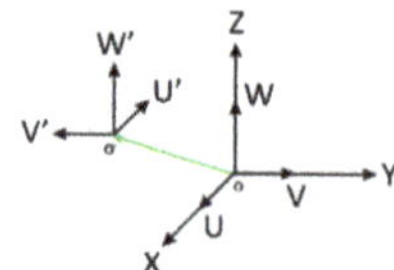

Figure 1. System "ouvw" rotated and translated with respect to "oxyz".

In both systems of reference, all the variables affecting the kinematics of the vehicle (i.e., velocity and acceleration) should be described, and the 3D rotation matrices were used for this purpose. These rotation matrices describe the counter clockwise rotations of the system around the x, y and z axes. The 3D matrix has been used with an interest of knowing the variation of the coordinates and not the variation of the vectors. In general, a solid can rotate in 3D space which can be described by three Euler angles: PhI (φ), Theta (θ) and Psi (ϕ). Each of these angles describes the rotation with respect to a coordinate axis: Phi—spinning with respect to "x", Theta and Psi—rotations around "y" and "z", respectively. A composition of rotations, considering that the product of the matrices is not commutative, is required for the representation of the orientation as described in Equation (3).

$$R = \begin{pmatrix} c\phi c\theta & s\phi c\phi + c\psi s\theta s\phi & s\psi s\phi + c\psi sc\phi \\ s\phi c\theta & c\psi c\theta + s\psi s\theta s\phi & -c\psi s\phi + s\psi s\theta c\phi \\ -s\theta & c\theta s\phi & c\theta c\phi \end{pmatrix} \tag{3}$$

where s and c symbolize the sine and cosine of the angle, respectively.

With rotation matrices, it is possible to project the speeds and linear accelerations from the secondary system to the primary system and vice versa if necessary. The use of angular velocities projection matrix given in Equation (4) is necessary for projecting the magnitudes of the angles [41].

$$S = \begin{pmatrix} 1 & sin\phi \times tan\theta & cos\phi \times tan\theta \\ 0 & cos\phi & -sin\phi \\ 0 & sin\phi/cos\theta & cos\phi/cos\theta \end{pmatrix} \tag{4}$$

Physical theories describe the way to predict the behavior of a moving rigid solid and these can be used to model the behavior of a quadcopter, as well. Generally, the movement of a rigid body can be complex, but the problem can be simplified by isolating different movements. A quadcopter is a machine that can follow very convoluted trajectories where it can turn around itself while making linear displacements and angular movements. It is possible to compare such a motion with a solid which follows a parabolic trajectory. Newton's second law describes the movement of the center of mass of a rigid body translation. If the center of mass is fixed, preventing its translation, it is possible to appreciate that the solid is only rotating. Then, the motion of a solid can be analyzed as the composition of the translational movement of its center of mass with respect to a system of reference and the rotation about an axis that passes through the center of mass. The Newton–Euler formulation describes the dynamics of this process and has been commonly used in this type of dynamic models [11,12,19,20].

To formulate a mathematical model of a quadcopter, it is essential to account for all the aspects related to the forces and torques acting on the vehicle. The model can be developed in two parts as dynamics of linear and angular movements. The main aim is to find the dynamic model tied to the primary system of reference, because it is the platform to control the quadcopter. However, initially, it is required to start from the mobile model as it is the system that experiences all the forces and torques.

2.1.1. Dynamics of Linear Movements

The dynamics of linear movements represent all the forces acting upon the quadcopter (i.e., the lifting forces provided by the engines and the gravitational force). The main idea is to simplify the problem allocating the forces to their respective system of coordinates. Hence, the lifting forces were assigned to the secondary system while the gravitational force was allocated to the primary system. These forces were allocated in this manner, as the quadcopter would be controlled from the primary system. The gravitational force is a vector pointing downwards (regardless of the vehicle movement), which is already known, and hence, it was allocated to the primary system. Furthermore, the lifting forces also act on the vehicle's body and hence were assigned to the secondary system. After allocating the forces, the expressions for the primary system were obtained by using the rotational 3D matrix and then the forces acting on both systems were determined. The gravitational and supporting forces on the secondary system after projecting on the primary system are given by Equations (5) and (6).

$$G_e = \begin{pmatrix} 0 & 0 & W_{gh} \end{pmatrix}' \tag{5}$$

where G_e: Gravitational force matrix on the primary system;
W_{gh}: Weight force vector on the primary system.

$$L_e = \begin{pmatrix} L_T \times (sin\phi sin\psi + cos\phi cos\psi sin\theta) \\ -L_T \times (cos\psi sin\phi - cos\phi sin\psi sin\theta) \\ L_T \times (cos\phi cos\theta) \end{pmatrix} \tag{6}$$

where L_e: Lift force matrix on the primary system;
L_T: Total lift force module on the primary system.

After analyzing the forces, the linear dynamic model on the earth system can be presented as given in Equations (7) and (8):

$$\sum F_e = m \times a \tag{7}$$

where F_e is the total force on the primary system (earth), m is the total mass and a is the acceleration due to the forces acting on the primary system.

$$\sum F_e = \begin{pmatrix} L_T \times (sin\phi sin\psi + cos\phi cos\psi sin\theta) \\ -L_T \times (cos\psi sin\phi - cos\phi sin\psi sin\theta) \\ L_T \times (cos\phi cos\theta) - g \times m \end{pmatrix} \tag{8}$$

Then, the expressions (9)–(11) were obtained to describe the accelerations in the primary reference system:

$$\ddot{x} = \frac{L_T \times (sin\phi sin\psi + cos\phi cos\psi sin\theta)}{m} \tag{9}$$

$$\ddot{y} = \frac{-L_T \times (cos\psi s\phi - cos\phi sin\psi sin\theta)}{m} \tag{10}$$

$$\ddot{z} = \frac{L_T \times (cos\psi cos\theta - g \times m)}{m} \tag{11}$$

where $\ddot{x}, \ddot{y}, \ddot{z}$ are the linear accelerations on the primary system.

2.1.2. Dynamics of Angular Movements

To obtain the dynamics of angular movements, the Euler formulation was used as presented in Equations (12) and (13):

$$\sum \tau_{extb} = I \times \ddot{\eta}_b + \dot{\eta}_b \times (I \times \dot{\eta}_b) \tag{12}$$

where $\tau_{extb} = \begin{pmatrix} \tau_\phi & \tau_\sigma & \tau_\psi \end{pmatrix}'$ is the external torque experienced by the secondary system, I is the inertia matrix, $\dot{\eta}_b$ is the angular velocities vector on the secondary system and $\ddot{\eta}_b$ is the angular accelerations vector on the secondary system.

Usually, a quadcopter is an aircraft with four engines. A pair of motors rotate in one direction while the other pair rotates in the opposite direction. Due to the "×" configuration used in this work, two rotors are used to generate the required torque on each axis, and these torques can be described by Equation (13):

$$\begin{pmatrix} \tau_\phi \\ \tau_\sigma \\ \tau_\psi \end{pmatrix} = \begin{pmatrix} \Delta L_{21,43} \times l \\ \Delta L_{42,13} \times l \\ \Delta D_{14,23} \times l \end{pmatrix} \tag{13}$$

where ΔL and ΔD are the increase in the lift and drag forces due to the torque increment of the two pairs of motors, respectively, and l is the arm length.

If the motor speeds are exactly the same, the sum of these velocities would be zero and then it can be said that the quadcopter is stable. Conversely, if the speeds of the motors are not the same, a torque due to the gyroscopic effect occurs [41], and this torque is described by Equation (14):

$$Gy_b = I_{TP} \times \ddot{\eta}_b \times \begin{pmatrix} -q & p & 0 \end{pmatrix}' \tag{14}$$

Then, the angular accelerations in the secondary reference system were obtained and are given by Equations (15)–(17):

$$\dot{p} = \frac{I_{yy} \times \dot{q} \times \dot{r} - I_{zz} \times \dot{q} \times \dot{r} - I_{TP} \times \dot{q} \times \Omega^2 + \tau_\phi}{I_{xx}} \tag{15}$$

$$\dot{q} = \frac{I_{TP} \times \dot{p} \times \Omega^2 + \tau_\sigma - I_{xx} \times \dot{p} \times \dot{r} + I_{zz} \times \dot{p} \times \dot{r}}{I_{yy}} \tag{16}$$

$$\dot{r} = \frac{\tau_\psi + I_{xx} \times \dot{p} \times \dot{q} - I_{yy} \times \dot{p} \times \dot{q}}{I_{zz}} \tag{17}$$

where $\ddot{p}, \ddot{q}$ and $\ddot{r}$ are the angular accelerations on the secondary system, I_{TP} is the total inertia of the propellers and Ω is the rotor's angular velocity.

By making the projection using the S matrix, the angular accelerations in the primary system can be obtained:

$$\ddot{\eta}_e = S \times \ddot{\eta}_b \tag{18}$$

where $\ddot{\eta}_e$ and $\ddot{\eta}_b$ are the angular acceleration vectors on the primary and secondary systems, respectively, and S is the angular acceleration matrix.

The expressions describing the angular accelerations in the primary reference system are given by Equations (19)–(21):

$$\ddot{\phi} = \frac{I_{TP} \times \dot{q} \times \Omega^2}{I_{xx}} + \frac{\tau_\phi}{I_{xx}} + \frac{(I_{yy} - I_{zz}) \times \dot{q} \times \dot{r}}{I_{xx}} + \frac{sin\phi \times tan\theta \times (-I_{TP} \times \dot{p} \times \Omega^2 + \tau_\theta + (I_{zz} - I_{xx}) \times \dot{p} \times \dot{r})}{I_{yy}} \tag{19}$$

$$\ddot{\theta} = \frac{cos\phi \times \left(-I_{TP} \times \dot{p} \times \Omega^2 + \tau_\theta + \left(I_{zz} - I_{yy}\right) \times \dot{p} \times \dot{r}\right)}{I_{yy}} + \frac{sin\phi \times \tau_\psi}{I_{zz}} \tag{20}$$

$$\ddot{\psi} = \frac{sin\phi \times \left(-I_{TP} \times \dot{p} \times \Omega^2 + \tau_\theta + \left(I_{zz} - I_{yy}\right) \times \dot{p} \times \dot{r}\right)}{I_{yy} \times cos\phi} - \frac{cos\phi \times \tau_\psi}{I_{zz} \times cos\phi} \tag{21}$$

where $\ddot{\phi}, \ddot{\theta}$ and $\ddot{\psi}$ are the angular accelerations on the primary system.

Once the mathematical model of the quadcopter was obtained, it was necessary to determine the specifications (i.e., the basic mass and size data) and parameters (i.e., the moment of inertia and other coefficients contained in the equations of the model). As per the previous studies reported in the literature, a quadcopter can be designed with different configurations: "+", "H" or "×". The "+" configuration presents a greater ease of calculating the moment of inertia of the engines which are located on the coordinate axes, but the movements are generated by a single motor. However, in this study, the "×" configuration was chosen to leverage the power of thrust to be offered by a couple of motors. It was assumed that this would allow the aerobatic maneuvers to be performed more easily. For calculating the moment of inertia, the components of the vehicle (i.e., motors, blades, etc.) were assumed as cylinders. Due to the "×" configuration (as shown in Figure 2a), no component coincides with the main axis; thus, in comparison with the "+" configuration, the calculations are relatively more complex.

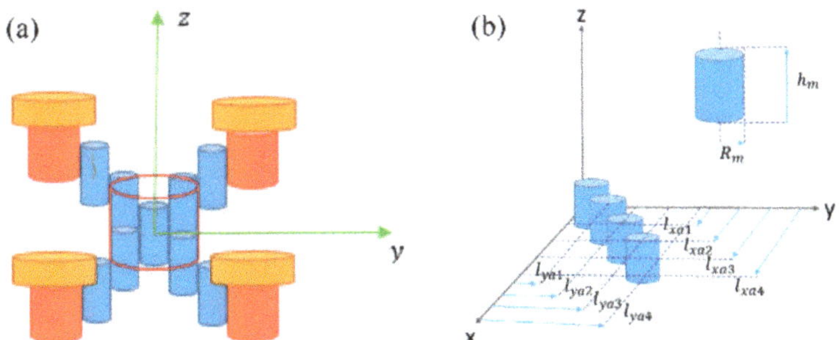

Figure 2. Inertial model of the quadcopter: (**a**) full model, (**b**) arms.

Of the components, the chassis presented difficulties in calculating the inertia. With the "+" configuration, it is possible to assume that the frame is composed of two cross cylinders that coincide with the main axis, whereas with the "×" configuration, none of these cylinders are coincided. Hence, to obtain the moment of inertia of the chassis, it was assumed that the arms are composed of multiple concatenated cylinders as shown in Figure 2. This was the simplest solution, because of its resemblance to the calculations used to find the inertia of the engines.

2.2. Design of the Controllers

Once the equations describing the accelerations of the system were obtained (i.e, Equations (15)–(17) and (19)–(21)), the Laplace transformation was applied to obtain the system in the frequency domain where the controllers will be implemented. The Equations (22)–(27) describe the six degrees of freedom of the quadcopter:

$$x(s) = \frac{L_T \times (sin\phi sin\psi + cos\phi cos\psi sin\theta)}{m \times s^2} \tag{22}$$

$$y(s) = \frac{-L_T \times (cos\psi sin\phi - cos\phi sin\psi sin\theta)}{m \times s^2} \tag{23}$$

$$z(s) = \frac{L_T \times (cos\psi cos\theta - g \times m)}{m \times s^2} \tag{24}$$

$$\phi(s) = \frac{I_{TP} \times q \times \Omega^2 + \tau_\phi + (I_{yy} - I_{zz}) \times q \times r}{I_{xx} \times s^2} + \frac{sin\phi \times tan\theta \times (-I_{TP} \times p \times \Omega^2 + \tau_\theta + (I_{zz} - I_{xx}) \times p \times r)}{I_{yy} \times s^2} \tag{25}$$

$$\theta(s) = \frac{cos\phi \times (-I_{TP} \times \dot{p} \times \Omega^2 + \tau_\theta + (I_{zz} - I_{yy}) \times \dot{p} \times r)}{I_{yy} \times s^2} + \frac{sin\phi \times \tau_\psi}{I_{zz} \times s^2} \tag{26}$$

$$\psi(s) = \frac{sin\phi \times (-I_{TP} \times p \times \Omega^2 + \tau_\theta + (I_{zz} - I_{yy}) \times p \times r)}{I_{yy} \times cos\phi \times s^2} - \frac{cos\phi \times \tau_\psi}{I_{zz} \times cos\phi \times s^2} \tag{27}$$

For the design of the control system, the non-linear dynamic system was reduced to a linear system by assuming that the quadcopter works with very small angles [42] (i.e., approximating the sine values to zero and the cosine values to one). Furthermore, since the angles are assumed to be very small, the angular speeds experienced by the quadcopter's body would also be approximated to be equal to zero. With these assumptions, a simplified model of the quadcopter was obtained and given by the Equations (28)–(33).

$$x(s) = 0 \tag{28}$$

$$y(s) = 0 \tag{29}$$

$$z(s) = \frac{1}{s^2} \times \left(\frac{L_T}{m} - g \right) \tag{30}$$

$$\phi(s) = \frac{\tau_\phi}{s^2 \times I_{xx}} \tag{31}$$

$$\theta(s) = \frac{\tau_\theta}{s^2 \times I_{yy}} \tag{32}$$

$$\psi(s) = -\frac{\tau_\psi}{s^2 \times I_{zz}} \tag{33}$$

To verify the performance, both the non-linear and the simplified models were simulated with unit step inputs. The results showed very similar behaviors between both models, and this confirms the accuracy of the simplified linear model. Based on the obtained model, two types of controllers were designed: a PID controller and two Fuzzy-PD controllers (i.e., Mandami and Sugeno). One of the major aims of this study is to verify whether a PID controller is sufficient to perform the basic maneuvers of a quadcopter, and to determine when the implementation of a non-linear controller is required. Subsequently, a GA is employed to improve the controllers' capabilities and performance.

2.2.1. PID Controller

First, the simplified linear model was implemented with a PID controller, and it makes the corrective actions by comparing the plant's current input and output states. The PID's proportional component maintains the stability of the quadcopter and the integral component achieves the precision, while the derivative component controls the speed. The respective gains (i.e., K_p, K_i and K_d) must be tuned for better control performance, and initially, the tuner offered by MATLAB Simulink was used for this purpose (i.e., the tuning was performed experimentally). Figure 3 illustrates the parallel PID control scheme with a filter coefficient which was implemented in the general circuit.

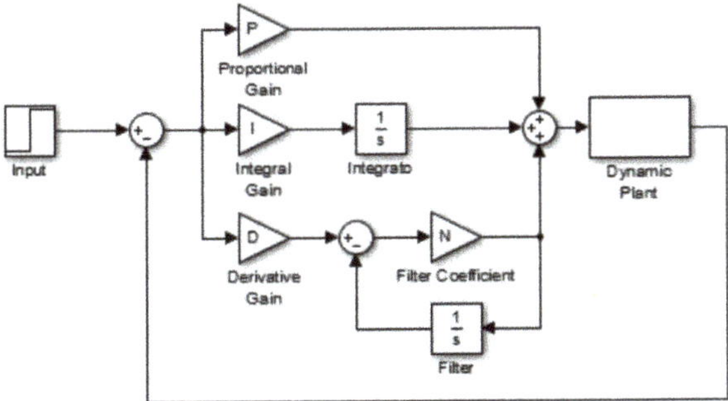

Figure 3. A schematic of the PID control scheme.

The tuning was carried out through the MATLAB optimization tool (see Figure 4). The optimization tool adjusts the inputs to achieve the desired output signal based on the pre-defined conditions. The simulation was carried out until it found the best possible combination of parameters that could match the output signal to the desired signal.

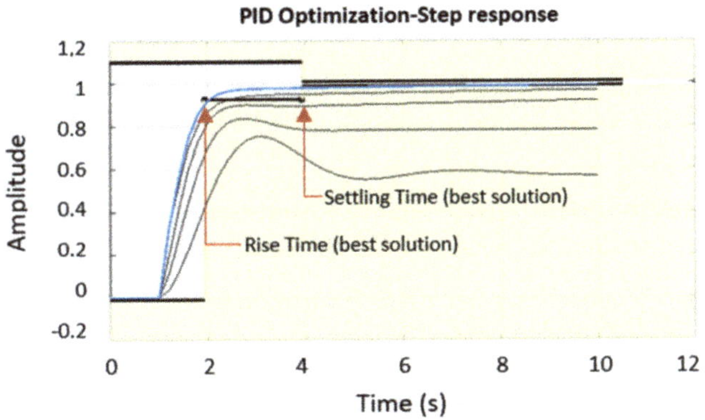

Figure 4. Tuning of the PID controller using the MATLAB optimization tool.

The PID controller responded accurately when the model was subjected to the trajectories with little variations. However, under stringent conditions such as sudden turns or sudden changes in altitude, these controllers began to lose the reference due to the non-linearities present in the model. Because of this limitation, the quadcopter was not able to make extreme turns or to perform acrobatic maneuvers. Each time a simulation was carried out, the model deviated from its linearity conditions, and hence, the PID controller required re-tuning of its gains. Therefore, the implementation of a supportive algorithm that provides the model with the necessary information on the changes in key parameters (or to update the gains of the controller) should be useful in achieving the desired stability.

2.2.2. Fuzzy-PD Controllers

Fuzzy controllers are characterized by non-linear systems. The values/information which govern the system are introduced by an expert as a series of rules or conditions and thus, the system can deal with information that is not clear (i.e., information with which the controller works will not be exact and/or numerical). The basis of rules introduced by the expert is of the antecedent-consequent type and the range covered by a membership function is labelled with a linguistic value that is usually a word or an adjective. Then, the linguistic variables are associated to a set of terms, defined in the same universe of discourse. The number of fuzzy sets determines the complexity of the controller, and these have a linguistic meaning such as "negative", "positive" or "zero". After defining the rule-base and the membership functions, the controller's operation includes three stages: fuzzification, inference and defuzzification. Fuzzification introduces input data, and they are processed to calculate the degree of membership that they will have on the controller. At the inference stage, the degree of membership in the input data is considered, and a decision is taken in the output space. Overall, the decisions are based on the scope of knowledge defined by the expert.

In this work, the fuzzy controller design was performed with the MATLAB Fuzzy Toolbox, which provides a simple and intuitive interface. Mamdani-type [43] and Sugeno-type [44] inference mechanisms were used to design these controllers. The design was carried out by analyzing the input and output signals of the previously designed PID controller discussed in Section 2.2.1. The behavior of the proportional, derivative and integral components of the PID controller was separately observed and the results were recorded for each time instant. The main idea was to understand the operation of the PID controller to set the initial platform of the fuzzy control.

Initially, the objective was to design a Fuzzy-PID controller, but due to certain difficulties, it was decided to design a Fuzzy-PD controller. The Fuzzy-PID required an extensive rule base (due to the number of inputs, the error, its derivative and integral), which would mean a higher computational cost. Finally, it was concluded that the problem with the Fuzzy-PID was due to the integral gain, and hence, it was decided to suppress this input. As a result, it was possible to simplify the rule base and in turn reduce the computational cost. If the Fuzzy-PD controller could work well with proportional and derivative gains, then the integral part could be added at a later stage to improve the performance. In this way, the Fuzzy-PD + I controller would be able to provide all the advantages of a parallel PID controller (the advantage offered by the integral part is given by its ability to consider past information). Once the Fuzzy-PD controller was designed, the integral action was added and it was observed that the error continued to be integrated indefinitely (i.e., the response had the "windup" effect). Once the problem was analyzed, two solutions were proposed: implementing an "anti-windup" or working only with the Fuzzy-PD controller. Finally, it was decided to eliminate the integral gain, and a Fuzzy-PD system as shown in Figure 5 was used.

Figure 5. Fuzzy–PD controller.

After selecting the controller arrangement, the two Fuzzy-PD controllers (i.e., Mamdani and Sugeno) were designed.

Mamdani Fuzzy-PD Controller

The first step of designing the Fuzzy-PD controller was to define the membership functions. The range of the controller and the linguistic labels were estimated accordingly. At first, several types of membership functions were created to explore the combination of labels that could offer the maximum possible accuracy between the inputs and outputs (i.e., the best possible control surface). It is known that a few rules mean a sharp surface (where changes occur abruptly) and more rules mean a smooth surface (where changes occur smoothly). The selected range of the controller and the linguistic labels are shown in Figure 6.

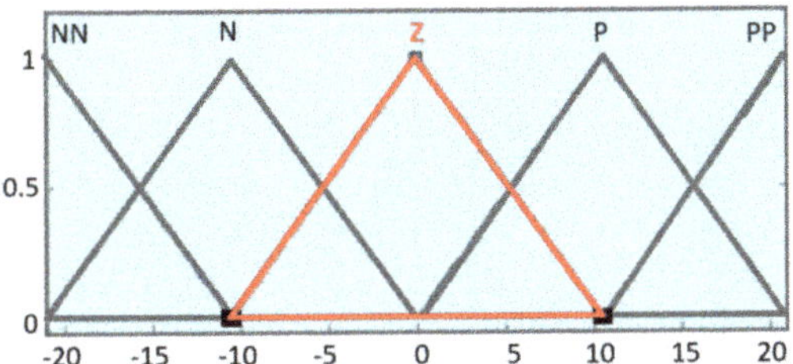

Figure 6. Output membership function for the Mamdani Fuzzy–PD controller.

The next step was to create a table of possible combinations of inputs and their potential outputs (i.e., the rule base). Moreover, this table must fulfil the control law of a PD controller as well which is given by a continuous-time transfer function presented in Equation (34):

$$U(s) = K_p + K_d \times s \tag{34}$$

This means that when $K_p = -K_d \times s$, the control signal should be $U(s) = 0$. With this logic, the identification of the outputs was made as illustrated in Figure 7a. Then, the rule base for the Fuzzy-PD was implemented using MATLAB. Nine rules were defined in total, and the corresponding control surface is shown in Figure 7b.

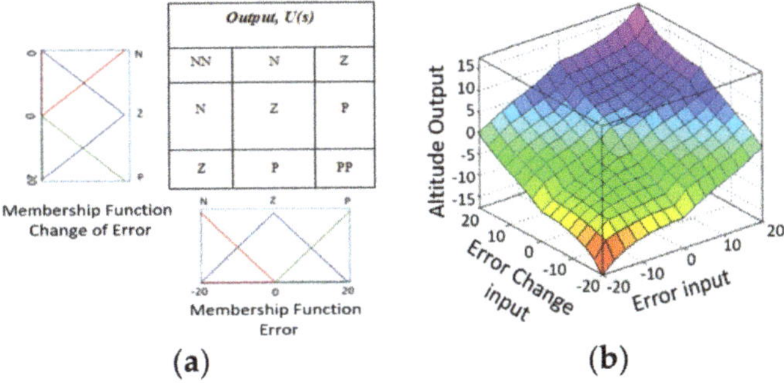

Figure 7. (**a**) Fuzzy outputs identification, (**b**) Mamdani Fuzzy–PD control surface.

Sugeno Fuzzy-PD Controller

The second Fuzzy-PD was designed by using the Sugeno methodology [44]. In the design, it was necessary to describe two membership functions for each of the inputs, as well as an output membership function. The membership functions of the inputs contain the error (i.e., the error between the input and the response of the plant to the control signal) and its change (i.e., the derivate of the error). In the design of the membership functions, sigmoidal functions have been used due to their principal characteristics and smoothness. This means that changes in the membership functions of a certain set do not happen drastically. Figure 8 shows that the climb segment for a membership function is a second order curve which changes from concavity at a given point, and once it reaches the maximum value (i.e., 1 or −1), it remains at that value. The ranges covered were from −1 to 1 for the inputs and from 0 to 1 for the output (see Figure 8). After describing the membership functions, the rule base was constructed. Due to the simplicity of the membership functions, the establishment of the rule base was simple, and it contained only four rules.

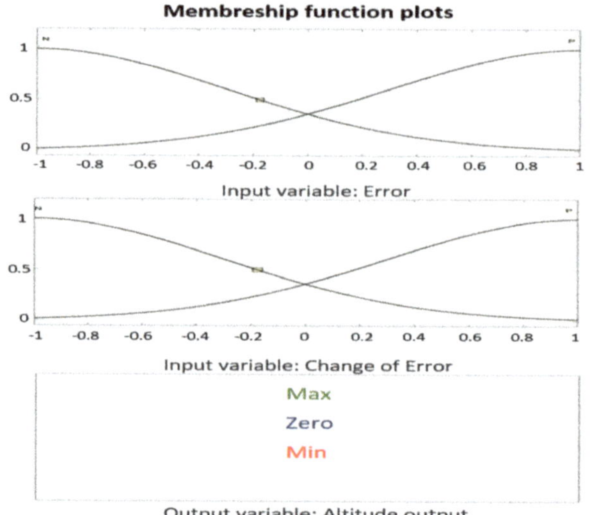

Figure 8. Sugeno membership functions.

Figure 9 shows the control surface obtained for the Fuzzy-PD controller with the Sugeno type fuzzy system. Moreover, Figure 9 shows that the surface is non-linear and offers a gradual evolution without abrupt changes. Achieving a smooth surface was the initial requirement considered.

Error	Connector	dError	Output
Neg	And	Neg	Min
Neg	And	Pos	Zero
Pos	And	Neg	Zero
Pos	And	Pos	Max

(a)

(b)

Figure 9. (**a**) Sugeno Fuzzy–PD rule base, (**b**) Sugeno Fuzzy–PD control surface.

Similar to the PID controller, the tuning of the gains for both controllers was performed through experimentation using the MATLAB optimization tool. The experimental tuning was realized by introducing a unit step input and executing the required actions. The gain allocated to the output of the Fuzzy-PD controller (K_f) is a scale factor which was the same gain used for the previously designed PID controller, as well. The tuning of the fuzzy controllers is an arduous task that consumes lots of time. However, the time here was substantially reduced thanks to the optimization tool in MATLAB.

Simulations were performed with both the Mamdani type and Sugeno type Fuzzy-PD controllers and excellent results were achieved compared to the PID controller. Moreover, it was observed that the Sugeno type Fuzzy-PD controller showed the worst performance in terms of tracking complicated trajectories (i.e., acrobatic maneuvers), as will be shown in Section 4. By re-tuning, it was possible to improve its performance, and it was concluded that the Sugeno Fuzzy-PD controller requires an adaptive adjustment of the gains to offer the most optimal response. On the other hand, the Mamdani Fuzzy-PD controller showed excellent performance in the majority of the tests (see Section 4).

2.3. Genetic Algorithm

After evaluating the performance of both Fuzzy-PD controllers, it was realized that the Sugeno type controller required continuous modifications of the gains for achieving better performance in terms of quadcopter stabilization. Hence, a GA was implemented as an adaptive mechanism to minimize the errors by providing the most appropriate gains in each possible situation.

GA originates from the proposition of using natural evolution as an optimization procedure, characterized by the basic operations of selection, crossing and mutation [45]. To evaluate the operations mentioned, it is necessary that the information to be optimized is encoded using the binary system (i.e., 0 and 1). In addition, this representation must be finite, each being an individual that makes up a population. Therefore, the algorithm is a search that uses such operations and begins by raising a family of individuals which are selected and then consider the most optimal to cross-do between them. However, to avoid falling into local minima, mutation is used. The form of selection can be made in a logical way using the criterion of choosing the most optimum individual. In addition to the operations cited above, the transition from one generation to another consists of a last element: replacement. Replacement is a procedure that is used to calculate a new generation of the above and its descendants. This is achieved by creating a space for offspring in the population by removing the parents from it.

GA Code Explanation

The elaboration of the GA is carried out following the methodology presented in a previous work [45]. First, the objective function is defined, which is the function of error. The error is a function of the reference signal, the plant and the expression of the controller, which in turn is a function of the gains. Then, the input parameters of the controller must be able to minimize the error.

Another important aspect is representation. It is necessary to encode the decision variables in a binary string. In the case of having three gains, sixteen bits must be used to represent a single variable (K_p, K_i or K_d). The assignment of the binary string to a real number is performed by the following expression presented in Equation (35):

$$Var = dec\_val \times \frac{up\_rang - low\_rang}{\left(2^{bit\_length} - 1\right) + low\_rang} \tag{35}$$

Thus, the total length of a chromosome is 48 bits: 16 for K_p, 16 for K_i and 16 for K_d.

The next step is to create an initial population. It is stipulated that in each generation, the size of the population is 20. The initial population is generated randomly, thus obtaining an initial population of 960 (i.e., 20 × 48 bits).

Once the initial population is generated, the first step is the evaluation, that is to calculate the fitness value of each member. This step has three sub-steps:

1. Conversion of the genotype of a chromosome to its phenotype by converting the binary chains into the corresponding real values.
2. Evaluation of the objective function by obtaining its value.
3. Conversion of the value of the objective function into an adjustment value where the fitness of each chromosome is equated to the objective function (i.e., evaluated for each chromosome) minus the maximum value obtained for the objective function. In this way, the chromosomes of a better fit are determined. In this work, these were the chromosomes with the lowest value.

Once the evaluation is complete, a new population is created based on the current generation. In this part, the reproduction, selection and crossing and mutation operators are used.

1. Reproduction: It can live and have offspring in the second population to the chromosomes that showed the best fit (i.e., the minimum value).

2. Selection and Crossing: The crossing is based on the calculation of the cumulative probability. This is carried out to decide which chromosomes will be selected for the crossover. This has three steps: calculating the total fit, calculating the probability of each chromosome and finally calculating the cumulative probability. The crossing method used was the Xover point, where a cut-off point is randomly selected and the parents in the right-hand side are exchanged to generate an offspring.

3. Mutation: The mutation is performed after the crossing by altering one or more genes with a probability equal to the mutation rate. In this case, the probability of mutation is 0.01. A population is generated as a function of the initial value (i.e., 20) and the number of bits (i.e., 48) and the last bit is changed from 1 to 0, and vice versa. The elite chromosomes of the above population are not subject to mutation, so after the mutation, they are restored. After an iteration of the GA, a new population is created. The procedure is repeated until there is no improvement in the best member of each generation after 20 iterations.

2.4. Autotuned Controllers

For tuning the parameters with the GA, only the PID and Sugeno Fuzzy-PD controllers were considered. The Mamdani Fuzzy-PD controller was left out, as it showed excellent performance without needing an adaptive adjustment of the gains (see Section 4). Initially, an attempt was made to cover the problem from a mathematical point of view. It was attempted to implement a GA considering the mathematical expression of the error. The idea was to make the GA deliver the gains through parallel computing with the Simulink model. This would be possible for the PID controller since it is relatively easier to obtain the function of the error depending on the parameters, but it would be very complex for the Fuzzy-PD controller. Furthermore, such a system would require the same computation twice, which entails a high computational cost. It was later verified that the expression of the error would not be necessary, because the algorithm can work directly with the error value obtained via the Simulink block diagram. In this way, it is possible to treat the error function like a black box.

2.4.1. Autotuned PID Using a GA

Later, it was observed that the PID controller, despite its good performance, presented some limitations when working under complex conditions. The quadcopter model presents several plants that are interconnected or interlaced, whose links create additional information or variations that were not initially seen and therefore not considered. This fact drove the PID controller to its limits by losing the reference in the most severe conditions. It was found that by modifying the gains for each of these situations separately, the PID controller could tackle the problem successfully. Therefore, a GA could help to achieve the required alterations to the parameters for each new situation.

Once the GA was designed, its implementation was achieved using a "MATLAB Function" block (see Figure 10). The new control system has four gains: K_p, K_i, K_d and K_o. The last parameter (K_o) is a scale factor located at the entrance of the plant. It was necessary to frequently modify K_o so the initial GA design was sufficient to meet the needs.

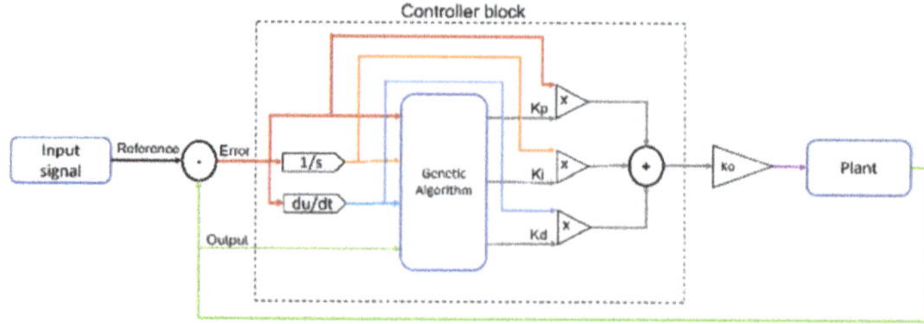

Figure 10. Autotuned GA–PID.

After the GA was implemented, the operational limits of the parameters were found, and the system was simulated showing promising results. The autotuned PID control system was able to pass the most demanding tests, and its level of performance is discussed in detail in Section 4.

2.4.2. Autotuned Sugeno Fuzzy-PD Using a GA

The fuzzy controller showed excellent performance and was able to stabilize the quadcopter in many adverse conditions. However, the designed controller presented static gains and hence, certain limitations were observed, which could have been avoided by the implementation of a GA. The designed autotuned GA-Sugeno Fuzzy-PD control system presents three gains: K_p, K_d and K_f. The parameter K_f corresponds to the output of the fuzzy controller and is responsible for scaling the signal to the plant. Following the steps of the autotuned PID design, an algorithm was implemented to control the K_p and K_d parameters, leaving K_f static. It was observed that the control system was not properly performing with a static K_f. Unlike the autotuned PID, it was necessary to have an adaptable gain for the output of the Fuzzy-PD controller (see Figure 11). This difference is due to the fact that the design of the Fuzzy-PD was not based on imitating the behavior of the traditional PID controller, and therefore, the controller has different requirements, even though the objective was the same. After designing the Fuzzy-PD system, the final step was to explore the parameter limits (explained in the next section). The results indicated that the autotuned GA-Sugeno Fuzzy-PD control system had excellent performance, which is discussed in detail in Section 4.

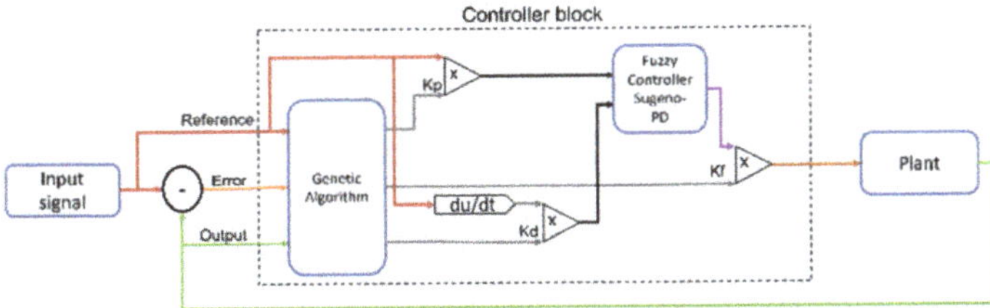

Figure 11. Autotuned GA–Sugeno Fuzzy–PD.

2.4.3. Parameter Work Limits

Once the GA was implemented for both controllers, the working limits of the parameters were obtained. Working limits ensure that the control system works accurately for system boundary conditions. It has been proven that despite the effort of the GA to minimize the error, the system will not be able to achieve stability if the working limits of the parameters are not properly adjusted. Finding the working limits of the parameters has been a task of pure experimentation and has taken a significant amount of time, since the system can be regulated in several ways. On the one hand, the GA-PID system did not present many difficulties in setting the limits. Initially, a step function was introduced to the system and the range of K_p was increased while increasing the static gain of the output (K_o). Once the controller started to respond and reached the reference signal, the range of K_d was increased and finally, K_p was increased for fine adjustment. It should be noted that the output gain determined the response speed. On the other hand, the autotuned GA-Sugeno Fuzzy-PD system presented many difficulties, as this system was made more complex by the possibility of modifying the limits of the membership functions to adjust the system. Initially, a Fuzzy-PID controller was designed that worked accurately in the complete model (showing the same limitations due to static gains). When implementing the GA and trying to find the working ranges of the parameters, many difficulties were observed that prevented the controller from performing its function. During each simulation, each parameter responded in a unique way by shifting the input signals away from the bounds of the membership functions of the controller. It was tried to adapt the functions that are associated to the input signals, but the controller showed a high sensitivity to the output, and some variation was reflected in the output, showing jumps and punctual losses of the reference. After analyzing the problem, it was decided that part of the faults could be due to the integral component, which had the "windup" effect. Initially, it was suggested to implement an "anti-windup", but finally, it was decided to design a Fuzzy-PD controller, as it would allow the elimination of the cited parameter (which was also presenting problems), resulting in a less complex system. The adjustment of the autotuned GA-Sugeno Fuzzy-PD ranks was less complex. Initially, it was proposed not to adjust the membership functions and only to configure the parameter ranges. The ranges for a unit step signal were set and the system responded accurately, but it could not respond to the maximum inputs (i.e., 360 degrees). This would prevent the quadcopter from making extreme turns and is therefore not acceptable. The adjustment was then made using the same parameters multiplied by other parameters until the system started responding to the 360-degree step input. Once this was achieved, the system began to accurately respond to all sorts of entries. It should be noted that the ranges of the parameters are very sensitive and even a minimal modification could make the system stop working accurately. However, we have found the optimal ranges that make the system a robust system.

3. 3D Simulator Development

A 3D simulator was developed to observe the behavior of the quadcopter in an intuitive way which would enable further analysis of the proposed controllers' performance and to explore the possible errors. Initially, the controllers were developed and the simulation results were used to fine-tune the controllers to achieve satisfactory performance. After achieving satisfactory results, the simulator was designed using MATLAB.

A dynamic model offers the possibility of obtaining movements performed in x, y and z planes. Therefore, it is possible to place the quadcopter in 3D space and to draw and examine the paths described by the vehicle. Here, defining of a point describing the path was possible, but it was impossible to see the rotation of the vehicle on its own axis while it was moving. The fundamental idea was then to consider the data that gave the dynamic model (i.e., linear and angular displacements) and provide these to the program written in MATLAB for plotting the points that draw the quadcopter's movement in real time. The main path described by the quadcopter was taken as the crucial point and this means that this point would be the main coordinate to follow and plot the vehicle's movement. This point coincided with the physical center of the model and it allowed the plotting of the vectors corresponding to the four arms that were traced with respect to the central point. However, the arms had an additional difficulty as they moved around the central point according to the maneuvers assigned. Hence, the coordinates of the arms would be four free points in the space and are related to each other like two cross vectors. Therefore, the only option was to use the Euler angles to know their orientation. To obtain the coordinates of the points that would indicate the orientation of the vehicle, arrays of homogeneous transformations were defined, and these provide the rotations and translations of the new points of orientation with respect to the central point.

Upon the completion of the design of the simulator, the corresponding characteristics were considered as necessary (i.e., displaying the path/s, the visualization of the speed of the engine in revolutions per minute (rpm)). To obtain the speed of the rotors, it was necessary to design a new circuit, and the sole purpose of this circuit was to separate the control signals to obtain the speeds of each engine. First, it was necessary to know that the inputs to the controllers are the speed changes between the pairs of rotors and then the controllers offer the corrected signals in their outputs. Then, the controllers' signals are introduced into the plants of the dynamic model in order to apply corrective actions. With this, it was possible to obtain each motor's speed in rpm. Table 1 shows the relevant information on the allocation of speed rates (only for a couple of engines in each plant), which were considered in the design of the circuit for the separation of signals.

Table 1. The information on control signal separation.

Inputs		Angular Velocity
Altitude	Σw_i	$\Omega_{mi} = \frac{\Sigma \Omega_i}{4}$
Roll	$\Delta w_{21,43}$	$w_1 = w_2 = \Omega_{mi} \pm \frac{\Delta \Omega_{21,43}}{2}$
Pitch	$\Delta w_{42,13}$	$w_1 = w_2 = \Omega_{mi} \pm \frac{\Delta \Omega_{42,13}}{2}$
Yaw	$\Delta w_{41,23}$	$w_2 = w_3 = \Omega_{mi} \pm \frac{\Delta \Omega_{41,23}}{2}$

4. Results and Discussion

This section discusses the results obtained during the simulations. Only the path results where the controllers have stopped working correctly will be shown, as all the controllers showed the same results for the paths that did not involve an acrobatic quadcopter maneuver.

The PID as well as both the Mamdani Fuzzy-PD and Sugeno Fuzzy-PD controllers were subjected to various tests to verify their limits of performance. The PID controller presented an excellent performance in terms of following easy trajectories, but significant differences were observed compared to the Fuzzy-PD controllers in terms of precision when describing complex trajectories. Figures 12 and 13 show the situation where the PID controller starts to fail. We can appreciate the controllers' response to the same step input with a pulse of 5^0 for a period of 0.5 s (i.e., a disturbance) that simulates a burst of air.

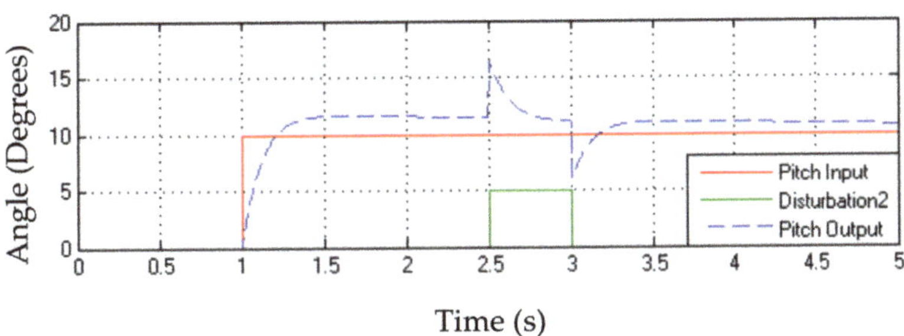

Figure 12. PID response to a step of 10° with a disturbance of 5°.

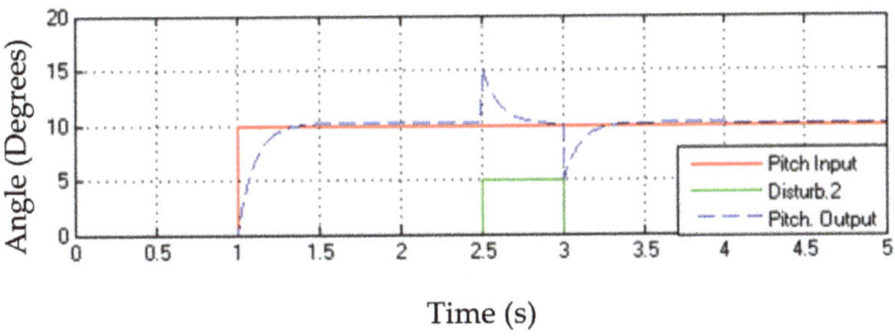

Figure 13. Fuzzy–PD response (the same for Mandami and Sugeno types) to a step of 10° with a disturbance of 5°.

It can be concluded that the PID controller could perform with the non-linear model, only with small angles and when no more than two inputs were introduced to the model. This is because when the system is working with simple paths, the dynamic model is not susceptible to many variations (i.e., the system becomes a little more static or constant over time and this condition was not unfavorable for the PID controller). On the other hand, it can be observed that the Fuzzy-PD controller proved to reach the reference with slightly less settling time than that of the PID controller. Such a difference in the settling time may be due to the adjustment of the gains. Tests with the non-linear model showed that the Fuzzy-PD controller is well adapted to all kinds of reference signals, but it showed difficulties when following acrobatic trajectories. Therefore, both the controllers required the implementation of a GA to improve their capabilities.

Figures 14–18 illustrate the simulation results of the five control systems designed, PID, Mandami Fuzzy-PD, Sugeno Fuzzy-PD, autotuned GA-PID and autotuned GA-Sugeno Fuzzy-PD, performing on a very complex track. From these figures, it is possible to appreciate the benefits of the implementation of a GA with both controllers. The details relevant to these figures are given in Table 2.

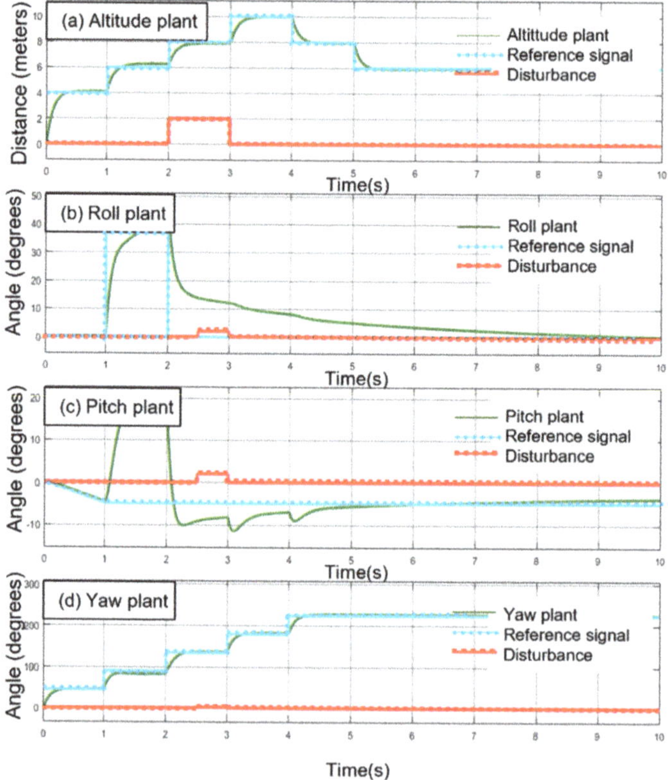

Figure 14. PID simulation results. (**a**) Altitude plant, (**b**) roll plant, (**c**) pitch plant, (**d**) yaw plant.

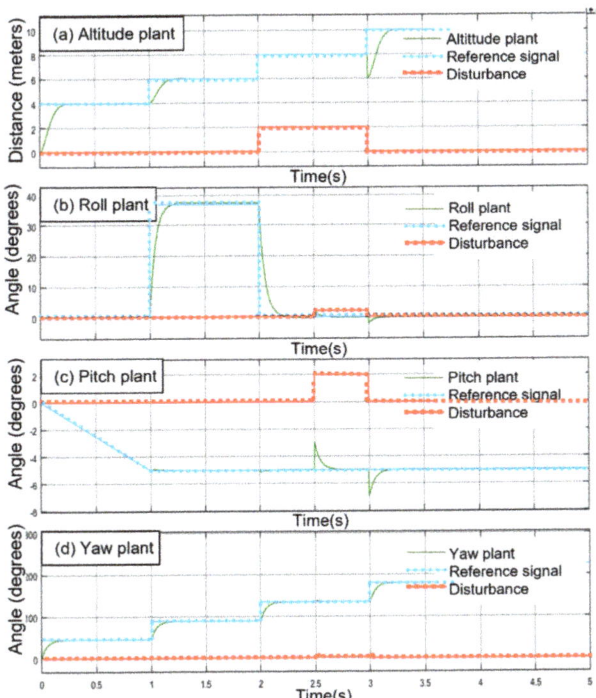

Figure 15. Mamdani Fuzzy–PD simulation results. (**a**) Altitude plant, (**b**) roll plant, (**c**) pitch plant, (**d**) yaw plant.

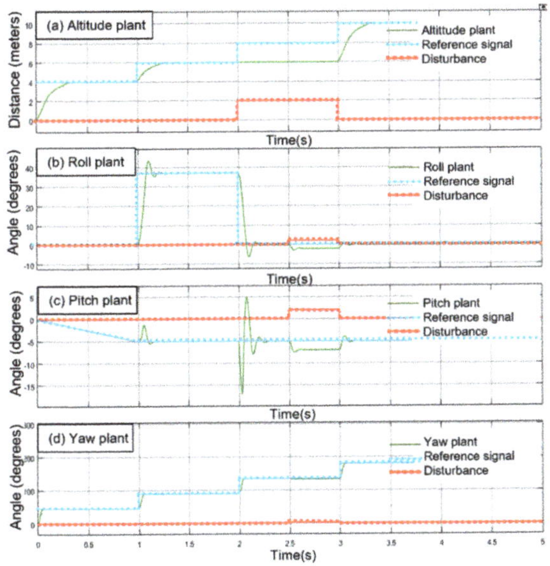

Figure 16. Sugeno Fuzzy–PD simulation results. (**a**) Altitude plant, (**b**) roll plant, (**c**) pitch plant, (**d**) yaw plant.

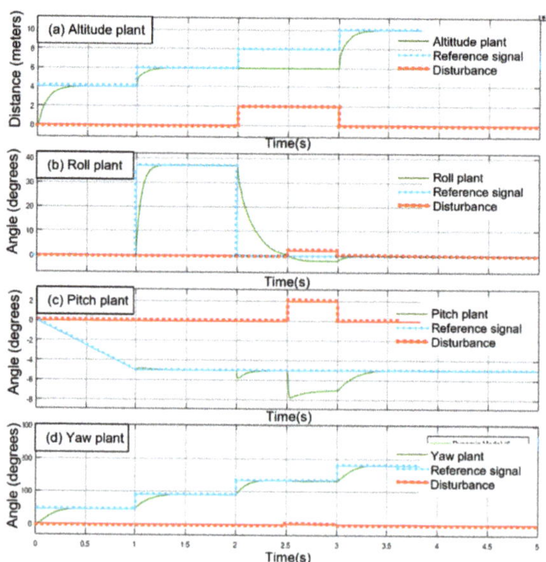

Figure 17. Autotuned GA–PID simulation results. (**a**) Altitude plant, (**b**) roll plant, (**c**) pitch plant, (**d**) yaw plant.

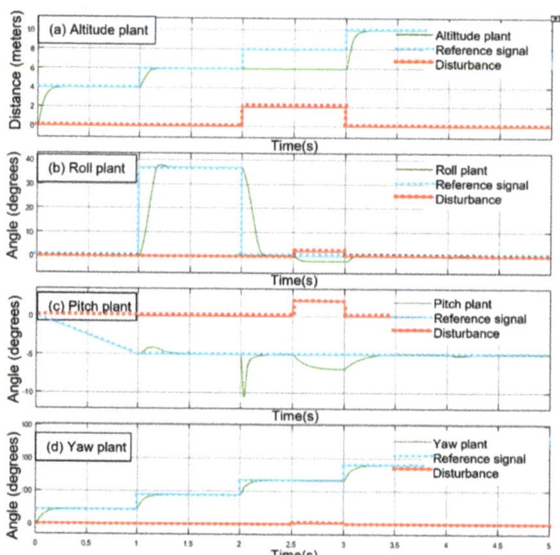

Figure 18. Autotuned GA–Sugeno Fuzzy–PD. (**a**) Altitude plant, (**b**) roll plant, (**c**) pitch plant, (**d**) yaw plant.

Table 2. Details of Fuzzy-PD simulation results (Figures 14–17).

Graph Label	Plant		Colors
(a)	Altitude	-	Green flat line: Plant response
(b)	Roll	-	Blue with circles: Reference signal (Trajectory)
(c)	Pitch	-	Red with squares: Disturbances
(d)	Yaw		

Figure 14 indicates the results of the four PID controllers. A satisfactory performance can be observed when the plants are subjected to conditions of minor variations. However, as the complexity of the trajectories increases, the controllers begin to fail by losing the references and then follow undesired trajectories. As shown in Figure 14b,c, the roll and pitch actions are conflicting, and this is because they are sharing variables and for this reason both can feel the modifications that exist between them. As a result, the stabilization of the roll and pitch plants (see Figure 14b,c) are achieved approximately at $t = 9$ s and $t = 6$ s, respectively. By means of experimentation, it has been verified that the PID controllers manage to stabilize the system in all cases, but these demand the variation of the gains in each simulation.

Figure 15 shows the simulation results achieved from the Mamdani Fuzzy-PD controllers. At first glance, a noticeable improvement can be observed when following the path in comparison with the PID controllers. This is due to the ability of the fuzzy controllers to work under non-linear conditions. These controllers show promising results without needing a GA. What is striking is the response to the disturbance at the altitude plant. As is shown in Figure 15a, it seems that the controller does not show any response when a disturbance occurs, and it is able to follow the reference with a high degree of accuracy. It should be stated that a good design of a fuzzy controller is more than enough to cover the control demands that the model of a quadcopter requires.

Figure 16 shows the results obtained for the Sugeno Fuzzy-PD controller. It is observed that the performance of this controller is better than that of the PID controller and manages to follow the reference well. However, it shows problems when dealing with acrobatic trajectories in the roll and pitch plants at $t = 2$ s (see Figure 16b,c). Hence, it was decided to reduce the speed of response of the controller, but after some tests, it could be seen that it did not respond to abrupt changes in an efficient manner. For this reason, this configuration gave large overshoots when sudden changes in the trajectory occurred. Therefore, as with the PID controller, the gains have been re-tuned, and the results have found to be better, and hence it was concluded that these controllers also required automatic tuning of the gains.

Figure 17 shows the results of the autotuned GA-PID controller. It can be observed that the algorithm works efficiently by providing the appropriate parameters to the PID controller. The controller was subjected to highly demanding trajectories and the results have been excellent in all tests. In the pitch graph shown in Figure 17c, there is still a small error at $t = 2$ s which was effectively corrected. Moreover, the controller of each plant was able to respond successfully to the disturbances.

Figure 18 shows the results of the autotuned GA-Sugeno Fuzzy-PD controller. The first point that stands out is the pitch plant error, as shown in Figure 18c at $t = 2$ s. In this controller the error is bigger than that shown in Figure 17c (autotuned GA-PID). This is because the controller's response is faster than the autotuned GA-PID and for this reason, there is an overshoot. The response curve of the GA-Sugeno Fuzzy-PD makes it faster than the GA-PID for large changes in the quadcopter's trajectory, but the PID is faster for small variations. This is clear from the details presented in Table 3 as the GA-PID is faster in most cases with the applied unit inputs (i.e., step, ramp and parabolic). On the other hand, it can be noticed that the GA-Sugeno Fuzzy-PD presents no overshoot. This could be due to the fact that the precision of the controller increases with the path changes with low demands.

Then, the precision of this controller would be compromised in the face of large changes in the plant where a higher response speed would be required, giving rise to the overshoot, as can be observed in Figure 18c at t = 2 s. The autotuned GA-Sugeno Fuzzy-PD controller was subjected to several tests with different input values, and all have been successful. It can be concluded that, owing to the implemented GA, the system is now able to adapt to the most adverse conditions, such as spins on the axis, loops or other acrobatic trajectories.

Table 3. Comparison of the steady-state and transient response of the controllers.

Controller	Plant	Static Response (Steady-State Error)			Transient Response			
		Step Input	Ramp Input	Parabolic Input	Rise Time (s)	Settling Time (s)	Peak Time (s)	Overshoot
PID	Altitude	-8.7×10^{-3}	4.8×10^{-3}	2.2×10^{-2}	0.207	0.430	0.501	0.000
	Roll	-5.9×10^{-3}	4.1×10^{-3}	2.1×10^{-2}	0.204	0.428	0.501	0.000
	Pitch	-1.3×10^{-2}	3.7×10^{-3}	2.1×10^{-2}	0.208	0.432	0.501	0.000
	Yaw	-7.3×10^{-3}	1.1×10^{-3}	2.1×10^{-2}	0.205	0.428	0.501	0.000
Mamdani Fuzzy-PD	Altitude	-7.6×10^{-6}	-7.6×10^{-6}	-1.9×10^{-4}	0.135	0.323	0.452	0.188
	Roll	-2.9×10^{-3}	-5.9×10^{-8}	4.1×10^{-8}	0.170	0.421	2.491	0.000
	Pitch	-4.4×10^{-9}	-9.5×10^{-9}	1.2×10^{-6}	0.171	0.422	2.487	0.000
	Yaw	-3.5×10^{-10}	1.3×10^{-9}	4.6×10^{-6}	0.099	0.281	1.802	0.000
Sugeno Fuzzy-PD	Altitude	1.9×10^{-3}	2.1×10^{-2}	2.5×10^{-2}	0.230	0.141	0.201	0.002
	Roll	1.9×10^{-6}	1.7×10^{-7}	2.1×10^{-5}	0.022	0.273	0.153	30.033
	Pitch	-9.6×10^{-7}	-1.7×10^{-7}	2.0×10^{-5}	0.021	0.273	0.153	30.860
	Yaw	3.5×10^{-10}	2.1×10^{-9}	6.1×10^{-7}	0.024	0.142	0.218	0.002
Autotuned GA-PID	Altitude	-2.4×10^{-3}	-2.6×10^{-7}	1.9×10^{-5}	0.206	0.458	1.573	0.048
	Roll	-1.7×10^{-3}	-7.5×10^{-3}	6.5×10^{-6}	0.103	0.282	0.629	0.042
	Pitch	-1.5×10^{-3}	-8.4×10^{-7}	6.2×10^{-6}	0.103	0.282	0.744	0.054
	Yaw	-1.7×10^{-3}	-8.0×10^{-5}	2.1×10^{-4}	0.197	0.392	0.505	0.713
Autotuned GA-Sugeno Fuzzy-PD	Altitude	2.1×10^{-6}	2.2×10^{-6}	5.5×10^{-5}	0.088	0.249	1.905	0.000
	Roll	-2.0×10^{-7}	4.9×10^{-7}	5.9×10^{-5}	0.105	0.277	4.504	0.000
	Pitch	-1.3×10^{-7}	-4.9×10^{-7}	5.7×10^{-5}	0.239	0.528	2.640	0.000
	Yaw	-5.9×10^{-11}	1.2×10^{-9}	3.4×10^{-6}	0.176	0.397	4.940	0.000

A quantitative comparison of the steady-state and transient response of the developed controllers is presented in Table 3. The results shown in Table 3 further confirm the results illustrated in Figures 14–18. It is clear that the autotuned GA-Sugeno Fuzzy-PD controller has the best overall performance in terms of the steady-state response to step, ramp and parabolic inputs. The Mamdani Fuzzy-PD controller shows excellent steady-state performance even without a GA to tune its parameters. Moreover, it can be seen that the performance of the traditional PID controller has been improved through the implementation of a GA. In terms of the transient performance, the Sugeno Fuzzy-PD controller shows the overall fastest performance and consequently has a high overshoot in the roll and pitch plants. This problem has been overcome in the autotuned GA-Sugeno Fuzzy-PD controller, and even though its response is a bit slower, it provides the best overall performance under both steady-state and transient conditions.

The quadcopter's behavior with the proposed control system (as illustrated by the 3D simulator) is presented in Figures 19 and 20. The speed of each motor is indicated in rpm. The screw trajectory shown in Figure 19 shows the quadcopter's response when an undefined turn is introduced into the yaw axis. It is possible to appreciate that the vehicle realizes a displacement of elevation while turning on its own axis. This is because initially, the quadcopter is in the rest position (i.e., in the ground at zero meters elevation) and when the simulation starts, the motors increase their speed until the vehicle takes off, in this case, up to a 1m height in the "Z" axis. This fact can be appreciated by observing the central trajectory described with red circles where the circles are initially separated as the vehicle is moving up. When it reaches its steady state, the vehicle stops and the red circles are accumulated in the same position. Thanks to this, it is possible to observe the rotation of the vehicle on its own axis, which describes a screw maneuver.

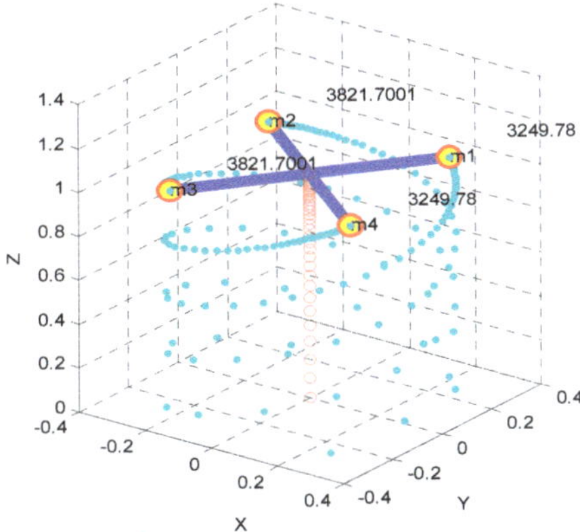

Figure 19. The quadcopter's operation along a spiral path.

Figure 20 illustrates the simulation of the quadcopter by making a movement along the "X" axis. Figure 20a shows the three-dimensional image of another trajectory executed by the vehicle, from its takeoff at the origin (X = 0, Y = 0, Z = 0) until the beginning of the movement in the negative direction of the "X" axis (i.e., the backward movement).

Figure 20b shows a top view where the rpm of each motor is observed in more detail. It can be seen that the displacement of the vehicle is made by the thrust of engines 1 and 3 (i.e., 5141.91 rpm), while engines 2 and 4 are responsible for keeping the vehicle at a constant height (i.e., 3712.47 rpm).

After several simulations with the simulator, it was verified that the responses of the dynamic model are compatible with the characteristics that were attributed to the quadcopter during its design stage. It means that a stationary state is obtained at about 3500 rpm and the thrust is realized using two motors (i.e., "×" configuration), while the twists and turns originate correctly.

It is evident from the results that the PID controllers were able to provide satisfactory performance when following simple paths, but failed when performing complex acrobatic maneuvers. Fuzzy-PD controllers outperformed the PID controllers in this regard, but still, the Sugeno Fuzzy-PD controller exhibited problems when dealing with acrobatic trajectories. The autotuned GA-PID and autotuned GA-Sugeno Fuzzy-PD controllers introduced in this study show that the implementation of a GA could enhance the performance of the traditional PID and Sugeno Fuzzy-PD controllers, which allowed the quadcopter to accurately follow complex acrobatic trajectories. Furthermore, it is clear that the Mamdani Fuzzy-PD controller was able to show excellent performance even without implementing a GA for tuning its parameters.

The proposed approach of implementing a GA for tuning the controller parameters has the benefit of improving the performance of traditional controllers, enabling the quadcopter to follow complex trajectories. For applications where the quadcopter has to follow simple trajectories only, a traditional PID controller can be equipped with a GA to improve its performance rather than using a more complex fuzzy controller. For quadcopter applications with more complex trajectories Fuzzy-PD controllers are more appropriate and their performance can further be enhanced by the implementation of a GA. However, this has several shortcomings, such as the time and effort needed to develop a fuzzy rule-based model and the GA implementation. Obtaining the working limits of the parameters is also a challenging task.

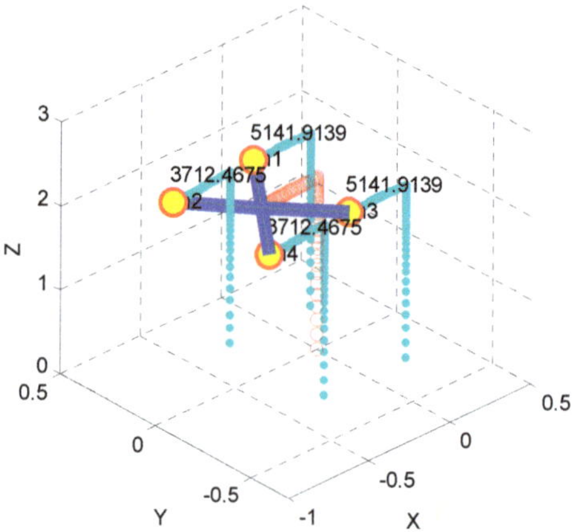

(**a**) A 3D view of the quadcopter performing a backward movement.

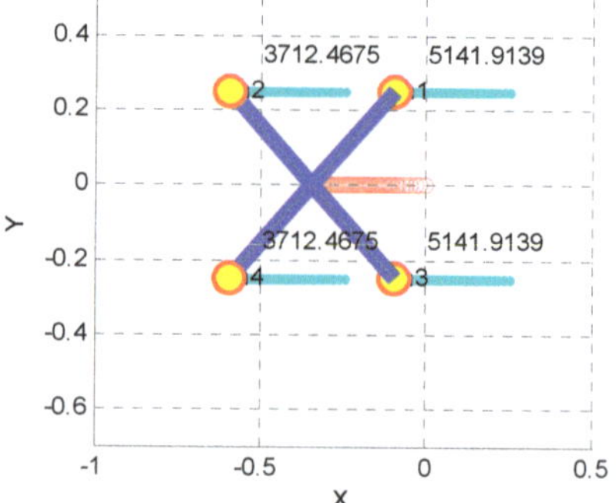

(**b**) Top view of the quadcopter performing a backward movement.

Figure 20. The quadcopter executing a backward movement.

5. Conclusions and Recommendations

Five controller types were designed for a quadcopter and their performance was evaluated in simulation: a PID, a Mamdani Fuzzy-PD, a Sugeno Fuzzy-PD, an autotuned GA-PID and an autotuned GA-Sugeno Fuzzy-PD. A paths simulator was also designed to visualize the quadcopter's behavior at different paths.

Through the simulations carried out, it has been verified that the stabilization of a quadcopter poses a challenge from the point of view of control engineering. Despite the effort to obtain the optimal parameters of the controllers, only the Mamdani Fuzzy-PD controller was able to face the most adverse conditions. This controller showed the best results with most of the difficult trajectories, and it did not require the implementation of a GA to improve its performance.

The time taken to tune the PID and Fuzzy-PD controllers was substantially reduced by the use of the MATLAB optimization tool. Moreover, it should be noted that the scale factor of the plant (placed in the output of the controllers) played a fundamental role in determining the speed of response of the plant. In other words, it was possible to achieve a high degree of sensitivity. The scale factor should be large enough to work with small gains, and in this way, the simulation and the adjustment process become a bit simpler and more intuitive.

With respect to the fuzzy controllers' design, it is recommended to work within the limits from 0 to 1. In this way, it is easier to analyze the system boundaries by analyzing the inputs and thus, to achieve the desired results more quickly. It has been verified that the performance of the controllers was improved when a GA was implemented. It was possible to observe this by subjecting the vehicle to the most demanding conditions (i.e., difficult trajectories) where the PID and Fuzzy-PD controllers required a re-adjustment of the gains for such conditions.

Other conclusions related to this work can be drawn as follows:

- For the quadcopter design, the "×" configuration was chosen for achieving better propulsion for the execution of acrobatic maneuvers. However, this choice led to complex calculations of the moments of inertia. On the other hand, the "+" configuration would be simpler in mathematical design and would provide a perfect rotation around its own axis due to the rotors acting in the same axis, where the vehicles with "×" configuration may fall slightly downward in this case.
- Even though the PID is a type of linear control system, it can be used with non-linear systems whenever working with a range of low values (i.e., in this case small angles and simple paths) and a high degree of accuracy is not required, given the conditions in indoor flights. However, the Fuzzy-PD controllers showed effectiveness with large angles and complex paths.
- The development of the PID could be improved by adding an anti-windup module to avoid the saturation. If so, the PID controller could have offered slightly better responses to non-linear conditions.
- The simulator proved to be a very useful tool in visualizing the paths and indicated when the dynamic model was not working properly. Hence, it was possible to investigate the issues in calculations to obtain an accurate dynamic model.

We would like to comment that the project is not closed yet, and that the dynamic model is currently being reviewed for a better definition with the aim of approaching more towards the real model.

In the future, novel solutions will be presented by the implementation of new controllers based on neural networks as well as a more intensive analysis of what has been achieved to date. Finally, we would like to emphasize that the use of a GA has exceeded expectations and it has proven to be a vital tool in achieving quadcopter stability.

Author Contributions: Conceptualization, I.S.L. and C.A.; methodology, I.S.L.; software, I.S.L.; validation, I.S.L., C.A. and Y.S.P.; formal analysis, I.S.L. and C.A.; investigation, I.S.L.; resources, I.S.L. and C.A.; data curation, I.S.L.; writing—original draft preparation, I.S.L., C.A. and Y.S.P.; writing—review and editing, I.S.L., C.A. and Y.S.P.; visualization, I.S.L.; supervision, C.A. and Y.S.P.; project administration, I.S.L., C.A. and Y.S.P. All authors have read and agreed to the published version of the manuscript.

Funding: This research received no external funding.

Abbreviations

Abbreviations

UAV	Unmanned Aerial Vehicle			
PI	Proportional-Integral			
GA	Genetic Algorithm		SPARC	Self-evolving Parameter-free Fu
MIMO	Multiple-Input Multiple-Output		MSE	Mean Square Error
PID	Proportional-Integral-Derivative		FLC	Fuzzy Logic Controller
P	Proportional		ABC	Artificial Bee Colony
ITAE	Integral Time Absolute Error		RPM	Revolutions per Minute
LQR	Linear Quadratic Regulator			
PD	Proportional-Derivative			
LQR-I	Linear Quadratic Regulator with Integrators			

References

1. Hoffmann, G.; Huang, H.; Waslander, S.; Tomlin, C. Quadrotor Helicopter Flight Dynamics and Control: Theory and Experiment. In Proceedings of the AIAA Guidance, Navigation and Control Conference and Exhibit, Hilton Head, South Carolina, 20–23 August 2007. [CrossRef]
2. Huang, H.; Hoffmann, G.M.; Waslander, S.L.; Tomlin, C.J. Aerodynamics and control of autonomous quadrotor helicopters in aggressive maneuvering. In Proceedings of the 2009 IEEE International Conference on Robotics and Automation, Kobe, Japan, 12–17 May 2009; pp. 3277–3282. [CrossRef]
3. Bolandi, H.; Rezaei, M.; Mohsenipour, R.; Nemati, H.; Smailzadeh, S.M. Attitude Control of a Quadrotor with Optimized PID Controller. *ICA* **2013**, *04*, 335–342. [CrossRef]
4. Sahul, M.P.V.; Chander, V.N.; Kurian, T. A novel method on Disturbance Rejection PID Controller for Quadcopter based on Optimization algorithm. *IFAC Proc. Vol.* **2014**, *47*, 192–199. [CrossRef]
5. Ahmad, F.; Kumar, P.; Bhandari, A.; Patil, P.P. Simulation of the Quadcopter Dynamics with LQR based Control. *Mater. Today Proc.* **2020**, *24*, 326–332. [CrossRef]
6. Nguyen, N.T.; Prodan, I.; Stoican, F.; Lefèvre, L. Reliable nonlinear control for quadcopter trajectory tracking through differential flatness. *IFAC-PapersOnLine* **2017**, *50*, 6971–6976. [CrossRef]
7. Al-Mahturi, A.; Santoso, F.; Garratt, M.A.; Anavatti, S.G. Nonlinear Altitude Control of a Quadcopter Drone Using Interval Type-2 Fuzzy Logic. In Proceedings of the 2018 IEEE Symposium Series on Computational Intelligence (SSCI), Bangalore, India, 18–21 November 2018; pp. 236–241. [CrossRef]
8. Ferdaus, M.M.; Hady, M.A.; Pratama, M.; Kandath, H.; Anavatti, S.G. RedPAC: A Simple Evolving Neuro-Fuzzy-based Intelligent Control Framework for Quadcopter. In Proceedings of the 2019 IEEE International Conference on Fuzzy Systems (FUZZ-IEEE), New Orleans, LA, USA, 18–21 June 2019; pp. 1–7. [CrossRef]
9. Amir, M.Y.; Abbass, V. Modeling of Quadrotor Helicopter Dynamics. In Proceedings of the 2008 International Conference on Smart Manufacturing Application, Goyang-si, Korea, 9–11 April 2008; pp. 100–105. [CrossRef]
10. Pounds, P.; Mahony, R.; Corke, P. Modelling and control of a large quadrotor robot. *Control. Eng. Pract.* **2010**, *18*, 691–699. [CrossRef]
11. Morar, I.-R.; Nascu, I. Model simplification of an unmanned aerial vehicle. In Proceedings of the 2012 IEEE International Conference on Automation, Quality and Testing, Robotics, Cluj-Napoca, Romania, 24–27 May 2012; pp. 591–596. [CrossRef]
12. Sa, R.C.; Barreto, G.A.; de Araujo, A.L.C.; Varela, A.T. Design and construction of a quadrotor-type unmanned aerial vehicle: Preliminary results. In Proceedings of the 2012 Workshop on Engineering Applications, Bogota, Colombia, 2–4 May 2012; pp. 1–6. [CrossRef]
13. Fernando, H.C.T.E.; de Silva, A.T.A.; de Zoysa, M.D.C.; Dilshan, K.A.D.C.; Munasinghe, S.R. Modelling, simulation and implementation of a quadrotor UAV. In Proceedings of the 2013 IEEE 8th International Conference on Industrial and Information Systems, Peradeniya, Sri Lanka, 17–20 December 2013; pp. 207–212. [CrossRef]
14. Barve, J.; Patel, K. Modelling, Simulation and Altitude-Range-Analysis of Quad-copter UAV. *IFAC Proc. Vol.* **2014**, *47*, 1126–1130. [CrossRef]
15. Hull, D.G. *Fundamentals of Airplane Flight Mechanics*; Springer: Berlin/Heidelberg, Germany; London, UK, 2007.
16. Alkamachi, A.; Erçelebi, E. Modelling and Genetic Algorithm Based-PID Control of H-Shaped Racing Quadcopter. *Arab. J. Sci. Eng.* **2017**, *42*, 2777–2786. [CrossRef]
17. Patrascu, M.; Hanchevici, A.B.; Dumitrache, I. Tuning of PID Controllers for Non-Linear MIMO Systems Using Genetic Algorithms. *IFAC Proc. Vol.* **2011**, *44*, 12644–12649. [CrossRef]
18. Lim, H.; Park, J.; Lee, D.; Kim, H.J. Build Your Own Quadrotor: Open-Source Projects on Unmanned Aerial Vehicles. *IEEE Robot. Automat. Mag.* **2012**, *19*, 33–45. [CrossRef]
19. Argentim, L.M.; Rezende, W.C.; Santos, P.E.; Aguiar, R.A. PID, LQR and LQR-PID on a quadcopter platform. In Proceedings of the 2013 International Conference on Informatics, Electronics and Vision (ICIEV), Dhaka, Bangladesh, 17–18 May 2013; pp. 1–6. [CrossRef]
20. Boudjit, K.; Larbes, C. Control and stabilization applied to micro quadrotor AR.Drone. In Proceedings of the 3rd International Conference on Application and Theory of Automation in Command and Control Systems—ATACCS'13, Naples, Italy, 28–30 May 2013; p. 122. [CrossRef]
21. Thu, K.M.; Gavrilov, A.I. Designing and Modeling of Quadcopter Control System Using L1 Adaptive Control. *Procedia Comput. Sci.* **2017**, *103*, 528–535. [CrossRef]
22. Hovakimyan, N.; Cao, C. *L1 Adaptive Control Theory: Guaranteed Robustness with Fast Adaptation*; Society for Industrial and Applied Mathematics: Philadelphia, PA, USA, 2010.
23. Njinwoua, B.J.; Wouwer, A.V. Cascade attitude control of a quadcopter in presence of motor asymmetry. *FAC-PapersOnLine* **2018**, *51*, 113–118. [CrossRef]
24. Cedro, L.; Wieczorkowski, K. Optimizing PID controller gains to model the performance of a quadcopter. *Transp. Res. Procedia* **2019**, *40*, 156–169. [CrossRef]
25. Paredes, J.; Sharma, P.; Ha, B.; Lanchares, M.; Atkins, E.; Gaskell, P.; Kolmanovsky, I. Development, implementation, and experimental outdoor evaluation of quadcopter controllers for computationally limited embedded systems. *Annu. Rev. Control.* **2021**, S1367578821000420. [CrossRef]

26. Hoffmann, F. Evolutionary algorithms for fuzzy control system design. *Proc. IEEE* **2001**, *89*, 1318–1333. [CrossRef]
27. Zulfatman, Z.; Rahmat, M.F. Application of self-tuning fuzzy pid controller on industrial hydraulic actuator using system identification approach. *Int. J. Smart Sens. Intell. Syst.* **2009**, *2*, 246–261. [CrossRef]
28. Santos, M.; López, V.; Morata, F. Intelligent Fuzzy Controller of a Quadrotor. In Proceedings of the 2010 IEEE International Conference on Intelligent Systems and Knowledge Engineering, Hangzhou, China, 15–16 November 2010; pp. 141–146. [CrossRef]
29. Sydney, N.; Smyth, B.; Paley, D.A. NSydney; Smyth, B.; Paley, D.A. Dynamic control of autonomous quadrotor flight in an estimated wind field. In Proceedings of the 52nd IEEE Conference on Decision and Control, Firenze, Italy, 10–13 December 2013; pp. 3609–3616. [CrossRef]
30. Fu, C.; Olivares-Mendez, M.A.; Campoy, P.; Suarez-Fernandez, R. UAS see-and-avoid strategy using a fuzzy logic controller optimized by Cross-Entropy in Scaling Factors and Membership Functions. In Proceedings of the 2013 International Conference on Unmanned Aircraft Systems (ICUAS), Atlanta, GA, USA, 28–31 May 2013; pp. 532–541. [CrossRef]
31. Gautam, D.; Ha, C. Control of a Quadrotor Using a Smart Self-Tuning Fuzzy PID Controller. *Int. J. Adv. Robot. Syst.* **2013**, *10*, 380. [CrossRef]
32. Fatan, M.; Sefidgari, B.L.; Barenji, A.V. An adaptive neuro PID for controlling the altitude of quadcopter robot. In Proceedings of the 2013 18th International Conference on Methods & Models in Automation & Robotics (MMAR), Miedzyzdroje, Poland, 26–29 August 2013; pp. 662–665. [CrossRef]
33. Benavidez, P.; Lambert, J.; Jaimes, A.; Jamshidi, M. Landing of an Ardrone 2.0 quadcopter on a mobile base using fuzzy logic. In Proceedings of the 2014 World Automation Congress (WAC), Waikoloa, HI, USA, 3–7 August 2014; pp. 803–812. [CrossRef]
34. Larrazabal, J.M.; Peñas, M.S. Intelligent rudder control of an unmanned surface vessel. *Expert Syst. Appl.* **2016**, *55*, 106–117. [CrossRef]
35. Garcia-Aunon, P.; Peñas, M.S.; García, J.M.D. Parameter selection based on fuzzy logic to improve UAV path-following algorithm. *J. Appl. Log.* **2017**, *24*, 62–75. [CrossRef]
36. Domingos, D.; Camargo, G.; Gomide, F. Autonomous Fuzzy Control and Navigation of Quadcopters. *IFAC-PapersOnLine* **2016**, *49*, 73–78. [CrossRef]
37. Yazid, E.; Garratt, M.; Santoso, F. Position control of a quadcopter drone using evolutionary algorithms-based self-tuning for first-order Takagi–Sugeno–Kang fuzzy logic autopilots. *Appl. Soft Comput.* **2019**, *78*, 373–392. [CrossRef]
38. Pham, T.H.; Ichalal, D.; Mammar, S. LPV and Nonlinear-based control of an Autonomous Quadcopter under variations of mass and moment of inertia. *IFAC-PapersOnLine* **2019**, *52*, 176–183. [CrossRef]
39. Ali, M.Z.; Ahmed, A.; Afridi, H.K. Control System Analysis and Design of Quadcopter in the Presence of Unmodelled Dynamics and Disturbances. *IFAC-PapersOnLine* **2020**, *53*, 8840–8846. [CrossRef]
40. Guerrero-Sánchez, M.E.; Lozano, R.; Castillo, P.; Hernández-González, O.; García-Beltrán, C.D.; Valencia-Palomo, G. Nonlinear control strategies for a UAV carrying a load with swing attenuation. *Appl. Math. Model.* **2021**, *91*, 709–722. [CrossRef]
41. Bresciani, T. Modelling, Identification and Control of a Quadrotor Helicopter. Master's Thesis, Lund University, Lund, Sweden, 2008.
42. Meriam, J.L.; Kraige, L.G. *Engineering Mechanics: Dynamics*, 6th ed.; Wiley: Hoboken, NJ, USA, 2006.
43. Mamdani, E.H.; Assilian, S. An Experiment in Linguistic Synthesis with a Fuzzy Logic Controller. *Int. J. Man-Mach. Stud.* **1975**, *7*, 1–13. [CrossRef]
44. Takagi, T.; Sugeno, M. Fuzzy Identification of Systems and Its Applications to Modeling and Control. *IEEE Trans. Syst. Man Cybern.* **1985**, *SMC-15*, 116–132. [CrossRef]
45. Sivanandam, S.N.; Deepa, S.N. *Introduction to Genetic Algorithms*; Springer: Berlin/Heidelberg, Germany; New York, NY, USA, 2007.

MDPI

St. Alban-Anlage 66

4052 Basel

Switzerland

Tel. +41 61 683 77 34

Fax +41 61 302 89 18

www.mdpi.com

Electronics Editorial Office

E-mail: electronics@mdpi.com

www.mdpi.com/journal/electronics